Lecture Notes in Mathematics

Edited by J.-M. Morel, F. Takens and B. Teissier

Editorial Policy
for the publication of monographs

1. Lecture Notes aim to report new developments in all areas of mathematics and their applications- quickly, informally and at a high level. Mathematical texts analysing new developments in modelling and numerical simulation are welcome.

   Monograph manuscripts should be reasonably self-contained and rounded off. Thus they may, and often will, present not only results of the author but also related work by other people. They may be based on specialised lecture courses. Furthermore, the manuscripts should provide sufficient motivation, examples and applications. This clearly distinguishes Lecture Notes from journal articles or technical reports which normally are very concise. Articles intended for a journal but too long to be accepted by most journals, usually do not have this "lecture notes" character. For similar reasons it is unusual for doctoral theses to be accepted for the Lecture Notes series, though habilitation theses may be appropriate.

2. Manuscripts should be submitted (preferably in duplicate) either to Springer's mathematics editorial in Heidelberg, or to one of the series editors (with a copy to Springer). In general, manuscripts will be sent out to 2 external referees for evaluation. If a decision cannot yet be reached on the basis of the first 2 reports, further referees may be contacted: The author will be informed of this. A final decision to publish can be made only on the basis of the complete manuscript, however a refereeing process leading to a preliminary decision can be based on a pre-final or incomplete manuscript. The strict minimum amount of material that will be considered should include a detailed outline describing the planned contents of each chapter, a bibliography   and several sample chapters.

   Authors should be aware that incomplete or insufficiently close to final manuscripts almost always result in longer refereeing times and nevertheless unclear referees' recommendations, making further refereeing of a final draft necessary.

   Authors should also be aware that parallel submission of their manuscript to another publisher while under consideration for LNM will in general lead to immediate rejection.

3. Manuscripts should in general be submitted in English. Final manuscripts should contain at least 100 pages of mathematical text and should always include

   - a table of contents;
   - an informative introduction, with adequate motivation and perhaps some historical remarks: it should be accessible to a reader not intimately familiar with the topic treated;
   - a subject index: as a rule this is genuinely helpful for the reader.

Continued on inside back-cover

Lecture Notes in Mathematics 1856

Editors:
J.-M. Morel, Cachan
F. Takens, Groningen
B. Teissier, Paris

Subseries:
Fondazione C.I.M.E., Firenze
Adviser: Pietro Zecca

K. Back   T.R. Bielecki   C. Hipp
S. Peng   W. Schachermayer

Stochastic Methods in Finance

Lectures given at the
C.I.M.E.-E.M.S. Summer School
held in Bressanone/Brixen, Italy,
July 6–12, 2003

Editors: M. Frittelli
W. Runggaldier

Fondazione
C.I.M.E.

Editors and Authors

Kerry Back
Mays Business School
Department of Finance
310C Wehner Bldg.
College Station, TX 77879-4218, USA
e-mail: back@olin.wustl.edu

Tomasz R. Bielecki
Department of Applied Mathematics
Illinois Inst. of Technology
10 West 32nd Street
Chicago, IL 60616, USA
e-mail: bielecki@iit.edu

Marco Frittelli
Dipartimento di Matematica per le Decisioni
Universitá degli Studi di Firenze
via Cesare Lombroso 6/17
50134 Firenze, Italy
e-mail: marco.frittelli@dmd.unifi.it

Christian Hipp
Institute for Finance, Banking and Insurance
University of Karlsruhe
Kronenstr. 34
76133 Karlsruhe, Germany
e-mail: christian.hipp@wiwi.uni-karlsruhe.de

Shige Peng
Institute of Mathematics
Shandong University
250100 Jinan
People's Republic of China
e-mail: peng@sdu.edu.cn

Wolfgang J. Runggaldier
Dipartimento di Matematica Pura ed Applicata
Universutá degli Studi di Padova
via Belzoni 7
35100 Padova, Italy
e-mail: runggal@math.unipd.it

Walter Schachermayer
Financial and Actuarial Mathematics
Vienna University of Technology
Wiedner Hauptstrasse 8/105-1
1040 Vienna, Austria
e-mail: wschach@fam.tuwien.ac.at

Library of Congress Control Number: 2004114748

Mathematics Subject Classification (2000):
60G99, 60-06, 91-06, 91B06, 91B16, 91B24, 91B28, 91B30, 91B70, 93-06, 93E11, 93E20

ISSN 0075-8434
ISBN 3-540-22953-1 Springer-Verlag Berlin Heidelberg New York
DOI: 10.1007/b100122

Springer is a part of Springer Science + Business Media
springeronline.com
© Springer-Verlag Berlin Heidelberg 2004
Printed in Germany

Typesetting: Camera-ready TeX output by the authors

41/3142/ du - 543210 - Printed on acid-free paper

Preface

A considerable part of the vast development in Mathematical Finance over the last two decades was determined by the application of stochastic methods. These were therefore chosen as the focus of the 2003 School on "Stochastic Methods in Finance". The growing interest of the mathematical community in this field was also reflected by the extraordinarily high number of applications for the CIME-EMS School. It was attended by 115 scientists and researchers, selected from among over 200 applicants. The attendees came from all continents: 85 were Europeans, among them 35 Italians.

The aim of the School was to provide a broad and accurate knowledge of some of the most up-to-date and relevant topics in Mathematical Finance. Particular attention was devoted to the investigation of innovative methods from stochastic analysis that play a fundamental role in mathematical modeling in finance or insurance: the theory of stochastic processes, optimal and stochastic control, stochastic differential equations, convex analysis and duality theory.

The outstanding and internationally renowned lecturers have themselves contributed in an essential way to the development of the theory and techniques that constituted the subjects of the lectures. The financial origin and motivation of the mathematical analysis were presented in a rigorous manner and this facilitated the understanding of the interface between mathematics and finance. Great emphasis was also placed on the importance and efficiency of mathematical instruments for the formalization and resolution of financial problems. Moreover, the direct financial origin of the development of some theories now of remarkable importance in mathematics emerged with clarity. The selection of the five topics of the CIME Course was not an easy task because of the wide spectrum of recent developments in Mathematical Finance. Although other topics could have been proposed, we are confident that the choice made covers some of the areas of greatest current interest.

We now propose a brief guided tour through the topics chosen and through the methodologies that modern financial mathematics has elaborated to unveil *Risk* beneath its different masks.

We begin the tour with expected utility maximization in continuous-time stochastic markets: this classical problem, which can be traced back to the seminal works by Merton, received a renewed impulse in the middle of the 1980's, when the so-called duality approach to the problem was first developed. Over the past twenty years, the theory constantly improved, until the general case of semimartingale stochastic models was finally tackled with great success. This prompted us to dedicate one series of lectures to this traditional as well as very innovative topic:

"Utility Maximization in Incomplete Markets", Prof. Walter Schachermayer, Technical University of Vienna.
This course was mainly focused on the maximization of the expected utility from terminal wealth in incomplete markets. A part of the course was dedicated to the presentation of the stochastic model of the market, with particular attention to the formulation of the condition of No Arbitrage. Some results of convex analysis and duality theory were also introduced and explained, as they are needed for the formulation of the dual problem with respect to the set of equivalent martingale measures. Then some recent results of this classical problem were presented in the general context of semi-martingale financial models.

The importance of the above-mentioned analysis of the utility maximization problem is also revealed in the theory of asset pricing in incomplete markets, where the agent's preferences have again to be given serious consideration, since *Risk* cannot be completely hedged. Different notions of "utility-based" prices have been introduced in the literature since the middle of the 1990's. These concepts determine pricing rules which are often non-linear outside the set of marketed claims. Depending on the utility function selected, these pricing kernels share many properties with non-linear valuations: this bordered on the realm of risk measures and capital requirements. Coherent or convex risk measures have been studied intensively in the last eight years but only very recently have risk measures been considered in a dynamic context. The theory of non-linear expectations is very appropriate for dealing with the genuinely dynamic aspects of the measures of *Risk*. This leads to the next topic:

"Nonlinear expectations, nonlinear evaluations and risk measures", Prof. Shige Peng, Shandong University.
In this course the theory of the so-called " g-expectations" was developed, with particular attention to the following topics: backward stochastic differential equations, F-expectation, g-martingales and theorems of decomposition of E-supermartingales. Applications to the theory of risk measures in a dynamic context were suggested, with particular emphasis on the issues of time consistency of the dynamic risk measures.

Among the many forms of *Risk* considered in finance, credit risk has received major attention in recent years. This is due to its theoretical relevance but certainly also to its practical implications among the multitude of investors. Credit risk is the risk faced by one party as a result of the possible decline in the creditworthiness of the counterpart or of a third party. An overview of the current state of the art was given in the following series of lectures:

"Stochastic methods in credit risk modeling: valuation and hedging", Prof. Tomasz Bielecki, Illinois Institute of Technology.
A broad review of the recent methodologies for the management of credit risk was presented in this course: structural models, intensity-based models, modeling of dependent defaults and migrations, defaultable term structures, copula based models. For each model the main mathematical tools have been described in detail, with particular emphasis on the theory of martingales, stochastic control, Markov chains. The written contribution to this volume involves, in addition to the lecturer, two co-authors, they too are among the most prominent current experts in the field.

The notion of *Risk* is not limited to finance, but has a traditional and dominating place also in insurance. For some time the two fields have evolved independently of one another, but recently they are increasingly interacting and this is reflected also in the financial reality, where insurance companies are entering the financial market and viceversa. It was therefore natural to have a series of lectures also on insurance risk and on the techniques to control it.

"Financial control methods applied in insurance", Prof. Christian Hipp, University of Karlsruhe.
The methodologies developed in modern mathematical finance have also met with wide use in the applications to the control and the management of the specific risk of insurance companies. In particular, the course showed how the theory of stochastic control and stochastic optimization can be used effectively and how it can be integrated with the classical insurance and risk theory.

Last but not least we come to the topic of partial and asymmetric information that doubtlessly is a possible source of *Risk*, but has considerable importance in itself since evidently the information is neither complete nor equally shared among the agents. Frequently debated also by economists, this topic was analyzed in the lectures:

"Partial and asymmetric information", Prof. Kerry Back, University of St. Louis.
In the context of economic equilibrium, a survey of incomplete and asymmetric information (or insider trading) models was presented. First, a review of filtering theory and stochastic control was introduced. In the second part of the course some work on incomplete information models was analyzed, focusing on Markov chain models. The last part was concerned with asymmetric information models, with particular emphasis on the Kyle model and extensions thereof.

As editors of these Lecture Notes we would like to thank the many persons and Institutions that contributed to the success of the school. It is our pleasure to thank the members of the CIME (Centro Internazionale Matematico Estivo) Scientific Committee for their invitation to organize the School; the Director, Prof. Pietro Zecca, and the Secretary, Prof. Elvira Mascolo, for their efficient support during the organization. We were particularly pleased by the fact that the European Mathematical Society (EMS) chose to co-sponsor this CIME-School as one of its two Summer Schools for 2003 and that it provided additional financial support through UNESCO-Roste.

Our special thanks go to the lecturers for their early preparation of the material to be distributed to the participants, for their excellent performance in teaching the courses and their stimulating scientific contributions. All the participants contributed to the creation of an exceptionally friendly atmosphere which also characterized the various social events organized in the beautiful environment around the School. We would like to thank the Town Council of Bressanone/Brixen for additional financial and organizational support; the Director and the staff of the Cusanus Academy in Bressanone/Brixen for their kind hospitality and efficiency as well as all those who helped us in the realization of this event.

This volume collects the texts of the five series of lectures presented at the Summer School. They are arranged in alphabetic order according to the name of the lecturer.

Firenze and Padova, March 2004

Marco Frittelli and Wolfgang J. Runggaldier

CIME's activity is supported by:

Istituto Nazionale di Alta Matematica "F. Severi":
Ministero dell'Istruzione, dell'Università e della Ricerca;
Ministero degli Affari Esteri - Direzione Generale per la Promozione e la Cooperazione - Ufficio V;
E. U. under the Training and Mobility of Researchers Programme and UNESCO-ROSTE, Venice Office

Contents

Incomplete and Asymmetric Information in Asset Pricing Theory

Kerry Back 1

1 Filtering Theory 1
   1.1 Kalman-Bucy Filter 3
   1.2 Two-State Markov Chain................................. 4
2 Incomplete Information 5
   2.1 Seminal Work.. 5
   2.2 Markov Chain Models of Production Economies 6
   2.3 Markov Chain Models of Pure Exchange Economies 7
   2.4 Heterogeneous Beliefs 11
3 Asymmetric Information 12
   3.1 Anticipative Information 12
   3.2 Rational Expectations Models 13
   3.3 Kyle Model... 16
   3.4 Continuous-Time Kyle Model 18
   3.5 Multiple Informed Traders in the Kyle Model 20
References 23

Modeling and Valuation of Credit Risk

Tomasz R. Bielecki, Monique Jeanblanc, Marek Rutkowski 27

1 Introduction 27
2 Structural Approach 29
   2.1 Basic Assumptions 29
      Defaultable Claims 29
      Risk-Neutral Valuation Formula......................... 31
      Defaultable Zero-Coupon Bond 32
   2.2 Classic Structural Models 34
      Merton's Model 34
      Black and Cox Model................................... 37
   2.3 Stochastic Interest Rates 43

2.4   Credit Spreads: A Case Study 45
2.5   Comments on Structural Models 46
3 Intensity-Based Approach 47
3.1   Hazard Function 47
       Hazard Function of a Random Time 48
       Associated Martingales.................................. 49
       Change of a Probability Measure 50
       Martingale Hazard Function 53
       Defaultable Bonds: Deterministic Intensity 53
3.2   Hazard Processes... 55
       Hazard Process of a Random Time 56
       Valuation of Defaultable Claims 57
       Alternative Recovery Rules 59
       Defaultable Bonds: Stochastic Intensity 63
       Martingale Hazard Process 64
       Martingale Hypothesis 65
       Canonical Construction 67
       Kusuoka's Counter-Example 69
       Change of a Probability 70
       Statistical Probability................................... 72
       Change of a Numeraire 74
       Preprice of a Defaultable Claim 77
       Credit Default Swaption 79
       A Practical Example..................................... 82
3.3   Martingale Approach 84
       Standing Assumptions 85
       Valuation of Defaultable Claims 85
       Martingale Approach under (H.1) 87
3.4   Further Developments 88
       Default-Adjusted Martingale Measure...................... 88
       Hybrid Models 89
       Unified Approach 90
3.5   Comments on Intensity-Based Models...................... 90
4 Dependent Defaults and Credit Migrations 91
4.1   Basket Credit Derivatives 92
       The ith-to-Default Contingent Claims 92
       Case of Two Entities.................................... 93
4.2   Conditionally Independent Defaults........................ 94
       Canonical Construction 94
       Independent Default Times 95
       Signed Intensities 96
       Valuation of FDC and LDC 96
       General Valuation Formula 97
       Default Swap of Basket Type 98

4.3 Copula-Based Approaches 99
    Direct Application......................................100
    Indirect Application 100
    Simplified Version 102
4.4 Jarrow and Yu Model....................................103
    Construction and Properties of the Model 103
    Bond Valuation 105
4.5 Extension of the Jarrow and Yu Model.....................106
    Kusuoka's Construction 107
    Interpretation of Intensities.............................108
    Bond Valuation 108
4.6 Dependent Intensities of Credit Migrations 109
    Extension of Kusuoka's Construction 109
4.7 Dynamics of Dependent Credit Ratings 112
4.8 Defaultable Term Structure..............................113
    Standing Assumptions 113
    Credit Migration Process................................116
    Defaultable Term Structure..............................117
    Premia for Interest Rate and Credit Event Risks 119
    Defaultable Coupon Bond 120
    Examples of Credit Derivatives............................121
4.9 Concluding Remarks....................................122
References 123

Stochastic Control with Application in Insurance
Christian Hipp 127
1 Preface..127
2 Introduction Into Insurance Risk 128
  2.1 The Lundberg Risk Model...............................128
  2.2 Alternatives 129
  2.3 Ruin Probability 129
  2.4 Asymptotic Behavior For Ruin Probabilities 131
3 Possible Control Variables and Stochastic Control 132
  3.1 Possible Control Variables 132
      Investment, One Risky Asset 132
      Investment, Two or More Risky Assets.....................133
      Proportional Reinsurance 134
      Unlimited XL Reinsurance 134
      XL-Reinsurance 135
      Premium Control.......................................135
      Control of New Business 135
  3.2 Stochastic Control......................................136
      Objective Functions 136
      Infinitesimal Generators.................................137
      Hamilton-Jacobi-Bellman Equations 139

    Verification Argument 141
    Steps for Solution 143
4 Optimal Investment for Insurers 143
  4.1 HJB and its Handy Form 143
  4.2 Existence of a Solution 145
  4.3 Exponential Claim Sizes 145
  4.4 Two or More Risky Assets 147
5 Optimal Reinsurance and Optimal New Business 148
  5.1 Optimal Proportional Reinsurance 150
  5.2 Optimal Unlimited XL Reinsurance 151
  5.3 Optimal XL Reinsurance 152
  5.4 Optimal New Business 153
6 Asymptotic Behavior for Value Function and Strategies 154
  6.1 Optimal Investment: Exponential Claims 154
  6.2 Optimal Investment: Small Claims 154
  6.3 Optimal Investment: Large Claims 155
  6.4 Optimal Reinsurance 156
7 A Control Problem with Constraint: Dividends and Ruin 157
  7.1 A Simple Insurance Model with Dividend Payments 157
  7.2 Modified HJB Equation 158
  7.3 Numerical Example and Conjectures 159
  7.4 Earlier and Further Work 161
8 Conclusions 162
References 163

**Nonlinear Expectations, Nonlinear Evaluations and Risk
Measures**
Shige Peng 165
1 Introduction 165
  1.1 Searching the Mechanism of Evaluations of Risky Assets 165
  1.2 Axiomatic Assumptions for Evaluations of Derivatives 166
    General Situations: $\mathcal{F}_t^X$–Consistent Nonlinear Evaluations 166
    $\mathcal{F}_t^X$–Consistent Nonlinear Expectations 167
  1.3 Organization of the Lecture 168
2 Brownian Filtration Consistent Evaluations and Expectations 169
  2.1 Main Notations and Definitions 169
  2.2 $\mathcal{F}_t$–Consistent Nonlinear Expectations 171
  2.3 $\mathcal{F}_t$–Consistent Nonlinear Evaluations 173
3 Backward Stochastic Differential Equations: g–Evaluations and
  g–Expectations 176
  3.1 BSDE: Existence, Uniqueness and Basic Estimates 176
  3.2 1–Dimensional BSDE 182
    Comparison Theorem 183
    Backward Stochastic Monotone Semigroups and g–Evaluations . 186
    Example: Black–Scholes Evaluations 188

g–Expectations 189
Upcrossing Inequality of $\mathcal{E}^g$–Supermartingales and Optional
    Sampling Inequality 193
3.3 A Monotonic Limit Theorem of BSDE 199
3.4 g–Martingales and (Nonlinear) g–Supermartingale
    Decomposition Theorem 201
4 Finding the Mechanism: Is an $\mathcal{F}$–Expectation a g–Expectation? 204
4.1 $\mathcal{E}^\mu$-Dominated $\mathcal{F}$-Expectations 204
4.2 $\mathcal{F}_t$-Consistent Martingales 207
4.3 BSDE under $\mathcal{F}_t$–Consistent Nonlinear Expectations 210
4.4 Decomposition Theorem for $\mathcal{E}$-Supermartingales 213
4.5 Representation Theorem
    of an $\mathcal{F}$–Expectation by a g–Expectation 216
4.6 How to Test and Find g? 219
4.7 A General Situation: $\mathcal{F}_t$–Evaluation Representation Theorem ... 220
5 Dynamic Risk Measures 221
6 Numerical Solution of BSDEs: Euler's Approximation 222
7 Appendix 224
7.1 Martingale Representation Theorem 224
7.2 A Monotonic Limit Theorem of Itô's Processes 226
7.3 Optional Stopping Theorem for $\mathcal{E}^g$–Supermartingale 232
References 238
References on BSDE and Nonlinear Expectations 240

Utility Maximisation in Incomplete Markets
Walter Schachermayer 255
1 Problem Setting 255
2 Models on Finite Probability Spaces 259
2.1 Utility Maximization 266
    The complete Case (Arrow) 266
    The Incomplete Case 272
3 The General Case 277
3.1 The Reasonable Asymptotic Elasticity Condition 277
3.2 Existence Theorems 281
References 289

Incomplete and Asymmetric Information in Asset Pricing Theory

Kerry Back

John M. Olin School of Business
Washington University in St. Louis
St. Louis, MO 63130
back@olin.wustl.edu

These notes could equally well be entitled "Applications of Filtering in Financial Theory." They constitute a selective survey of incomplete and asymmetric information models. The study of asymmetric information, which emphasizes differences in information, means that we will be concerned with equilibrium theory and how the less informed agents learn in equilibrium from the more informed agents. The study of incomplete information is also most interesting in the context of economic equilibrium.

Excellent surveys of incomplete information models in finance [48] and of asymmetric information models [10] have recently been published. In these notes, I will not attempt to repeat these comprehensive surveys but instead will give a more selective review.

The first part of this article provides a review of filtering theory, in particular establishing the notation to be used in the later parts. The second part reviews some work on incomplete information models, focusing on recent work using simple Markov chain models to model the behavior of the market portfolio. The last part reviews asymmetric information models, focusing on the Kyle model and extensions thereof.

1 Filtering Theory

Let us start with a brief review of filtering theory, as exposited in [33]. Note first that engineers and economists tend to use the term "signal" differently. Engineers take the viewpoint of the transmitter, who sends a "signal," which is then to be estimated (or "filtered") from a noisy observation. Economists tend to take the viewpoint of the receiver, who observes a "signal" and then uses it to estimate some other variable. To avoid confusion, I will try to avoid the term, but when I use it (in the last part of the chapter), it will be in the sense of economists.

We work on a finite time horizon $[0, T]$ and a complete probability space $(\Omega, \mathcal{A}, P)$. The problem is to estimate a process X from the observations of another process Y. In general, one considers estimating the conditional expectation $E[f(X_t)|\mathcal{F}_t^Y]$, where $\{\mathcal{F}_t^Y\}$ is the the filtration generated by Y augmented by the P–null sets in $\mathcal{A}$, and f is a real-valued function satisfying some minimal regularity conditions but otherwise arbitrary. By estimating $E[f(X_t)|\mathcal{F}_t^Y]$ for arbitrary f, one can obtain the distribution of X_t conditional on $\mathcal{F}_t^Y$.

For any process θ, we will use the conventional notation $\hat{\theta}_t$ to denote $E[\theta_t|\mathcal{F}_t^Y]$. More precisely, $\hat{\theta}_t$ denotes for each t a version of $E[\theta_t|\mathcal{F}_t^Y]$ chosen so that the resulting process $(t, \omega) \mapsto \hat{\theta}_t(\omega)$ is jointly measurable.

Let W be an n–dimensional Wiener process on its own filtration and define $\mathcal{F}_t$ to be the σ–field generated by $(X_s, W_s; s \leq t)$ augmented by the P–null sets in $\mathcal{A}$. We assume for each t that $\mathcal{F}_t$ is independent of the σ–field generated by $(W_v - W_u; t \leq u \leq v \leq T)$, which simply means that the future changes in the Wiener process cannot be foretold by X. Henceforth, we will assume that all processes are $\{\mathcal{F}_t\}$–adapted.

The Wiener process W creates the noise that must be filtered from the observation process. Specifically, assume the observation process Y satisfies

$$dY_t = h_t\, dt + dW_t; \quad Y_0 = 0 \tag{1}$$

where h is a jointly measurable $\Re^n$–valued process satisfying $E \int_0^T \|h_t\|^2\, dt < \infty$.

Assume X takes values in some complete separable metric space, define $f_t = f(X_t)$, and assume

$$df_t = g_t\, dt + dM_t, \tag{2}$$

for some jointly measurable process g and right-continuous martingale M such that $E \int_0^T |g_t|^2 dt < \infty$. If X is given as the solution of a stochastic differential equation and f is smooth, the processes g and M can of course be computed from Itô's formula. We assume further that $E[f_t^2] < \infty$ for each t and $E \int_0^T \|f_t h_t\|^2\, dt < \infty$.

The "innovation process" is defined as

$$\begin{aligned} dZ_t &= dY_t - \hat{h}_t\, dt \\ &= (h_t - \hat{h}_t)\, dt + dW_t \end{aligned} \tag{3}$$

with $Z_0 = 0$. The differential dZ is interpreted as the innovation or "surprise" in the variable Y, which consists of two parts, one being the error in the estimation of the drift h_t and the other being the random change dW.

The main results of filtering theory, due to Fujisaka, Kallianpur, and Kunita [22], are the following.

1) The innovation process Z is an $\{\mathcal{F}_t^Y\}$–Brownian Motion.

2) For any separable L^2–bounded $\{\mathcal{F}_t^Y\}$–martingale H, there exists a jointly measurable $\{\mathcal{F}_t^Y\}$–adapted $\Re^n$–valued process ϕ such that $E \int_0^T \| \phi_t \|^2 \, dt < \infty$, and

$$dH_t = \sum_{i=1}^{n} \phi_t^i \, dZ_t^i.$$

3) There exist jointly measurable adapted processes α^i such that $d[M, W^i]_t = \alpha_t^i \, dt$, for $i = 1, \ldots, N$.

4) $\hat{f}$ evolves as

$$d\hat{f}_t = \hat{g}_t \, dt + \left(\widehat{fh}_t - \hat{f}_t \hat{h}_t + \hat{\alpha}_t \right)' dZ_t, \tag{4}$$

where $\widehat{fh}_t$ denotes $E[f_t h_t | \mathcal{F}_t^Y]$.

Part (1) means in particular that Z is a martingale; thus the innovations dZ are indeed "unpredictable." Given that it is a martingale, the fact that it is a Brownian motion follows from Levy's theorem and the fact, which follows immediately from (3), that the covariations are $d\langle Z^i, Z^j \rangle = dt$ if $i = j$ and 0 otherwise. Part (2) means that the process Z "spans" the $\{\mathcal{F}_t^Y\}$–martingales (which would follow from $\{\mathcal{F}_t^Y\} = \{\mathcal{F}_t^Z\}$, though this condition does not hold in general). Part (3) means that the square-bracket processes are absolutely continuous, though in our applications we will assume M and the W^i are independent, implying $\alpha^i = 0$ for all i.

Part (4) is the filtering formula. The estimate $\hat{f}$ is updated because f is expected to change (which is obviously captured by the term $\hat{g}_t \, dt$) and because new information from dZ is available to estimate f. The observation process Y (or equivalently the innovation process Z) is useful for estimating f due to two factors. One is the possibility of correlation between the martingales W and M. This is reflected in the term $\hat{\alpha}_t \, dZ_t$. The other factor is the correlation between f and the drift h_t of Y. This is reflected in the term $(\widehat{fh}_t - \hat{f}_t \hat{h}_t) \, dZ_t$. Note that $\widehat{fh}_t - \hat{f}_t \hat{h}_t$ is the covariance of f_t and h_t, conditional on $\mathcal{F}_t^Y$. The formula (4) generalizes the linear prediction formula

$$\hat{x} = \bar{x} + \frac{\mathrm{cov}(x, y)}{\mathrm{var}(y)} (y - \bar{y}),$$

which yields $\hat{x} = E[x|y]$ when x and y are joint normal.

We consider two applications.

1.1 Kalman-Bucy Filter

Assume X_0 is distributed normally with variance σ^2 and

$$dX_t = aX_t \, dt + dB_t,$$
$$dY_t = cX_t \, dt + dW_t,$$

where B and W are independent real-valued Brownian motions that are independent of X_0. In this case, the distribution of X_t conditional on $\mathcal{F}_t^Y$ is normal with deterministic variance Σ_t. Moreover,

$$d\hat{X}_t = a\hat{X}_t\,dt + c\Sigma_t\,dZ_t, \tag{5}$$

where the innovation process Z is given by

$$dZ_t = dY_t - c\hat{X}_t\,dt. \tag{6}$$

Furthermore,

$$\Sigma_t = \frac{\gamma\alpha e^{\lambda t} - \beta}{\gamma e^{\lambda t} + 1}, \tag{7}$$

where α and $-\beta$ are the two roots of the quadratic equation $1 + 2ax - c^2x^2 = 0$, with both α and β positive, $\lambda = c^2(\alpha + \beta)$ and $\gamma = (\sigma^2 + \beta)/(\alpha - \sigma^2)$. One can consult, e.g., [33] or [41] for the derivation of these results from the general filtering results cited above. In the multivariate case, an equation of the form (5) also holds, where Σ_t is the covariance matrix of X_t conditional on $\mathcal{F}_t^Y$. In this circumstance, the covariance matrix evolves deterministically and satisfies an ordinary differential equation of the Riccati type, but there is in general no closed-form solution of the differential equation.

1.2 Two-State Markov Chain

A very simple model that lies outside the Gaussian family is a two-state Markov chain. There is no loss of generality in taking the states to be 0 and 1, and it is convenient to do so. Consider the Markov chain X satisfying

$$dX_t = (1 - X_{t-})\,dN_t^0 - X_{t-}\,dN_t^1, \tag{8}$$

where $X_{t-} \equiv \lim_{s \uparrow t} X_s$ and the N^i are independent Poisson processes with parameters λ^i that are independent of X_0. This means that X stays in each state an exponentially distributed amount of time, with the exponential distribution determining the transition from state i to state j having parameter λ^i. This fits in our earlier framework as

$$dX_t = g_t\,dt + dM_t,$$

where

$$g_t = (1 - X_{t-})\lambda^0 - X_{t-}\lambda^1, \text{ and}$$
$$dM_t = (1 - X_{t-})\,dM_t^0 - X_{t-}\,dM_t^1,$$

with M^i being the martingale $M_t^i = N_t^i - \lambda^i t$.
     Assume

$$dY_t = h(X_{t-})\,dt + dW_t, \tag{9}$$

where W is an n–dimensional Brownian motion independent of the N^i and X_0. Thus, the drift vector of Y is $h(0)$ or $h(1)$ depending on the state X_{t-}. In terms of our earlier notation, $h_t = h(X_{t-})$.

Write π_t for $\hat{X}_t$. This is the conditional probability that $X_t = 1$. The general filtering formula (4) implies[1]

$$d\pi_t = \left[(1 - \pi_t)\lambda^0 - \pi_t\lambda^1\right] dt + \pi_t(1 - \pi_t)\left[h(1) - h(0)\right]' dZ_t, \qquad (10)$$

where the innovation process Z is given by

$$dZ_t = dY_t - \left[(1 - \pi_t)h(0) + \pi_t h(1)\right] dt. \qquad (11)$$

This is a special case of the results on Markov chain filtering due to Wonham [47].

Note the similarity of (10) with the Kalman-Bucy filter (5): $h(1) - h(0)$ is the vector c in the equation

$$
\begin{aligned}
dY_t &= h(X_{t-}) \, dt + dW_t \\
&= \left[(1 - X_{t-})h(0) + X_{t-}h(1)\right] dt + dW_t \\
&= h(0) \, dt + cX_{t-} \, dt + dW_t,
\end{aligned}
$$

and $\pi_t(1 - \pi_t)$ is the variance of X_t conditional on $\mathcal{F}_t^Y$.

2 Incomplete Information

2.1 Seminal Work

Early work in portfolio choice and market equilibrium under incomplete information includes [16], [19], and [23]. These papers analyze models of the following sort. The instantaneous rate of return on an asset is given by

$$
\begin{aligned}
\tfrac{dS}{S} &= \mu_t \, dt + \sigma \, dW, \quad where \\
d\mu_t &= \kappa(\theta - \mu_t) \, dt + \phi \, dB
\end{aligned}
$$

and W and B are Brownian motions with a constant correlation coefficient ρ, and where μ_0 is normally distributed and independent of W and B. It is assumed that investors observe S but not μ; i.e., their filtration is the filtration generated by S (augmented by the P–null sets). The innovation process is

$$dZ = \frac{\mu_t - \hat{\mu}_t}{\sigma} \, dt + dW,$$

which is an $\{\mathcal{F}_t^S\}$–Brownian motion. Moreover, we can write

---

[1] Note that (4) implies π is continuous and then from bounded convergence we have $\pi_t = E\left[X_{t-}|\mathcal{F}_t^Y\right]$, so $\hat{g}_t = (1 - \pi_t)\lambda^0 - \pi_t\lambda^1$.

$$\frac{dS}{S} = \hat{\mu}_t \, dt + \sigma \, dZ. \tag{12}$$

Because $\hat{\mu}$ is observable (adapted to $\{\mathcal{F}_t^S\}$), this is equivalent to a standard complete information model, and the portfolio choice theory of Merton applies to (12). This is a particular application of the separation principle for optimal control under incomplete information, and in fact the primary contribution of these early papers was to highlight the role of the separation principle.

These early models were interpreted as equilibrium models by assuming the returns are the returns of physical investment technologies having constant returns to scale, as in the Cox-Ingersoll-Ross model [12]. In other words, the assets are in infinitely elastic supply. We will call such an economy a "production economy," though obviously it is a very special type of production economy. In this case, there are no market clearing conditions to be satisfied. Equilibrium is determined by the optimal investments and consumption of the agents. Given an equilibrium, prices of other zero net supply assets can be determined—for example, term structure models can be developed. However, the set of such models that can be generated by assuming incomplete information is the same as the set that can be generated with complete information, given the equivalence of (12) with complete information models. In particular, the Kalman-Bucy filtering equations imply particular dynamics for $\hat{\mu}$, but one could equally well assume the same dynamics for μ and assume μ is observable.

2.2 Markov Chain Models of Production Economies

In Gaussian models (with Gaussian priors) the conditional covariance matrix of the unobserved variables is deterministic. This means that there is no real linkage between Gaussian incomplete information models and the well-documented phenomenon of stochastic volatility. Detemple observes in [17] that, within a model that is otherwise Gaussian, stochastic volatility can be generated by assuming non-Gaussian priors. However, more recent work has focused on Markov chain models.

David in [13] and [14] studies an economy in which the assets are in infinitely elastic supply, assuming a two-state Markov chain for which the transition time from each state is exponentially distributed as in Section 1.2. In David's model, there are two assets ($i = 0, 1$), with

$$\frac{dS^i}{S^i} = \mu^i(X_{t-}) \, dt + \sigma^i \, dW^i,$$

where W^0 and W^1 are independent Brownian motions, $X_t \in \{0, 1\}$, and $\mu^0(x) = \mu^1(1-x)$. Set $\mu_a = \mu^0(0)$ and $\mu_b = \mu^0(1)$. Then when $X_{t-} = 0$, the growth rates of the assets are μ_a for asset 0 and μ_b for asset 1, and the growth rates of the assets are reversed when $X_{t-} = 1$. With complete information in this economy, the investment opportunity set is independent of X_{t-}. However, with incomplete information, investors do not know for certain which asset is most productive. Suppose, for example, that $\mu_a > \mu_b$. Then asset 0 is most productive in state 0 and asset 1 is most productive in state 1. The filtering equation for the model is (10), with observation process $Y = (Y^0, Y^1)$, where

$$dY_t^i = \frac{d \log S_t^i}{\sigma^i} = \left(\frac{\mu^i(X_{t-})}{\sigma^i} - \frac{\sigma^i}{2} \right) dt + dW^i.$$

In terms of the innovation processes (the following equations actually define the innovation processes), we have

$$\frac{dS^0}{S^0} = \left[(1 - \pi_t)\mu_a + \pi_t \mu_b \right] dt + \sigma^0 \, dZ^0,$$
$$\frac{dS^1}{S^1} = \left[\pi_t \mu_a + (1 - \pi_t)\mu_b \right] dt + \sigma^1 \, dZ^1.$$

As in [16], [19] and [23], this is equivalent to a complete information model in which the expected rates of return of the assets are stochastic with particular dynamics given by the filtering equations, but the volatilities of assets are constant.

David focuses on the volatility of the market portfolio, assuming a representative investor with power utility. The weights of the two assets in the market portfolio will depend on π_t (e.g., asset 0 will be weighted more highly when π_t is small, because this means a greater belief that the expected return of asset 0 is $\mu_a > \mu_b$). Assume for example that $\sigma^1 = \sigma^2$. Then, due to diversification, the instantaneous volatility of the market portfolio will be smallest when the assets are equally weighted, which will be the case when $\pi_t = 1/2$, and the volatility will be higher when π_t is near 0 or 1. Therefore, the market portfolio will have a stochastic volatility. Using simulation evidence, David shows that the return on the market portfolio in the model can be consistent with the following stylized facts regarding asset returns.

1) Excess kurtosis: the tails of asset return distributions are "too fat" to be consistent with normality.
2) Skewness: large negative returns occur more frequently than large positive returns.
3) Covariation between returns and changes in conditional variances: large negative returns are associated with a greater increase in the conditional variance than are large positive returns.

2.3 Markov Chain Models of Pure Exchange Economies

Arguably, a more interesting context in which to study incomplete information is an economy of the type studied by Lucas in [40], in which the assets are in fixed supply. This is a "pure exchange" economy, in which the essential economic problem is to allocate consumption of the asset dividends. In this case, the prices and returns of the assets are determined in equilibrium by the market-clearing conditions and hence will be affected fundamentally by the nature of information.

David and Veronesi (see [44], [45] and [15]) study models of this type and discuss various issues regarding the volatility and expected return of the market portfolio. Their models are variations on the following basic model. Assume there is a single asset, with supply normalized to one, which pays dividends at rate D. Assume

$$\frac{dD_t}{D_t} = \alpha_D(X_{t-}) \, dt + \sigma_D \, dW^1, \tag{13}$$

where X is a two-state Markov chain with switching between states occurring at exponentially distributed times, as in Section 1.2. Here W^1 is a real-valued Brownian motion independent of X_0. Investors observe the dividend rate D but do not observe the state X_{t-}, which determines the growth rate of dividends. We may also assume investors observe a process

$$dH_t = \alpha_H(X_{t-}) \, dt + \sigma_H \, dW^2, \tag{14}$$

where W^2 is a real-valued Brownian motion independent of W^1 and X_0. The process H summarizes any other information investors may have about the state of the economy.

The filtering equations for this model are the same as those described earlier, where we set

$$Y = \left(\frac{\log D}{\sigma_D}, \frac{H}{\sigma_H}\right) \quad \text{and} \quad \mu = \left(\frac{\alpha_D - \sigma_D^2/2}{\sigma_D}, \frac{\alpha_H}{\sigma_H}\right).$$

In terms of the innovation process $Z = (Z^1, Z^2)$, we have

$$\frac{dD_t}{D_t} = \left[\pi_t \alpha_D(1) + (1 - \pi_t)\alpha_D(0)\right] dt + \sigma_D \, dZ^1, \tag{15}$$

$$dH = \left[\pi_t \alpha_H(1) + (1 - \pi_t)\alpha_H(0)\right] dt + \sigma_H \, dZ^2, \tag{16}$$

and the conditional probability π_t evolves as

$$d\pi_t = \left[(1 - \pi_t)\lambda^0 - \pi_t \lambda^1\right] dt$$
$$+ \pi_t(1 - \pi_t)\left[\frac{\alpha_D(1) - \alpha_D(0)}{\sigma_D} \, dZ^1 + \frac{\alpha_H(1) - \alpha_H(0)}{\sigma_H} \, dZ^2\right]. \tag{17}$$

Note that (15) and (17) form a Markovian system in which the growth rate of dividends is stochastic. From here, the analysis is entirely standard. It is assumed that there is a representative investor[2] who is infinitely-lived and who maximizes the expected discounted utility of consumption $u(c_t)$, with discount rate δ. The representative investor must consume the aggregate dividend in.

---

[2] For the construction of a representative investor, see for example [20].

equilibrium, and the price of the asset is determined by his marginal rate of substitution. Specifically, the asset price at time t must be

$$S_t = E\left[\int_t^\infty \frac{e^{-\delta(s-t)}u'(D_s)}{u'(D_t)} D_s\, ds \middle| \pi_t, D_t\right]. \tag{18}$$

In the case of logarithmic utility, we obtain $S_t = D_t/\delta$, so the asset return is given by

$$\frac{dS_t}{S_t} = \left[\pi_t \alpha_D(1) + (1 - \pi_t)\alpha_D(0)\right] dt + \sigma_D\, dZ^1.$$

This is essentially the same as the early models on incomplete information, because we have simply specified the expected return

$$\pi_t \alpha_D(1) + (1 - \pi_t)\alpha_D(0)$$

as a particular stochastic process.

The case of power utility $u(c) = c^\gamma/\gamma$ is more interesting. Note that for $s \geq t$ we have from (13) that

$$D_s^\gamma = D_t^\gamma e^{\gamma \int_t^s \left\{[\alpha_D(X_{a-})-\sigma_D^2/2]\,da + \sigma_D\, dW_a^1\right\}}.$$

Using this, equation (18) yields

$$\begin{aligned}
S_t &= D_t^{1-\gamma} E\left[\int_t^\infty e^{-\delta(s-t)} D_s^\gamma\, ds \middle| \pi_t, D_t\right] \\
&= D_t^{1-\gamma} \left\{(1-\pi_t)E\left[\int_t^\infty e^{-\delta(s-t)} D_s^\gamma\, ds \middle| X_{t-} = 0, D_t\right] \right. \\
&\qquad\left. + \pi_t E\left[\int_t^\infty e^{-\delta(s-t)} D_s^\gamma\, ds \middle| X_{t-} = 1, D_t\right]\right\} \\
&= D_t \left\{(1-\pi_t)E\left[\int_t^\infty e^{-\delta(s-t)} e^{\gamma \int_t^s \left\{[\alpha_D(X_{a-})-\sigma_D^2/2]\,da + \sigma_D\, dW_a^1\right\}}\, ds \middle| X_{t-}=0\right] \right. \\
&\qquad\left. + \pi_t E\left[\int_t^\infty e^{-\delta(s-t)} e^{\gamma \int_t^s \left\{[\alpha_D(X_{a-})-\sigma_D^2/2]\,da + \sigma_D\, dW_a^1\right\}}\, ds \middle| X_{t-}=1\right]\right\}.
\end{aligned}$$

Due to the time-homogeneity of the Markovian system (15) and (17), the conditional expectations in the above are independent of the date t. Denoting the first expectation by C_0 and the second by C_1, we have

$$S_t = D_t\left\{(1 - \pi_t)C^0 + \pi_t C^1\right\}.$$

This implies

$$\frac{dS}{S} = \frac{dD}{D} + \frac{(C^1 - C^0)\,d\pi}{(1-\pi)C^0 + \pi C^1} + \frac{(C^1 - C^0)\,d\langle D, \pi \rangle}{D[(1-\pi)C^0 + \pi C^1]} \qquad (19)$$

$$= \text{something}\,dt + \sigma_D\,dZ^1$$

$$+ \left[\frac{(C^1 - C^0)\pi(1-\pi)}{(1-\pi)C^0 + \pi C^1} \right]$$

$$\times \left[\frac{\alpha_D(1) - \alpha_D(0)}{\sigma_D}\,dZ^1 + \frac{\alpha_S(1) - \alpha_S(0)}{\sigma_S}\,dZ^2 \right].$$

The factor

$$\frac{(C^1 - C^0)\pi(1-\pi)}{(1-\pi)C^0 + \pi C^1} \qquad (20)$$

introduces stochastic volatility. Thus, stochastic volatility can arise in a model in which the volatility of dividends is constant.

There are obviously other ways than incomplete information to introduce a stochastic growth rate of dividends in a Markovian model similar to (15) and (17). However, this approach leads to a very sensible connection between investors' uncertainty about the state of the economy and the volatility of assets. Note that the factor $\pi_t(1 - \pi_t)$ in the numerator of (20) is the conditional variance of X_t—it is largest when π_t is near $1/2$, when investors are most uncertain about the state of the economy, and smallest when π_t is near zero or one, which is when investors are most confident about the state of the economy. Thus, the volatility of the asset is linked to investors' confidence about future economic growth.

Veronesi actually assumes in [44] that the level of dividends (rather than the logarithm of dividends) follows an Ornstein-Uhlenbeck process as in (13) and he assumes the representative investor has negative exponential utility (i.e., he assumes constant *absolute* risk aversion rather than constant *relative* risk aversion). David and Veronesi study in [15] the model described here but assume the representative investor also has an endowment stream. They show that the model can generate a time-varying correlation between the return and volatility of the market portfolio (for example, sometimes the correlation may be positive and sometimes it may be negative) and use the model to generate an option pricing formula for options on the market portfolio. Time-varying correlation has been noted to be necessary to reconcile stochastic volatility models with market option prices. In the David-Veronesi model, it arises quite naturally. When investors believe they are in the high growth state (π_t is high), a low dividend realization will lead to both a negative return on the market and an increase in volatility, because it increases the uncertainty about the actual state (i.e., it increases the conditional variance $\pi_t(1 - \pi)_t$). Thus, volatility and returns are negatively correlated in this circumstance. In contrast, if investors believe they are in the low growth state (π_t is low), a low dividend realization will lead to a negative return and a decrease in volatility, because it reaffirms the belief that the state is low, decreasing the conditional variance $\pi_t(1 - \pi_t)$. Thus, volatility and returns are positively correlated in this circumstance.

In [45], Veronesi studies the above model but assuming there are n states of the world rather than just two. One way to express his model is to let the state variable X_t take values in $\{1, \ldots, n\}$ with dynamics

$$dX_t = \sum_{i=1}^{n} (i - X_{t-}) \, dN_t^i,$$

where the N^i are independent Poisson processes with parameters λ^i. This means that X jumps to state i at each arrival date of the Poisson process N^i, independent of the prior state (in particular, X stays in state i if $X_{t-} = i$ and $\Delta N_t^i = 1$). The process $N \equiv \sum_{i=1}^{n} N^i$ is a Poisson process with parameter $\lambda \equiv \sum_{i=1}^{n} \lambda^i$. Conditional on $\Delta N_t = 1$, there is probability λ^i/λ that $\Delta N_t^i = 1$ and therefore probability λ^i/λ that $X_t = i$, independent of the prior state X_{t-}. Define $X_t^i = 1_{\{X_t = i\}}$. Then $E\left[X_t^i \mid \mathcal{F}_t^Y\right]$, which we will denote by π_t^i, is the probability that $X_t = i$ conditional on $\mathcal{F}_t^Y$. The distribution of X_t conditional on $\mathcal{F}_t^Y$ is clearly defined by the π_t^i. The process X_t^i is a two-state Markov chain with dynamics

$$dX_t^i = (1 - X_{t-}^i) \, dN_t^i - X_{t-}^i \, dN_t^{-i}, \tag{21}$$

where $N^{-i} \equiv \sum_{j \neq i} N_t^j$ is a Poisson process with parameter $\lambda^{-i} \equiv \sum_{j \neq i} \lambda^j$, because, if X^i is in state 0, it exits at an arrival time of N^i, and, if it is in state 1, it exits at an arrival time of N^{-i}. Equation (21) is of the same form as equation (8), and, therefore, the dynamics of π^i are given by the filtering equation (10) for two-state Markov chains. The resulting formula for the dynamics of the asset price S is a straightforward generalization of (19).

2.4 Heterogeneous Beliefs

Economists often assume that all agents have the same prior beliefs. A rationale for this assumption is given by Harsanyi in [29]. To some, this rationale seems less than compelling, motivating the analysis of heterogeneous prior beliefs. A good example is the Detemple-Murthy model [18]. This model is of a single-asset Lucas economy similar to the one described in the previous section (but with the unobservable dividend growth rate being driven by a Brownian motion instead of following a two-state Markov chain). Instead of assuming a representative investor, Detemple and Murthy assume there are two classes of investors with different beliefs about the initial value of the dividend growth rate. Finally, they assume each type of investor has logarithmic utility and the investors all have the same discount rate. The focus of their paper is the impact of margin requirements, which limit short sales of the asset and limit borrowing to buy the asset. This is an example of an issue that cannot be addressed in a representative investor model, because margin requirements are never binding in equilibrium on a representative investor, given that he simply holds the market portfolio in equilibrium. In a frictionless complete-markets economy one can always construct a representative investor, but that is not necessarily true in an economy with margin requirements or other frictions or incompleteness of markets. In the absence of a representative investor, it can be difficult to compute or characterize an equilibrium, but this task is considerably simplified by assuming logarithmic utility, because that implies investors are "myopic"—they hold the tangency portfolio and do not have hedging demands. However, if all investors have logarithmic utility, then heterogeneity must be introduced through some other mechanism than the utility function. The assumption of incomplete information and heterogeneous priors is a simple device for generating this heterogeneity among agents. Basak and Croitoru study in [8] the effect of introducing "arbitrageurs" (for example, financial intermediaries) in the model of Detemple and Murthy. Jouini and Napp discuss in [36] the existence of representative investors in markets with incomplete information and heterogeneous beliefs.

Another way to introduce heterogeneity of posterior beliefs is to assume investors have different views regarding the dynamical laws of economic processes. As an example, consider the economy with dividend process (13) and observation process (14). We might assume some investors believe the Brownian motions W^1 and W^2 are correlated while others believe they are independent, or more generally we may assume investors have different beliefs regarding the correlation coefficient. Scheinkman and Xiong study a similar model in [42], though in their model there are two assets. To each asset there corresponds a process D satisfying (13), though $D(t)$ is interpreted as the cumulative dividends paid between 0 and t instead of the rate of dividends at time t. To each asset there also corresponds an observation process of the form (14). There are two types of investors. One type thinks the observation process associated with the first asset has positive instantaneous correlation with its cumulative dividend process while the other type thinks the two Brownian motions are independent. The reverse is true for the second asset. Scheinkman and Xiong intepret this as "overconfidence," with each investor weighting the innovation process for one of the assets too highly when updating his beliefs. They link this form of overconfidence to speculative bubbles, the volume of trading, and the "excess volatility" puzzle.

3 Asymmetric Information

3.1 Anticipative Information

Recently, a literature has developed using the theory of enlargement of filtrations to study the topic of "insider trading." See [9], [25], [26], [31], [34], [38] and the references therein. One starts with asset prices of the usual form[3]

$$\frac{dS_t^i}{S_t^i} = \mu_t^i \, dt + \sigma_t^i \, dW_t^i, \tag{22}$$

on the horizon $[0, T]$ where the W^i are correlated Brownian motions on the filtered probability space $(\Omega, \mathcal{F}, \{\mathcal{F}_t\}, P)$. Then one supposes there is an $\mathcal{F}_T$-measurable random variable Y (with values in $\Re^k$ or some more general space) and an "insider" has access to the filtration $\{\mathcal{G}_t\}$, which is the usual augmentation of the filtration $\{\mathcal{F}_t \vee \sigma - (Y)\}$. By "access to the filtration," I mean that the insider is allowed to choose trading strategies that are $\{\mathcal{G}_t\}$–adapted.

Some interesting questions are (1) does the model make mathematical sense—i.e., are the price processes $\{\mathcal{G}_t\}$–semimartingales? (2) is there an arbitrage opportunity for the insider? (3) is the market complete for the insider? (4) how much additional utility can the insider earn from his advance knowledge of Y? (5) how would the insider value derivatives? .... For the answer to the first question, the essential reference is [32]. In [9], Baudoin describes the setup I have outlined here as the case of "strong information" and also introduces a concept of "weak information."

The study of anticipative information can be useful as a first step to developing an equilibrium model. Because the insider is assumed to take the price process (22) as given (unaffected by his portfolio choice) the equilibrium model would be of the "rational expectations" variety described in the next section. If one does not solve for an equilibrium, the assumed price dynamics could be quite arbitrary. Suppose for example that there is a constant riskless rate r and the advance information Y is the vector of asset prices S_T. Then there is an arbitrage opportunity for the insider unless

$$S_t^i = e^{-r(T-t)} S_T^i$$

for all i and t, which of course cannot be the case if the volatilities σ^i are nonzero. One might simply say that this is not an acceptable model and adopt hypotheses that exclude it. However, the rationale for excluding it must be a belief that exploitation of arbitrage opportunities tends to eliminate them. In other words, buying and selling by the insider would be expected to change market prices. This is true in general and not just in this specific example. The idea that market prices reflect in some way and to some extent the information of economic agents is a cornerstone of finance and of economics in general. In the remainder of this article, we will discuss equilibrium models of asymmetric information.

3.2 Rational Expectations Models

The term "rational expectations" means that agents understand the mapping from the information of various agents to the equilibrium price; thus they make

---

[3] Assume either that there are no dividends or that the S_i represent the prices of the portfolios in which dividends are reinvested in new shares.

correct inferences from prices (see [27]). The original rational expectations models were "competitive" models in the sense that agents were assumed to be "price takers," meaning that they assume their own actions have no effect on prices. Now the term is generally reserved for competitive models, and I will use it in that sense. We will examine strategic models, in which agents understand the impact of their actions on prices, in the next sections.

An important rational expectations model is that of Wang [46]. Wang studies a Lucas economy in which the dividend rate D_t of the asset has dynamics

$$dD_t = (\Pi_t - kD_t) \, dt + b_D \, dW, \tag{23}$$

where W is an $\Re^3$–valued Brownian motion. Moreover, it is assumed that

$$d\Pi_t = a_\Pi(\bar{\Pi} - \Pi_t) \, dt + b_\Pi \, dW \tag{24}$$

for a constant $\bar{\Pi}$. It is also assumed that there is a Cox-Ingersoll-Ross-type asset (i.e., one in infinitely elastic supply) that pays the constant rate of return r. There are two classes of investors, each having constant absolute risk aversion.

One class of investors (the "informed traders") observes D and Π. The other class (the "uninformed traders") observes only D. As described thus far, the model should admit a "fully revealing equilibrium," in which the uninformed traders could infer the value of Π_t from the equilibrium price of the asset. This equilibrium suffers from the "Grossman-Stiglitz paradox"—in reality it presumably costs some effort or money to become informed, but if prices are fully revealing, then no one would pay the cost of becoming informed; however, if no one is informed, prices cannot be fully revealing (and it would presumably be worthwhile in that case for someone to pay the cost of becoming informed). Wang avoids this outcome by the device introduced by Grossman and Stiglitz in [28]: he assumes the asset is subject to supply shocks that are unobserved by all traders. The noise introduced by the supply shocks prevents uninformed traders from inverting the price to compute the information Π_t of informed traders.[4] Specifically, Wang assumes the supply of the asset is $1 + \Theta_t$, where

$$d\Theta_t = -a_\Theta \Theta \, dt + b_\Theta \, dW. \tag{25}$$

The general method used to solve rational expectations models is still that described by Grossman in [27], even though Grossman did not assume there were supply shocks and obtained a fully revealing equilibrium. The trick is to consider an "artificial economy" in which traders are endowed with certain additional information. One computes an equilibrium of the artificial economy

---

[4] In fact, this type of mechanism was first introduced by Lucas [39], who assumes the money supply is unobservable in the short run, and hence real economic shocks cannot be distinguished from monetary shocks, leading to real effects of monetary policy in the short run.

and then shows that prices in this artificial economy reveal exactly the additional information traders were assumed to possess. Thus, the equilibrium of the artificial economy is an equilibrium of the actual economy in which traders make correct inferences from prices.

In Wang's artificial economy, the informed traders observe Θ as well as D and Π. The uninformed traders observe a linear combination of Θ and Π as well as D. In the equilibrium of the artificial economy, the price reveals the linear combination of Θ and Π, given knowledge of D. This implies that it reveals Θ to the informed traders, given that they are endowed with knowledge of Π and D. Therefore, the equilibrium of the artificial economy is an equilibrium of the actual economy.

Specifically, Wang conjectures that the equilibrium price S_t is a linear combination of D_t, Π_t, Θ_t and $\hat{\Pi}$, where $\hat{\Pi}$ denotes the expectation of Π conditional on the information of the uninformed traders. For this to make sense, one has to specify the filtration of the uninformed traders, and in the artificial economy it is specified as the filtration generated by D and a particular linear combination of Π and Θ. Let this linear combination be

$$H_t = \alpha \Pi_t + \beta \Theta_t \qquad (26)$$

Then the observation process of the uninformed traders in the artificial economy is $Y_t = (D, H)$ and the unobserved process they wish to estimate is Π. For the equilibrium of the artificial economy to be an equilibrium of the actual economy, we will need S_t to be a linear combination of D_t, H_t and $\hat{\Pi}_t$; i.e.,

$$S_t = \delta + \gamma D_t + \kappa H_t + \lambda \hat{\Pi}_t. \qquad (27)$$

Conditional on $\mathcal{F}_t^Y$, Π_t is normally distributed with mean $\hat{\Pi}_t$ and a deterministic variance. Wang derives an equilbrium in which S_t is a linear combination of D_t, Π_t, Θ_t and $\hat{\Pi}_t$ with time-invariant coefficients by focusing on the steady-state solution of the model. Specifically, he assumes the variance of Π_0 is the equilibrium point of the ordinary differential equation that the variance satisfies.

Given the specification of the price process (26)–(27) and the filtering formula, it is straightforward to calculate the demands of the two classes of traders. The market clearing equation is that the sum of the demands equals Θ_t. This is a linear equation that must hold for all values of D_t, Π_t, Θ_t and $\hat{\Pi}_t$. Imposing this condition gives the equilibrium values of α, β, δ, γ, κ and λ.

In addition to the usual issues regarding the expected return and volatility of the market portfolio, Wang is able to describe the portfolio behavior of the two classes of investors; in particular, uninformed traders tend to act as "trend chasers," buying the asset when its price increases, and informed traders act as "contrarians," selling the asset when its price increases.

3.3 Kyle Model

The price-taking assumption in rational expectations models is often problematic. In the extreme case, prices are fully revealing, and traders can form their demands as functions of the fully revealing prices, ignoring the information they possessed prior to observing prices. But, if traders all act independently of their own information, how can prices reveal information? Moreover, as mentioned earlier, full revelation of information by prices would eliminate the incentive to collect information in the first place.

The price-taking assumption is particularly problematic when information is possessed by only one or a few traders. Consider the case of a piece of information that is held by only a single trader. In general, the equilibrium price in a rational expectations model will reflect this information to some extent. Moreover, traders are assumed to make correct inferences from prices, so the trader is assumed to be aware that his information enters prices. But how can he anticipate that the price will reflect his private information, when he assumes that his actions do not affect the price? In [30], Hellwig describes this as "schizophrenia" on the part of traders.

These issues do not arise in strategic models, in which agents are assumed to recognize that their actions affect prices and it is only through their actions that private information becomes incorporated into prices. The most prominent model of strategic trading with asymmetric information is due to Kyle [35]. Kyle's model has been applied on many occasions, beginning with [1], to study various issues in market microstructure.

The Kyle model focuses on a single risky asset traded over the time period $[0, T]$. It is assumed that there is also a riskless asset, with the risk-free rate normalized to zero. Unlike models described previously in which the single risky asset is interpreted as the market portfolio, with the dividend of the asset equaling aggregate consumption, the Kyle model is not a model of the market portfolio. In fact, the risk of the asset is best interpreted as idiosyncratic, because investors are assumed to be risk neutral. As in [28] it is assumed that the supply of the asset is subject to random shocks, which we interpret as resulting from the trade of "noise traders." The noise traders trade for reasons that are unmodeled. For example, they may experience liquidity shocks (endowments of cash to be invested or desires for cash for consumption) and for that reason are often called "liquidity traders." In addition to one or more strategic traders and the noise traders, it is assumed that there are competitive risk-neutral "market makers," who are somewhat analogous to the uninformed traders in Wang's model. The market makers observe the net demands of the strategic traders and noise traders and compete to fill their demands. As a result of their competition (and their risk neutrality and the fact that the risk-free rate is zero), the transaction price is always the expectation of the asset value, conditional on the information of the market makers, i.e., conditional on the information in the history of orders.

It is assumed that the information asymmetry is erased by a public announcement at date T. Since this eliminates the "lemons problem," all positions can be liquidated at this announced value. Denote this value by v. From now on, we will adopt the normalization that $T = 1$. In the remainder of this section, we will describe the single-period model in [35], in which there is a single informed trader.

In this model, there is trading only at date 0, and consumption occurs at date 1. The asset value v is normally distributed with mean $\bar{v}$ and variance σ_v^2. The informed trader observes v and submits an order $x(v)$. Noise traders submit an order z that is independent of v and normally distributed with mean zero and variance σ_z^2. Market makers observe $y \equiv x + z$ and set the price equal to $p = E[v|x(v) + z]$. The informed trader wishes to maximize his expected profit, which is $E[x(v - p)]$. We search for a "linear equilibrium," in which the price is set as $p = \bar{v} + \lambda y$ and the insider's trade is $x = \eta(v - \bar{v})$, for constants λ and η. An equilibrium is defined by

1) Given $x = \eta(v - \bar{v})$, pricing satisfies Bayes' rule; i.e., $\bar{v} + \lambda y = E[v|y]$, and
2) Given $p = \bar{v} + \lambda y$, the insider's strategy is optimal; i.e., $\eta(v - \bar{v}) = \operatorname{argmax}_x E[x(v - \bar{v} - \lambda(x + z))]$.

Condition (1) implies

$$\lambda = \frac{\operatorname{cov}(v, y)}{\operatorname{var}(y)} = \frac{\eta \sigma_v^2}{\eta^2 \sigma_v^2 + \sigma_z^2},$$

and condition (2) implies

$$\eta = \frac{1}{2\lambda}.$$

The solution of these two equations is

$$\eta = \frac{\sigma_z}{\sigma_v} \quad \text{and} \quad \lambda = \frac{\sigma_v}{2\sigma_z}.$$

A slightly more general definition of a linear equilibrium would allow the constants in the affine pricing rule and trading strategy to be of general form, but it is easily seen that the equilibrium obtained here is unique within that class also.

Kyle defines the reciprocal of λ as the "depth" of the market. It measures the number of shares that can be traded causing only a unit change in the price. Of interest is the fact that the depth of the market is proportional to the amount of noise trading as measured by σ_v and inversely proportional to the amount of private information as measured by σ_v. Thus, markets are deeper in this model when uninformed trading is more prevalent and when the degree of information asymmetry is smaller.

Kyle analyzed a discrete-time multiperiod version of the model, assuming the variance of noise trades in each period is $\sigma_z^2 \Delta t$, where Δt is the length of each period. He showed that the equilibria converge to the equilibrium of a continuous-time model in which the noise trades arrive as a Brownian motion with volatility σ_z.

3.4 Continuous-Time Kyle Model

The continuous-time version of the model was formalized and generalized in
[2]. Subsequent generalizations appear in [3], [4], [5], [6], [7], [11], and [37].
In the continuous-time model, given that the risk-free rate is assumed to be
zero, the budget equation (self-financing condition) for the informed trader is
$dW = X\,dS$, where W denotes his wealth, X is the number of shares he holds,
and S is the price. Let $C = W - XS$ denote the amount of cash he holds.
Assuming X and S are continuous semimartingales (on the interval $[0,1)$ at
least) and applying Itô's formula to $W = XS + C$, we obtain

$$dW = dC + X\,dS + S\,dX + d\langle X, S\rangle,$$

so the budget equation implies

$$dC = -S\,dX - d\langle X, S\rangle.$$

It is common in the finance literature to write the differential of the sharp
bracket process $\langle X, S\rangle$ as $dS\,dX$. Adopting this notation, we can write

$$dC = -S\,dX - dS\,dX = -(S + dS)\,dX.$$

Thus, we can interpret the change in the cash position as equaling the cost
of shares purchased, where the number of shares purchased is dX and the
price paid is $S + dS$, which can be interpreted as the price prevailing at
the end of the infinitesimal period dt. This interpretation has nothing to do
with insider trading. We are simply interpreting the usual budget equation.
This intepetation is well understood and in fact is the motivation for the
continuous-time budget equation.

However, this application of Itô's formula (integration by parts) is useful
for analyzing the choice problem of the insider in the Kyle model. Specifically,
we are assuming the insider can sell his shares for the known value v at date
1. This will create a jump in his cash position at date 1 equal to vX_{1-} (where,
as usual, X_{t-} denotes $\lim_{s\uparrow t} X_s$). Normalizing both the number of shares he
owns at date 0 and his initial cash to be zero, his wealth at date 1 will equal

$$\begin{aligned}
C_1 &= vX_{1-} - \int_0^{1-} S_t\,dX_t - \int_0^{1-} d\langle X, S\rangle_t \\
&= \int_0^{1-} (v - S_t)\,dX_t - \langle X, S\rangle_{1-}, \\
&= \int_0^1 (v - S_t)\,dX_t - \langle X, S\rangle_{1-},
\end{aligned}$$

the last equality being a result of the equality $S_1 = v$.

In addition to the advance information about the asset value v, the other
distinctive characteristic of the insider's portfolio choice problem is that he
understands that market prices react to his trades. Specifically, we assume

$$dS_t = \phi_t \, dt + \lambda_t \, dY_t, \tag{28}$$

for some stochastic processes ϕ and $\lambda > 0$, where $Y = X + Z$ and Z is the Brownian motion of noise trades. This implies

$$\langle X, S \rangle_{1-} = \int_0^{1-} \lambda_t \, d\langle X, X \rangle_t + \int_0^{1-} \lambda_t \, d\langle X, Z \rangle_t.$$

In [2], it is shown that it is strictly suboptimal to have

$$\int_0^{1-} \lambda_t \, d\langle X, X \rangle_t > 0,$$

which implies that optimal X must be finite variation processes. This is quite different from the Merton model. However, in the Merton model, there is no term of the form $\int_0^{1-} \lambda_t \, d\langle X, X \rangle_t$, because Merton (and almost all subsequent authors) studied a price-taking investor.

Equilibrium requires that the insider's strategy X be optimal, given the pricing rule (28), and that the pricing rule satisfy $S_t = E[v|\mathcal{F}_t^Y]$. It turns out that in equilibrium the insider's strategy is absolutely continuous, so $dX_t = \theta_t \, dt$ for some stochastic process θ and the insider's final wealth is $\int_0^1 (v - S_t) \theta_t \, dt$. Moreover, in equilibrium, the observation process Y is an $\{\mathcal{F}_t^Y\}$-Brownian motion, which means that, up to scaling by $1/\sigma_z$, the observation process equals the innovation process.

Under the larger filtration of the insider, Y is a Brownian bridge. This is feasible because the insider controls Y via $dY = \theta_t \, dt + dZ$. The Brownian bridge terminates at a value dependent on v, and the Brownian motion/Brownian bridge distinction completely characterizes the information asymmetry in equilibrium. The market makers understand that Y is a Brownian bridge on the insider's filtration, but they do not know the value at which it will terminate. Integrating over the distribution of possible terminal values converts the Brownian bridge into a Brownian motion. Note the similarity with a model of anticipative information when the private signal of the insider is the vector W_T of terminal values of the Brownian motions in (22).

One point worth noting is that when the insider is risk-neutral, it is not actually necessary to assume he knows the value at which the asset can be liquidated at date 1. His expected profit from trading is the same whether v is the actual liquidation value or merely the conditional expectation of the liquidation value given his information at date 0. Likewise, the filtering problem of the market makers is the same when v simply denotes the expected value of the asset conditional on the insider's information. This equivalence does not hold when the insider is risk averse, because then the number of shares he wishes to hold at date 1 is affected by the remaining risk regarding the liquidation value. The continuous-time Kyle model with a single informed trader having negative exponential utility is analyzed in [7] and [11]. The equilibrium price in that case is of the form $S_t = H(t, U_t)$ where $U_t = \int_0^t \kappa(s) \, dY_s$ for a deterministic function κ (in the risk-neutral case, $\kappa = 1$).

3.5 Multiple Informed Traders in the Kyle Model

Here we will discuss the continuous-time Kyle model with multiple informed traders developed in [6]. Their work builds on the analysis in [21] of a discrete-time model with multiple traders. In the model of [6] – herafter BCW – there are N risk-neutral traders who observe signals y^i at date 0. The signals are assumed to be joint normally distributed with the liquidation value, and the joint distribution is assumed to be symmetric in the y^i. As noted at the end of the previous section, the interesting value is not really the liquidation value but rather the conditional expectation of the liquidation value, in this case conditional on all the signals of the traders. Denote this value by v. Because of the joint normality, v is an affine function of the y^i and, by affinely transforming the y^i, we can assume $v = \sum_{i=1}^{N} y^i$.

BCW search for a linear equilibrium. Defining $Y = Z + \sum_{i=1}^{N} X^i$, "linearity" means that the price evolves as

$$dS_t = \phi(t)\,dt + \lambda(t)\,dY_t, \tag{29}$$

for some deterministic functions ϕ and λ and trading strategies take the form $dX_t^i = \theta_t^i\,dt$, where

$$\theta_t^i = \alpha(t)S_t + \beta(t)y^i. \tag{30}$$

for some deterministic functions α and β.

Given trading strategies of this type, the observation process of market makers is

$$dY_t = N\alpha(t)S_t\,dt + \beta(t)v\,dt + dZ_t$$

which is equivalent (because S by definition is $\{\mathcal{F}_t^Y\}$–adapted) to observing a process with dynamics $\beta(t)v\,dt + dZ_t$, so estimation of v by the market makers is a simple Gaussian filtering problem as in Section 1.1. Let $\hat{v}$ denote the solution to this filtering problem.

Equilibrium requires $S = \hat{v}$. Equating the coefficients in the dynamics of $\hat{v}$ given by the Kalman-Bucy filtering equation (5) to the proposed linear dynamics (29) for S, it can easily be seen that we must have $\alpha = -\beta/N$ and $\phi = 0$. Thus, in any linear equilibrium,

$$dS_t = \lambda(t)\,dY_t, \quad \text{and} \tag{31}$$
$$dY_t = \beta(t)\big[v - \hat{v}_t\big]\,dt + dZ_t. \tag{32}$$

Equation (32) means that the observation process Y is (up to rescaling by $1/\sigma_z$) the innovation process for the market makers, and, as in the single-trader model, Y is an $\{\mathcal{F}_t^Y\}$–Brownian motion. Moreover, the Kalman-Bucy filtering theory implies that λ is a specific functional of β.

The novelty of this model, relative to Kyle's model with a single informed trader, is that each trader is trying to estimate the signals of others and each knows that others are trying to estimate his signal, etc. Denote trader i's estimate of v at time t by $\hat{v}_t^i$. This estimate is based on the signal y^i and on having observed the price up to date t. Due to the proposed linear dynamics for the price, observing the price allows the trader to infer Y and therefore, because he also knows X^i, he can infer $Z + X^{-i}$, where $X^{-i} \equiv \sum_{j \neq i} X^j$. Thus, trader i's observation process is

$$dZ_t + dX_t^{-i} = (N-1)\alpha(t)S_t \, dt + \beta(t)y^{-i} \, dt + dZ_t, \qquad (33)$$

where $y^{-i} \equiv \sum_{j \neq i} y^j$, and observing this is equivalent to observing a process with dynamics $\beta(t)y^{-i} \, dt + dZ_t$. Hence calculating $\widehat{y^{-i}}$ is again a simple Gaussian filtering problem, and trader i's estimate of the value is then given by $\hat{v}_t^i = y^i + \widehat{y^{-i}}$. The innovation process for trader i (up to scaling by $1/\sigma_z$) is W^i defined by

$$dW_t^i = \beta(t) \left[y^{-i} - \widehat{y_t^{-i}} \right] dt + dZ_t = \beta(t) \left[v - \hat{v}_t^i \right] dt + dZ_t. \qquad (34)$$

It is worthwhile to point out that the simplicity of the filtering problems is due to the assumption that each trader plays a strategy of the form (30). Given the results of the single-trader model, it might have been more natural to guess a strategy of the form $\theta_t^i = \eta(t) \left[\hat{v}_t^i - S_t \right]$. However, to start with such a guess would make the analysis of the filtering impossible. To compute $\hat{v}_t^i$, we would need to know the dynamics of $\hat{v}_t^j$ for all $j \neq i$, because these variables would appear in the observation process of trader i. However, to know the dynamics of $\hat{v}_t^j$, we would need to know the dynamics of $\hat{v}_t^i$, because this would appear in the observation process of trader j. This circularity is known in economics (cf. [43]) as the "forecasting the forecasts of others" problem. The circularity does not arise when trading strategies are specified as functions of signals rather than as functions of estimates. However, the existence of an equilibrium with strategies of the form (30) is something that requires verification. Foster and Viswanathan first showed that this approach works in the discrete-time version of the model they studied in [21], and BCW extend this to continuous time. Moreover, BCW show that in equilibrium it is indeed true that $\theta_t^i = \eta(t) \left[\hat{v}_t^i - S_t \right]$ for some function η, as I have suggested one might conjecture.

The control problems of the informed traders are not as simple as the filtering problems. Assuming absolutely continuous strategies, the objective function of trader i is

$$E \left[\int_0^1 (v - S_t)\theta_t^i \, dt \, \middle| \, y^i \right] = E \left[\int_0^1 (\hat{v}_t^i - S_t)\theta_t^i \, dt \right]. \qquad (35)$$

The trader's strategy does not influence his estimate $\hat{v}_t^i$ of the asset value. As mentioned above, the state variable $\hat{v}_t^i$ evolves as

$$d\hat{v}_t^i = \gamma(t) \, dW_t^i, \qquad (36)$$

where γ is a function that is given to us by filtering theory (as a functional of β) and W^i is the innovation process defined in (34). However, the trader's strategy does affect the price S_t as specified in (31). Note that equation (32) for dY must hold in equilibrium, but we cannot assume it here, because each trader has the option to deviate from his equilibrium strategy, and we must prove that such deviations are not optimal. We assume that all traders $j \neq i$ play strategies of the form (30). Thus, (31) implies

$$
dS_t = \lambda(t) \left[\theta_t^i \, dt + \sum_{j \neq i} \theta_t^j \, dt + dZ_t \right]
$$

$$
= \lambda(t) \left[\theta_t^i \, dt + \sum_{j \neq i} \left\{ \alpha(t) S_t \, dt + \beta(t) y^j \, dt \right\} + dZ_t \right]
$$

$$
= \lambda(t) \left[\theta_t^i \, dt + (N-1)\alpha(t) S(t) \, dt + \beta(t) y^{-i} \, dt + dZ_t \right].
$$

Now substituting from (34) we have

$$
dS_t = \lambda(t) \left[\theta_t^i \, dt + (N-1)\alpha(t) S(t) t \, dt + \beta(t) \widehat{y_t^{-i}} \, dt + dW_t^i \right]
$$
$$
= \lambda(t) \left[\theta_t^i \, dt + (N-1)\alpha(t) S(t) t \, dt + \beta(t) \{ \hat{v}_t^i - y^i \} \, dt + dW_t^i \right] \quad (37)
$$

The objective is to be maximized subject to the state dynamics (36) and (37) over all processes θ^i adapted to the trader's filtration. Note that the objective function (35) and state dynamics (36) and (37) define a Markovian control problem involving a single Brownian motion W^i.

A key characteristic of the control problem, as in the single-trader model, is that both the instantaneous reward $(\hat{v}_t^i - S_t)\theta_t^i$ and the state variable dynamics are linear in the control θ_t^i. This implies that the control problem has a certain degeneracy. In order for the HJB equation to be satisfied, the coefficient of θ_t^i in the maximization problem must be zero and the remaining terms in the problem must add to zero. Letting $J(t, S, \hat{v}^i)$ denote the value function, setting the coefficient of θ_t^i to be zero yields

$$
\hat{v}^i - S + \lambda \frac{\partial J}{\partial S} = 0. \quad (38)
$$

The condition that the remaining terms sum to zero is

$$
\frac{\partial J}{\partial t} + \lambda \{ (N-1)\alpha S + \beta[\hat{v}^i - y^i] \, dt \} \frac{\partial J}{\partial S}
$$
$$
+ \frac{\sigma_z^2}{2} \left(\lambda^2 \frac{\partial^2 J}{\partial S^2} + 2\lambda\gamma \frac{\partial^2 J}{\partial S \partial \hat{v}^i} + \gamma^2 \frac{\partial^2 J}{\partial (\hat{v}^i)^2} \right) = 0. \quad (39)
$$

Differentiating equation (26) in S yields a pde for $\partial J/\partial S$. However, the derivatives of $\partial J/\partial S$ can be calculated in terms of λ and its derivative from (8), and substituting these expressions into the pde for $\partial J/\partial S$ eliminates the derivatives of $\partial J/\partial S$ and reduces the pde to the following condition:

$$(S - \hat{v}^i)\frac{d}{dt}\left(\frac{1}{\lambda}\right) + \left(\frac{2N-1}{N}\hat{v}^i - y^i - \frac{2N-2}{N}S\right)\beta = 0. \tag{40}$$

Here we have imposed for the strategies θ^j ($j \neq i$) that $\alpha = -\beta/N$, as was noted is necessary for equilibrium. Equation (40) is a linear restriction on the state variables $(S, \hat{v}^i)$. The usual verification theorem shows indeed that a strategy is optimal if and only if it controls the state variables to satisfy this linear restriction at all times. Thus, because of the local linearity of the problem, the usual first-order condition from the HJB equation does not determine the optimal control, but the optimal control is determined by the HJB equation via this dimensionality reduction. The feasibility of controlling the state variables to satisfy this linear restriction depends of course on the fact that there is only a single Brownian motion driving both state variables.

To obtain a symmetric equilibrium, we need the strategy (30) assumed to be played by traders $j \neq i$ to be optimal for trader i also. Thus, we need this strategy to imply that equation (40) holds at all times. BCW show that there is a unique function β (with $\phi = 0$ in (29) and $\alpha = -\beta/N$ in (30) and with λ the functional of β implied by the Kalman-Bucy filtering theory) for which this is true. Specifically, they show that the equilibrium conditional variance of v given $\{\mathcal{F}_t^Y\}$ is obtained from the inverse of the incomplete gamma function, and the other components of the equilibrium are simple functions of this conditional variance.

An important characteristic of the equilibrium is that the depth of the market reduces to zero at the terminal date, due to a relatively large degree of asymmetric information remaining near the end of the trading period. This contrasts with the single-trader Kyle model with a normal distribution, in which the depth is constant over time and the asymmetric information disappears linearly in time. BCW also show that there is no linear equilibrium if the insiders' signals are perfectly correlated.

References

1. Admati, A., Pfleiderer, P.: A theory of intraday patterns: volume and price variability. Review of Financial Studies, **1**, 3–40 (1988)
2. Back, K.: Insider trading in continuous time. Review of Financial Studies, **5**, 387–409 (1992)
3. Back, K.: Asymmetric information and options. Review of Financial Studies, **6**, 435–472 (1993)
4. Back, K., Baruch, S. Information in securities markets: Kyle meets Glosten and Milgrom. Econometrica (forthcoming)
5. Back, K., Pedersen, H.: Long-lived information and intraday patterns. Journal of Financial Markets, **1**, 385–402 (1998)
6. Back, K., Cao, H., Willard, G.: Imperfect competition among informed traders. Journal of Finance, **55**, 2117–2155 (2000)
7. Baruch, S.: Insider trading and risk aversion. Journal of Financial Markets, **5**, 451–464 (2002)

8. Basak, S., Croitoru, B.: On the role of arbitrageurs in rational markets, preprint (2003)
9. Baudoin, F.: The financial value of weak information on a financial market. In: Paris-Princeton Lectures on Mathematical Finance 2002, Springer (2003)
10. Brunnermeier, M. K.: Asset Pricing under Asymmetric Information: Bubbles, Crashes, Technical Analysis, and Herding. Oxford University Press (2001)
11. Cho, K.-H.: Continuous auctions and insider trading: uniqueness and risk aversion. Finance and Stochastics **7**, 47–71 (2003)
12. Cox, J., Ingersoll, J., Ross, S.: An Intertemporal general equilibrium model of asset prices, Econometrica, **53**, 363–384 (1985)
13. David, A.: Business cycle risk and the equity premium, Ph.D. dissertation, University of California at Los Angeles (1993)
14. David, A: Fluctuating confidence in stock markets: implications for returns and volatility, Journal of Financial and Quantitative Analysis, **32**, 427–462 (1997)
15. David, A., Veronesi, P: Option prices with uncertain fundamentals: theory and evidence on the dynamics of implied volatilities, preprint (2002)
16. Detemple, J.: Asset pricing in a production economy with incomplete information, Journal of Finance, **41**, 383–391 (1986)
17. Detemple, J.: Further results on asset pricing with incomplete information, Journal of Economic dynamics and Control, **15**, 425–454 (1991)
18. Detemple, J., Murthy, S.: Equilibrium asset prices and no-arbitrage with portfolio constraints, Review of Financial Studies, **10**, 1133–1174 (1997)
19. Dothan, M. U., Feldman, D.: Equilibrium interest rates and multiperiod bonds in a partially observable economy, Journal of Finance, **41**, 369–382 (1986)
20. Duffie, J. D.: Dynamic Asset Pricing Theory, Princeton University Press (2001)
21. Foster, F. D., Viswanathan, S.: Strategic trading when agents forecast the forecasts of others, Journal of Finance, **51**, 1437–1478 (1996)
22. Fujisaki, M., Kallianpur, G., Kunita, H.: Stochastic differential equations for the non-linear filtering problem, Osaka Journal of Mathematics, **9**, 19–40 (1972)
23. Gennotte, G.: Optimal portfolio choice under incomplete information, Journal of Finance, **41**, 733–746 (1986)
24. Glosten, L., Milgrom, P.: Bid, ask, and transaction prices in a specialist market with heterogeneously informed traders, Journal of Financial Economics, **13**, 71–100 (1985)
25. Grorud, A., Pontier, M.: Insider trading in a continuous time market model, International Journal of Theoretical and Applied Finance, **1**, 331–347 (1988)
26. Grorud, A., Pontier, M.: Asymmetrical information and incomplete markets, International Journal of Theoretical and Applied Finance, **4**, 285–302 (2001)
27. Grossman, S.: An introduction to the theory of rational expectations under asymmetric information, Review of Economic Studies, **31**, 573–585 (1981)
28. Grossman, S., Stiglitz, J.: On the impossibility of informationally efficient markets, American Economic Review, **70**, 393–408 (1980)
29. Harsanyi, J.: Games with incomplete information played by 'Bayesian' players, part III: the basic probability distribution of the game, Management Science, **14**, 486–502 (1968)

30. Hellwig, M.: On the aggregation of information in competitive markets, Journal of Economic Theory, **26**, 279–312 (1980)
31. Imkeller, P.: Malliavin's calculus in insider models: additional utility and free lunches, Mathematical Finance, **13**, 153–169 (2003)
32. Jacod, J.: Groissement initial, hypothèse h' et theórème de Girsanov, in Séminaire de Calcul Stochastique 1982-1983, Lecture Notes in Mathematics 1118, 15–35, Springer (1985)
33. Kallianpur, G.: Stochastic Filtering Theory, Springer-Verlag (1980).
34. Karatzas, I., Pikovsky, I.: Anticipative portfolio optimization, Advances in Applied Probability, **28**, 1095–1122 (1996)
35. Kyle, A. S.: Continuous auctions and insider trading, Econometrica, **53**, 1315–1335 (1985)
36. Jouni, E., Napp, C.: Consensus consumer and intertemporal asset pricing under heterogeneous beliefs, preprint (2003)
37. Lasserre, G.: Asymmetric information and imperfect competition in a continuous-time multivariate security model, Finance and Stochastics (forthcoming)
38. León, J., Navarro, R., Nualart D.: An anticipating calculus approach to the utility maximization of an insider, Mathematical Finance, **13**, 171–185 (2003)
39. Lucas, R.: Expectations and the neutrality of money, Journal of Economic Theory, **4**, 103–124 (1972)
40. Lucas, R.: Asset prices in an exchange economy, Econometrica, **40**, 1429–1444. (1978)
41. Rogers, L. C. G., Williams, D.: Diffusions, Markov Processes and Martingales, Vol. 2, Cambridge University Press (2000)
42. Scheinkman, J., Xiong, W.: Speculative bubbles and overconfidence, Journal of Political Economy (forthcoming)
43. Townsend, R.: Forecasting the forecasts of others, Journal of Political Economy, **91**, 546-588 (1983)
44. Veronesi, P.: Stock market overreaction to bad news in good times: a rational expectations model, Review of Financial Studies, **12**, 975–1007 (1999)
45. Veronesi, P.: How does information quality affect stock returns? Journal of Finance, **55**, 807–837 (2000)
46. Wang, J.: A model of intertemporal asset prices under asymmetric information, Review of Economic Studies, **60**, 249–282 (1993)
47. Wonham, W.: Some applications of stochastic differential equations to optimal nonlinear filtering, SIAM Journal of Control, Series A, **2**, 347–369 (1965)
48. Ziegler, A.: Incomplete Information and Heterogeneous Beliefs in Continuous-Time Finance, Springer (2003)

Modeling and Valuation of Credit Risk

Tomasz R. Bielecki[1,*], Monique Jeanblanc[2], and Marek Rutkowski[3,**]

[1] Department of Applied Mathematics
Illinois Institute of Technology
Chicago, USA
bielecki@iit.edu
[2] Equipe d'Analyse et Probabilités
Université d'Évry-Val d'Essonne
Évry, France
[3] Faculty of Mathematics and Information Science
Warsaw University of Technology and
Institute of Mathematics of the Polish Academy of Sciences
Warszawa, Poland

1 Introduction

The goal of this work is to present a survey of recent developments in the area of mathematical modeling of *credit risk* and *credit derivatives*. Credit risk embedded in a financial transaction is the risk that at least one of the parties involved in the transaction will suffer a financial loss due to decline in the creditworthiness of the counter-party to the transaction, or perhaps of some third party. For example:

- A holder of a corporate bond bears a risk that the (market) value of the bond will decline due to decline in credit rating of the issuer.
- A bank may suffer a loss if a bank's debtor defaults on payment of the interest due and (or) the principal amount of the loan.
- A party involved in a trade of a credit derivative, such as a credit default swap (CDS), may suffer a loss if a reference credit event occurs.
- The market value of individual tranches constituting a collateralized debt obligation (CDO) may decline as a result of changes in the correlation between the default times of the underlying defaultable securities (i.e., of the collateral).

The most extensively studied form of credit risk is the *default risk* – that is, the risk that a counterparty in a financial contract will not fulfil a contractual commitment to meet her/his obligations stated in the contract. For

[*] The first author was supported in part by NSF Grant 0202851.
[**] The third author was supported by KBN Grant PBZ-KBN-016/P03/1999.

this reason, the main tool in the area of credit risk modeling is a judicious specification of the random time of default. A large part of the present text will be devoted to this issue, examined from different perspectives by various authors.

Our main goal is to present the most important mathematical tools that are used for the arbitrage valuation of defaultable claims, which are also known under the name of credit derivatives. We decided to examine the important issue of hedging credit risk in a separate work (see the forthcoming paper by Bielecki et al. (2004)).

These lecture notes are organized as follows. First, in Chapter 1, we provide a concise summary of the main developments within the so-called *structural approach* to modeling and valuation of credit risk. This was historically the first approach used in this area, and it goes back to the fundamental papers by Black and Scholes (1973) and Merton (1974). Since the main object to be modeled in the structural approach is the process representing the total value of the firm's assets (for instance, the issuer of a corporate bond), this methodology is frequently termed the *value-of-the-firm approach* in financial literature.

Chapter 2 is devoted to the *intensity-based approach*, which is also known as the *reduced-form approach*. This approach is purely probabilistic in nature and, technically speaking, it has a lot in common with the reliability theory. Since, typically, the value of the firm is not modeled, the specification of the default time is directly related to the likelihood of default event conditional on an information flow. More specifically, the default risk is reflected either by a deterministic default intensity function, or, more generally, by a stochastic intensity.

The final chapter provides an introduction to the area of modeling dependent credit migrations and defaults. Arguably, this is the most important and the most difficult research area with regard to credit risk and credit derivatives. We describe the case of conditionally independent default time, the copula-based approach, as well as the Jarrow and Yu (2001) approach to the modeling of dependent stochastic intensities. We conclude by summarizing one of the approaches that were recently developed for the purpose of modeling term structure of corporate interest rates.

Acknowledgments

Since this is a survey article, we do not provide here, with rare exceptions, the proofs of mathematical results that are presented in the text. For the demonstrations, the interested reader is referred to numerous original papers, as well as recent monographs, which are collected in the (non-exhaustive) list of references. Let us only mention, that the proofs of most results can be found in Bielecki and Rutkowski (2002) and Jeanblanc and Rutkowski (2000, 2001).

Finally, it should be acknowledged that some results (especially within the intensity-based approach presented in Chapter 2) were obtained independently by various authors, who worked under different sets of assumptions and within distinct setups, and thus we decided not to provide specific credentials in most cases. We hope that respective authors and the readers will be understanding in this regard.

2 Structural Approach

In this chapter, we present the *structural approach* to modeling credit risk (as already mentioned in the introduction, it is also known as the *value-of-the-firm approach*). This methodology directly refers to economic fundamentals, such as the capital structure of a company, in order to model credit events (a default event, in particular). As we shall see in what follows, the two major driving concepts in the structural modeling are: the total value of the firm's assets and the default triggering barrier.

2.1 Basic Assumptions

We fix a finite horizon date $T^* > 0$, and we suppose that the underlying probability space $(\Omega, \mathcal{F}, \mathbb{P})$, endowed with some (reference) filtration $\mathbb{F} = (\mathcal{F}_t)_{0 \le t \le T^*}$, is sufficiently rich to support the following objects:

- The *short-term interest rate* process r, and thus also a default-free term structure model.
- The *firm's value process* V, which is interpreted as a model for the total value of the firm's assets.
- The *barrier process* v, which will be used in the specification of the default time τ.
- The *promised contingent claim* X representing the firm's liabilities to be redeemed at maturity date $T \le T^*$.
- The process C, which models the *promised dividends,* i.e., the liabilities stream that is redeemed continuously or discretely over time to the holder of a defaultable claim.
- The *recovery claim* $\tilde{X}$ representing the recovery payoff received at time T, if default occurs prior to or at the claim's maturity date T.
- The *recovery process* Z, which specifies the recovery payoff at time of default, if it occurs prior to or at the maturity date T.

Defaultable Claims

Technical Assumptions

We postulate that the processes V, Z, C and v are progressively measurable with respect to the filtration $\mathbb{F}$, and that the random variables X and $\tilde{X}$ are $\mathcal{F}_T$-measurable. In addition, C is assumed to be a process of finite variation, with $C_0 = 0$. We assume without mentioning that all random objects introduced above satisfy suitable integrability conditions.

Probabilities $\mathbb{P}$ *and* $\mathbb{P}^*$

The probability $\mathbb{P}$ is assumed to represent the *real-world* (or *statistical*) probability, as opposed to the *martingale measure* (also known as the *risk-neutral probability*). The latter probability is denoted by $\mathbb{P}^*$ in what follows.

Default Time

Let us denote by τ the random time of default. It is essential to emphasize that the various approaches to valuing and hedging of defaultable securities differ between themselves with regard to the ways in which the default event – and thus also the default time τ – are modeled. In the structural approach, the default time τ will be typically defined in terms of the value process V and the barrier process v. We set

$$\tau = \inf\{t > 0 : t \in \mathcal{T} \text{ and } V_t \leq v_t\}$$

with the usual convention that the infimum over the empty set equals $+\infty$. The set $\mathcal{T}$ is assumed to be a Borel measurable subset of the time interval $[0, T]$ (or $[0, \infty)$ in the case of perpetual claims). In particular, depending on the model and the purpose we may have that $\mathcal{T} = \{T\}$ as in the classical Merton model, or that $\mathcal{T} = \{T_1, T_2, \ldots, T_n\}$ if default can only happen (or, rather, can be declared) at some discrete time instants, such as the coupon payment dates. In most cases we have either $\mathcal{T} = [0, T]$ or $\mathcal{T} = [0, \infty)$. In classic structural models, the default time τ is given by the formula:

$$\tau = \inf\{t > 0 : t \in [0, T] \text{ and } V_t \leq \bar{v}(t)\},$$

where $\bar{v} : [0, T] \to \mathbb{R}_+$ is some deterministic function, termed the *barrier*.

Predictability of Default Time

Typically, the random variable τ is defined in such a way that it is an $\mathbb{F}$-stopping time. Since the underlying filtration $\mathbb{F}$ in most structural models is generated by a standard Brownian motion, τ will be an $\mathbb{F}$-predictable stopping time (as any stopping time with respect to a Brownian filtration). The latter property means that within the framework of the structural approach there exists a sequence of increasing stopping times announcing the default time; in this sense, the default time can be forecasted with some degree of certainty.

In some structural models, the value process V is assumed to follow a jump diffusion, in which case the default time is not predicable with respect to the reference filtration, in general. Some other structural models are constructed so that the barrier process is not adapted to the reference filtration $\mathbb{F}$, neither it is adapted to some 'enlarged' filtration, denoted by $\mathbb{G}$ in the next chapter. Consequently, τ is not predictable with respect to $\mathbb{G}$ in these models. Also, in general, this will be the case if the value process is a discontinuous semimartingale.

Remarks. Later in this article, we shall discuss the so-called intensity-based approach to modeling credit risk. In this alternative approach, the default time will not be a predictable stopping time with respect to the 'enlarged' filtration, in general. In typical examples, the filtration $\mathbb{G}$ will encompass some Brownian filtration $\mathbb{F}$, but $\mathbb{G}$ will be strictly larger than $\mathbb{F}$. At the intuitive level, in the intensity-based approach the occurrence of the default event comes as a total surprise. For any date t, the default intensity γ_t will specify the conditional probability of the occurrence of default over an infinitesimally small time interval $[t, t + dt]$.

Recovery Rules

If default does not occur before or at time T, the promised claim X is paid in full at time T. Otherwise, depending on the market convention, either (1) the amount $\tilde{X}$ is paid at the maturity date T, or (2) the amount Z_τ is paid at time τ. As a matter of fact, in reality, the recovery payment may also be distributed over time. However, for the modeling purposes it suffices to consider recovery payment only at default time or at maturity, as other possibilities can be reduced to the above by means of forward or backward discounting. In the case when default occurs at maturity, i.e., on the event $\{\tau = T\}$, we postulate that only the recovery payment $\tilde{X}$ is paid. In a general setting, we consider simultaneously both kinds of recovery payoff, and thus a generic defaultable claim is formally defined as a quintuple $(X, C, \tilde{X}, Z, \tau)$.

Remarks. The above notation emphasizes the role of the default time τ in the definition of a generic defaultable claim. Within the structural framework it would be more appropriate to denote the defaultable claim as a sextuple $(X, C, \tilde{X}, Z, V, v)$, since τ is defined in terms of V and v.

Risk-Neutral Valuation Formula

Suppose that our financial market model is arbitrage-free, in the sense that there exists a *martingale measure (risk-neutral probability)* $\mathbb{P}^*$, meaning that price process of any tradeable security, which pays no coupons or dividends, becomes an $\mathbb{F}$-martingale under $\mathbb{P}^*$, when discounted by the *savings account* B, given as

$$B_t = \exp \left(\int_0^t r_u \, du \right).$$

We introduce the jump process $H_t = \mathbb{1}_{\{\tau \le t\}}$, and we denote by D the process that models all cash flows received by the owner of a defaultable claim. Let us denote

$$X^d(T) = X \mathbb{1}_{\{\tau > T\}} + \tilde{X} \mathbb{1}_{\{\tau \le T\}}.$$

Definition 2.1. The *dividend process* D of a defaultable contingent claim $(X, C, \tilde{X}, Z, \tau)$, which settles at time T, equals

$$D_t = X^d(T)\mathbb{1}_{\{t \geq T\}} + \int_{]0,t]} (1 - H_u) \, dC_u + \int_{]0,t]} Z_u \, dH_u.$$

It is apparent that D is a process of finite variation, and

$$\int_{]0,t]} (1 - H_u) \, dC_u = \int_{]0,t]} \mathbb{1}_{\{\tau > u\}} \, dC_u = C_{\tau-} \mathbb{1}_{\{\tau \leq t\}} + C_t \mathbb{1}_{\{\tau > t\}}.$$

Note that if default occurs at some date t, the promised dividend $C_t - C_{t-}$, which is due to be paid at this date, is not received by the holder of a default-able claim. Furthermore, if we set $\tau \wedge t = \min\{\tau, t\}$ then

$$\int_{]0,t]} Z_u \, dH_u = Z_{\tau \wedge t} \mathbb{1}_{\{\tau \leq t\}} = Z_\tau \mathbb{1}_{\{\tau \leq t\}}.$$

Remarks. In principle, the promised payoff X could be incorporated into the promised dividends process C. However, this would be inconvenient, since in practice the recovery rules concerning the promised dividends C and the promised claim X are different, in general. For instance, in the case of a defaultable coupon bond, it is frequently postulated that in case of default the future coupons are lost, but a strictly positive fraction of the face value is usually received by the bondholder.

We are in the position to define the ex-dividend price S_t of a defaultable claim. At any time t, the random variable S_t represents the current value of all future cash flows associated with a given defaultable claim.

Definition 2.2. For any date $t \in [0, T[$, the *ex-dividend price* of the default-able claim $(X, C, \tilde{X}, Z, \tau)$ is given as

$$S_t = B_t \, \mathbb{E}_{\mathbb{P}^*} \left(\int_{]t,T]} B_u^{-1} \, dD_u \,\Big|\, \mathcal{F}_t \right). \tag{1}$$

In addition, we always set $S_T = X^d(T)$. In the next chapter, we shall use the same definition of the price, but with the probability measure $\mathbb{P}^*$ substituted with $\mathbb{Q}^*$, and the filtration $\mathbb{F}$ replaced by $\mathbb{G}$.

It needs to be emphasized that we are not concerned here with the issue of completeness of our market. In particular, we are not concerned in this article whether the relevant pricing measures are unique or not. For the study of pricing and hedging of credit risk in incomplete markets we refer to Bielecki et al. (2004a) and (2004b).

Defaultable Zero-Coupon Bond

Assume that $C \equiv 0$, $Z \equiv 0$ and $X = L$ for some positive constant $L > 0$. Then the value process S represents the arbitrage price of a *defaultable zero-coupon bond* (also known as the *corporate discount bond*) with the face value L and recovery at maturity only. In general, the price $D(t, T)$ of such a bond equals

$$D(t, T) = B_t \, \mathbb{E}_{\mathbb{P}^*} \big(B_T^{-1} (L \mathbb{1}_{\{\tau > T\}} + \tilde{X} \mathbb{1}_{\{\tau \leq T\}}) \,\big|\, \mathcal{F}_t \big).$$

It is convenient to rewrite the last formula as follows:

$$D(t, T) = L B_t \, \mathbb{E}_{\mathbb{P}^*} \big(B_T^{-1} (\mathbb{1}_{\{\tau > T\}} + \delta(T) \mathbb{1}_{\{\tau \leq T\}}) \,\big|\, \mathcal{F}_t \big),$$

where the random variable $\delta(T) = \tilde{X}/L$ represents the so-called *recovery rate upon default*. It is natural to assume that $0 \leq \tilde{X} \leq L$ so that $\delta(T)$ satisfies $0 \leq \delta(T) \leq 1$. Alternatively, we may re-express the bond price as follows:

$$D(t, T) = L \Big(B(t, T) - B_t \, \mathbb{E}_{\mathbb{P}^*} \big(B_T^{-1} w(T) \mathbb{1}_{\{\tau \leq T\}} \,\big|\, \mathcal{F}_t \big) \Big),$$

where

$$B(t, T) = B_t \, \mathbb{E}_{\mathbb{P}^*} \big(B_T^{-1} \,\big|\, \mathcal{F}_t \big)$$

is the price of a unit default-free zero-coupon bond, and $w(T) = 1 - \delta(T)$ is the *writedown rate upon default*. Generally speaking, the time-t value of a corporate bond depends on the joint probability distribution under $\mathbb{P}^*$ of the three-dimensional random variable $(B_T, \delta(T), \tau)$ or, equivalently, $(B_T, w(T), \tau)$.

Example 2.1. Merton (1974) postulates that the recovery payoff upon default equals $\tilde{X} = V_T$, where the random variable V_T is the firm's value at maturity date T of a corporate bond. Consequently, the random recovery rate upon default equals $\delta(T) = V_T/L$, and the writedown rate upon default equals $w(T) = 1 - V_T/L$.

Expected Writedowns

For simplicity, we assume that the savings account B is non-random – that is, the short-term rate r is deterministic. Then the price of a default-free zero-coupon bond equals $B(t, T) = B_t B_T^{-1}$, and the price of a zero-coupon corporate bond satisfies

$$D(t, T) = L_t (1 - w^*(t, T)),$$

where $L_t = L B(t, T)$ is the present value of future liabilities, and $w^*(t, T)$ is the *conditional expected writedown rate* under $\mathbb{P}^*$. It is given by the following equality:

$$w^*(t, T) = \mathbb{E}_{\mathbb{P}^*} \big(w(T) \mathbb{1}_{\{\tau \leq T\}} \,\big|\, \mathcal{F}_t \big).$$

Notice that we may set $w(T) = 0$ on the event $\{\tau > T\}$.

The *conditional expected writedown rate upon default* equals, under $\mathbb{P}^*$,

$$w_t^* = \frac{\mathbb{E}_{\mathbb{P}^*} \big(w(T) \mathbb{1}_{\{\tau \leq T\}} \,\big|\, \mathcal{F}_t \big)}{\mathbb{P}^* \{\tau \leq T \,|\, \mathcal{F}_t\}} = \frac{w^*(t, T)}{p_t^*},$$

where $p_t^* = \mathbb{P}^*\{\tau \leq T \mid \mathcal{F}_t\}$ is the *conditional risk-neutral probability of default*. Finally, let $\delta_t^* = 1 - w_t^*$ be the *conditional expected recovery rate upon default* under $\mathbb{P}^*$. In terms of p_t^*, δ_t^* and p_t^*, we obtain

$$D(t,T) = L_t(1 - p_t^*) + L_t p_t^* \delta_t^* = L_t(1 - p_t^* w_t^*).$$

If the random variables $w(T)$ and τ are conditionally independent with respect to the σ-field $\mathcal{F}_t$ under $\mathbb{P}^*$, then we have $w_t^* = \mathbb{E}_{\mathbb{P}^*}(w(T) \mid \mathcal{F}_t)$.

Example 2.2. In practice, it is common to assume that the recovery rate is non-random. Let the recovery rate $\delta(T)$ be constant, specifically, $\delta(T) = \delta$ for some real number δ. In this case, the writedown rate $w(T) = w = 1 - \delta$ is non-random as well. Then $w^*(t,T) = w p_t^*$ and $w_t^* = w$ for every $0 \leq t \leq T$. Furthermore, the price of a defaultable bond has the following representation

$$D(t,T) = L_t(1 - p_t^*) + \delta L_t p_t^* = L_t(1 - w p_t^*).$$

We shall return to various recovery schemes later in the text.

2.2 Classic Structural Models

Classic structural models are based on the assumption that the risk-neutral dynamics of the value process of the assets of the firm V are given by the SDE:

$$dV_t = V_t\left((r - \kappa)\,dt + \sigma_V\,dW_t^*\right), \quad V_0 > 0,$$

where κ is the constant payout (dividend) ratio, and the process W^* is a standard Brownian motion under the martingale measure $\mathbb{P}^*$.

Merton's Model

We present here the classic model due to Merton (1974).

Basic assumptions. A firm has a single liability with promised terminal payoff L, interpreted as the zero-coupon bond with maturity T and face value $L > 0$. The ability of the firm to redeem its debt is determined by the total value V_T of firm's assets at time T. Default may occur at time T only, and the default event corresponds to the event $\{V_T < L\}$. Hence, the stopping time τ equals

$$\tau = T\mathbb{1}_{\{V_T < L\}} + \infty\mathbb{1}_{\{V_T \geq L\}}.$$

Moreover $C = 0$, $Z = 0$, and

$$X^d(T) = V_T\mathbb{1}_{\{V_T < L\}} + L\mathbb{1}_{\{V_T \geq L\}}$$

so that $\tilde{X} = V_T$. In other words, the payoff at maturity equals

$$D_T = \min(V_T, L) = L - \max(L - V_T, 0) = L - (L - V_T)^+.$$

The latter equality shows that the valuation of the corporate bond in Merton's setup is equivalent to the valuation of a European put option written on the firm's value with strike equal to the bond's face value. Let $D(t,T)$ be the price at time $t < T$ of the corporate bond. It is clear that the value $D(V_t)$ of the firm's debt equals

$$D(V_t) = D(t,T) = L\,B(t,T) - P_t,$$

where P_t is the price of a put option with strike L and expiration date T. It is apparent that the value $E(V_t)$ of the firm's equity at time t equals

$$E(V_t) = V_t - D(V_t) = V_t - LB(t,T) + P_t = C_t,$$

where C_t stands for the price at time t of a call option written on the firm's assets, with the strike price L and the exercise date T. To justify the last equality above, we may also observe that at time T we have

$$E(V_T) = V_T - D(V_T) = V_T - \min(V_T, L) = (V_T - L)^+.$$

We conclude that the firm's shareholders are in some sense the holders of a call option on the firm's assets.

Merton's Formula

Using the option-like features of a corporate bond, Merton (1974) derived a closed-form expression for its arbitrage price. Let N denote the standard Gaussian cumulative distribution function:

$$N(x) = \frac{1}{\sqrt{2\pi}} \int_{-\infty}^{x} e^{-u^2/2}\,du, \quad \forall x \in \mathbb{R}.$$

Proposition 2.1. *For every $0 \le t < T$ the value $D(t,T)$ of a corporate bond equals*

$$D(t,T) = V_t e^{-\kappa(T-t)} N\big(-d_+(V_t, T-t)\big) + L\,B(t,T)N\big(d_-(V_t, T-t)\big)$$

where

$$d_\pm(V_t, T-t) = \frac{\ln(V_t/L) + \big(r - \kappa \pm \frac{1}{2}\sigma_V^2\big)(T-t)}{\sigma_V\sqrt{T-t}}.$$

The unique replicating strategy for a defaultable bond involves holding at any time $0 \le t < T$ the $\phi_t^1 V_t$ units of cash invested in the firm's value and $\phi_t^2 B(t,T)$ units of cash invested in default-free bonds, where

$$\phi_t^1 = e^{-\kappa(T-t)} N\big(-d_+(V_t, T-t)\big)$$

and

$$\phi_t^2 = \frac{D(t,T) - \phi_t^1 V_t}{B(t,T)} = LN\big(d_-(V_t, T-t)\big).$$

Credit Spreads

For notational simplicity, we set $\kappa = 0$. Then Merton's formula becomes:

$$D(t,T) = LB(t,T)\big(N(d - \sigma_V\sqrt{T-t}) + \Gamma_t N(-d)\big),$$

where we denote $\Gamma_t = V_t/LB(t,T)$ and

$$d = d(V_t, T - t) = \frac{\ln(V_t/L) + (r + \sigma_V^2/2)(T - t)}{\sigma_V\sqrt{T-t}}.$$

Since $LB(t,T)$ represents the current value of the face value of the firm's debt, the quantity Γ_t can be seen as a proxy of the asset-to-debt ratio $V_t/D(t,T)$. It can be easily verified that the inequality $D(t,T) < LB(t,T)$ is valid. This property is equivalent to the positivity of the corresponding credit spread (see below).

Observe that in the present setup the continuously compounded yield $r(t,T)$ at time t on the T-maturity Treasury zero-coupon bond is constant, and equal to the short-term rate r. Indeed, we have

$$B(t,T) = e^{-r(t,T)(T-t)}B(t,T) = e^{-r(T-t)}.$$

Let us denote by $r^d(t,T)$ the continuously compounded yield on the corporate bond at time $t < T$, so that

$$D(t,T) = Le^{-r^d(t,T)(T-t)}.$$

From the last equality, it follows that

$$r^d(t,T) = -\frac{\ln D(t,T) - \ln L}{T - t}.$$

For $t < T$ the *credit spread* $S(t,T)$ is defined as the excess return on a defaultable bond:

$$S(t,T) = r^d(t,T) - r(t,T) = \frac{1}{T-t}\ln\frac{LB(t,T)}{D(t,T)}.$$

In Merton's model, we have

$$S(t,T) = -\frac{\ln\big(N(d - \sigma_V\sqrt{T-t}) + \Gamma_t N(-d)\big)}{T-t} > 0.$$

This agrees with the well-known fact that risky bonds have an expected return in excess of the risk-free interest rate. In other words, the yields on corporate bonds are higher than yields on Treasury bonds with matching notional amounts. Notice, however, when t tends to T, the credit spread in Merton's model tends either to infinity or to 0, depending on whether $V_T < L$ or $V_T > L$. Formally, if we define the *forward short spread at time T* as

$$FSS_T = \lim_{t \uparrow T} S(t, T)$$

then

$$FSS_T(\omega) = \begin{cases} 0, & \text{if } \omega \in \{V_T > L\}, \\ \infty, & \text{if } \omega \in \{V_T < L\}. \end{cases}$$

Black and Cox Model

By construction, Merton's model does not allow for a premature default, in the sense that the default may only occur at the maturity of the claim. Several authors put forward structural-type models in which this restrictive and unrealistic feature is relaxed. In most of these models, the time of default is given as the *first passage time* of the value process V to either a deterministic or a random barrier. In principle, the bond's default may thus occur at any time before or on the maturity date T. The challenge is to appropriately specify the lower threshold v, the recovery process Z, and to explicitly evaluate the conditional expectation that appears on the right-hand side of the risk-neutral valuation formula

$$S_t = B_t \, \mathbb{E}_{\mathbb{P}^*} \left(\int_{]t,T]} B_u^{-1} \, dD_u \, \Big| \, \mathcal{F}_t \right),$$

which is valid for $t \in [0, T[$. As one might easily guess, this is a non-trivial mathematical problem, in general. In addition, the practical problem of the lack of direct observations of the value process V largely limits the applicability of the first-passage-time models based on the value of the firm process V.

Corporate Zero-Coupon Bond

Black and Cox (1976) extend Merton's (1974) research in several directions, by taking into account such specific features of real-life debt contracts as: safety covenants, debt subordination, and restrictions on the sale of assets. Following Merton (1974), they assume that the firm's stockholders receive continuous dividend payments, which are proportional to the current value of firm's assets. Specifically, they postulate that

$$dV_t = V_t \big((r - \kappa) \, dt + \sigma_V \, dW_t^* \big), \quad V_0 > 0,$$

where the constant $\kappa \geq 0$ represents the payout ratio, and $\sigma_V > 0$ is the constant volatility. The short-term interest rate r is assumed to be constant.

Safety covenants. Safety covenants provide the firm's bondholders with the right to force the firm to bankruptcy or reorganization if the firm is doing poorly according to a set standard. The standard for a poor performance is set by Black and Cox in terms of a time-dependent deterministic barrier

$\bar{v}(t) = Ke^{-\gamma(T-t)}$, $t \in [0, T[$, for some constant $K > 0$. As soon as the value of firm's assets crosses this lower threshold, the bondholders take over the firm. Otherwise, default takes place at debt's maturity or not depending on whether $V_T < L$ or not.

Default time. Let us set

$$v_t = \begin{cases} \bar{v}(t), \text{ for } t < T, \\ L, \qquad\qquad \text{ for } t = T. \end{cases}$$

The default event occurs at the first time $t \in [0, T]$ at which the firm's value V_t falls below the level v_t, or the default event does not occur at all. The default time equals ($\inf \emptyset = +\infty$)

$$\tau = \inf \{ t \in [0, T] : V_t < v_t \}.$$

The recovery process Z and the recovery payoff $\tilde{X}$ are proportional to the value process: $Z \equiv \beta_2 V$ and $\tilde{X} = \beta_1 V_T$ for some constants $\beta_1, \beta_2 \in [0, 1]$. The case examined by Black and Cox (1976) corresponds to $\beta_1 = \beta_2 = 1$.

To summarize, we consider the following model:

$$X = L, \ C \equiv 0, \ Z \equiv \beta_2 V, \ \tilde{X} = \beta_1 V_T, \ \tau = \bar{\tau} \wedge \hat{\tau},$$

where the *early default time* $\bar{\tau}$ equals

$$\bar{\tau} = \inf \{ t \in [0, T) : V_t < \bar{v}(t) \}$$

and $\hat{\tau}$ stands for Merton's default time: $\hat{\tau} = T \mathbb{1}_{\{V_T < L\}} + \infty \mathbb{1}_{\{V_T \geq L\}}$.

Bond Valuation

Similarly as in Merton's model, it is assumed that the short term interest rate is deterministic and equal to a positive constant r. We postulate, in addition, that $\bar{v}(t) \leq LB(t, T)$ or, more explicitly,

$$Ke^{-\gamma(T-t)} \leq Le^{-r(T-t)}, \quad \forall t \in [0, T],$$

so that, in particular, $K \leq L$. This condition ensures that the payoff to the bondholder at the default time τ never exceeds the face value of debt, discounted at a risk-free rate.

PDE approach. Since the model for the value process V is given in terms of a Markovian diffusion, a suitable partial differential equation can be used to characterize the value process of the corporate bond. Let us write $D(t, T) = u(V_t, t)$. Then the pricing function $u = u(v, t)$ of a defaultable bond satisfies the following PDE:

$$u_t(v, t) + (r - \kappa)v u_v(v, t) + \tfrac{1}{2}\sigma_V^2 v^2 u_{vv}(v, t) - ru(v, t) = 0$$

on the domain

$$\{(v,t) \in \mathbb{R}_+ \times \mathbb{R}_+ : 0 < t < T, \, v > Ke^{-\gamma(T-t)}\},$$

with the boundary condition

$$u(Ke^{-\gamma(T-t)}, t) = \beta_2 Ke^{-\gamma(T-t)}$$

and the terminal condition $u(v,T) = \min(\beta_1 v, L)$.

Probabilistic approach. For any $t < T$ the price $D(t,T) = u(V_t, t)$ of a defaultable bond has the following probabilistic representation, on the set $\{\tau > t\} = \{\bar{\tau} > t\}$

$$D(t,T) = \mathbb{E}_{\mathbb{P}^*} \left(Le^{-r(T-t)} \mathbb{1}_{\{\bar{\tau} \geq T, \, V_T \geq L\}} \, \Big| \, \mathcal{F}_t \right)$$

$$+ \mathbb{E}_{\mathbb{P}^*} \left(\beta_1 V_T e^{-r(T-t)} \mathbb{1}_{\{\bar{\tau} \geq T, \, V_T < L\}} \, \Big| \, \mathcal{F}_t \right)$$

$$+ \mathbb{E}_{\mathbb{P}^*} \left(K\beta_2 e^{-\gamma(T-\bar{\tau})} e^{-r(\bar{\tau}-t)} \mathbb{1}_{\{t < \bar{\tau} < T\}} \, \Big| \, \mathcal{F}_t \right).$$

After default – that is, on the set $\{\tau \leq t\} = \{\bar{\tau} \leq t\}$, we clearly have

$$D(t,T) = \beta_2 \bar{v}(\tau) B^{-1}(\tau,T) B(t,T) = K\beta_2 e^{-\gamma(T-\tau)} e^{r(t-\tau)}.$$

To compute the expected values above, we observe that:

- the first two conditional expectations can be computed by using the formula for the conditional probability $\mathbb{P}^*\{V_s \geq x, \, \tau \geq s \, | \, \mathcal{F}_t\}$,
- to evaluate the third conditional expectation, it suffices employ the conditional probability law of the first passage time of the process V to the barrier $\bar{v}(t)$.

Black and Cox Formula

Before we state the bond valuation result due to Black and Cox (1976), we find it convenient to introduce some notation. We denote

$$\nu = r - \kappa - \tfrac{1}{2}\sigma_V^2,$$

$$\hat{\nu} = \nu - \gamma = r - \kappa - \gamma - \tfrac{1}{2}\sigma_V^2,$$

and $\hat{a} = \hat{\nu}\sigma_V^{-2}$. For the sake of brevity, in the statement of Proposition 2.2 we shall write σ instead of σ_V. As already mentioned, the probabilistic proof of this result is based on the knowledge of the probability law of the first passage time of the geometric (exponential) Brownian motion to an exponential barrier.

Proposition 2.2. *Assume that $\hat{\nu}^2 + 2\sigma^2(r - \gamma) > 0$. Prior to bond's default, that is: on the set $\{\tau > t\}$, the price process $D(t,T) = u(V_t, t)$ of a defaultable bond equals*

$$D(t,T) = LB(t,T)\big(N\big(h_1(V_t,T-t)\big) - R_t^{2\hat{a}}N\big(h_2(V_t,T-t)\big)\big)$$
$$+ \beta_1 V_t e^{-\kappa(T-t)}\big(N\big(h_3(V_t,T-t)\big) - N\big(h_4(V_t,T-t)\big)\big)$$
$$+ \beta_1 V_t e^{-\kappa(T-t)} R_t^{2\hat{a}+2}\big(N\big(h_5(V_t,T-t)\big) - N\big(h_6(V_t,T-t)\big)\big)$$
$$+ \beta_2 V_t\big(R_t^{\theta+\zeta}N\big(h_7(V_t,T-t)\big) + R_t^{\theta-\zeta}N\big(h_8(V_t,T-t)\big)\big),$$

where $R_t = \bar{v}(t)/V_t$, $\theta = \hat{a}+1$, $\zeta = \sigma^{-2}\sqrt{\hat{v}^2 + 2\sigma^2(r-\gamma)}$ and

$$h_1(V_t,T-t) = \frac{\ln(V_t/L) + \nu(T-t)}{\sigma\sqrt{T-t}},$$

$$h_2(V_t,T-t) = \frac{\ln \bar{v}^2(t) - \ln(LV_t) + \nu(T-t)}{\sigma\sqrt{T-t}},$$

$$h_3(V_t,T-t) = \frac{\ln(L/V_t) - (\nu+\sigma^2)(T-t)}{\sigma\sqrt{T-t}},$$

$$h_4(V_t,T-t) = \frac{\ln(K/V_t) - (\nu+\sigma^2)(T-t)}{\sigma\sqrt{T-t}},$$

$$h_5(V_t,T-t) = \frac{\ln \bar{v}^2(t) - \ln(LV_t) + (\nu+\sigma^2)(T-t)}{\sigma\sqrt{T-t}},$$

$$h_6(V_t,T-t) = \frac{\ln \bar{v}^2(t) - \ln(KV_t) + (\nu+\sigma^2)(T-t)}{\sigma\sqrt{T-t}},$$

$$h_7(V_t,T-t) = \frac{\ln(\bar{v}(t)/V_t) + \zeta\sigma^2(T-t)}{\sigma\sqrt{T-t}},$$

$$h_8(V_t,T-t) = \frac{\ln(\bar{v}(t)/V_t) - \zeta\sigma^2(T-t)}{\sigma\sqrt{T-t}}.$$

Special Cases

Assume that $\beta_1 = \beta_2 = 1$ and the barrier function $\bar{v}$ is such that $K = L$. Then necessarily $\gamma \geq r$. It can be checked that for $K = L$ we have $D(t,T) = D_1(t,T) + D_3(t,T)$ where:

$$D_1(t,T) = LB(t,T)\big(N\big(h_1(V_t,T-t)\big) - R_t^{2\hat{a}}N\big(h_2(V_t,T-t)\big)\big)$$

$$D_3(t,T) = V_t\big(R_t^{\theta+\zeta}N\big(h_7(V_t,T-t)\big) + R_t^{\theta-\zeta}N\big(h_8(V_t,T-t)\big)\big).$$

Case $\gamma = r$. If we also assume that $\gamma = r$ then $\zeta = -\sigma^{-2}\hat{v}$, and thus

$$V_t R_t^{\theta+\zeta} = LB(t,T), \quad V_t R_t^{\theta-\zeta} = V_t R_t^{2\hat{a}+1} = LB(t,T)R_t^{2\hat{a}}.$$

It is also easy to see that in this case

$$h_1(V_t,T-t) = \frac{\ln(V_t/L) + \nu(T-t)}{\sigma\sqrt{T-t}} = -h_7(V_t,T-t),$$

while

$$h_2(V_t, T - t) = \frac{\ln \bar{v}^2(t) - \ln(LV_t) + \nu(T - t)}{\sigma\sqrt{T - t}} = h_8(V_t, T - t).$$

We conclude that if $\bar{v}(t) = Le^{-r(T-t)} = LB(t, T)$ then $D(t, T) = LB(t, T)$. This result is quite intuitive. A corporate bond with a safety covenant represented by the barrier function, which equals the discounted value of the bond's face value, is equivalent to a default-free bond with the same face value and maturity.

Case $\gamma > r$. For $K = L$ and $\gamma > r$, it is natural to expect that $D(t, T)$ would be smaller than $LB(t, T)$. It is also possible to show that when γ tends to infinity (all other parameters being fixed), then the Black and Cox price converges to Merton's price.

Further Developments

The Black and Cox first-passage-time approach was later developed by, among others: Brennan and Schwartz (1977, 1980) – an analysis of convertible bonds, Kim et al. (1993) – a random barrier and random interest rates, Nielsen et al. (1993) – a random barrier and random interest rates, Leland (1994), Leland and Toft (1996) – a study of an optimal capital structure, bankruptcy costs and tax benefits, Longstaff and Schwartz (1995) – a constant barrier and random interest rates.

Optimal Capital Structure

We consider a firm that has an interest paying bonds outstanding. We assume that it is a consol bond, which pays continuously coupon rate c. Assume that $r > 0$ and the payout rate κ is equal to zero. This condition can be given a financial interpretation as the restriction on the sale of assets, as opposed to issuing of new equity. Equivalently, we may think about a situation in which the stockholders will make payments to the firm to cover the interest payments. However, they have the right to stop making payments at any time and either turn the firm over to the bondholders or pay them a lump payment of c/r per unit of the bond's notional amount.

Recall that we denote by $E(V_t)$ ($D(V_t)$, resp.) the value at time t of the firm equity (debt, resp.), hence the total value of the firm's assets satisfies $V_t = E(V_t) + D(V_t)$.

Black and Cox (1976) argue that there is a critical level of the value of the firm, denoted as v^*, below which no more equity can be sold. The critical value v^* will be chosen by stockholders, whose aim is to minimize the value of the bonds (equivalently, to maximize the value of the equity). Let us observe that v^* is nothing else than a constant default barrier in the problem under consideration; the optimal default time τ^* thus equals $\tau^* = \inf\{t \geq 0 : V_t \leq v^*\}$.

To find the value of v^*, let us first fix the bankruptcy level $\bar{v}$. The ODE for the pricing function $u^\infty = u^\infty(V)$ of a consol bond takes the following form (recall that $\sigma = \sigma_V$)

$$\tfrac{1}{2}V^2\sigma^2 u^\infty_{VV} + rV u^\infty_V + c - ru^\infty = 0,$$

subject to the lower boundary condition $u^\infty(\bar{v}) = \min(\bar{v}, c/r)$ and the upper boundary condition

$$\lim_{V \to \infty} u^\infty_V(V) = 0.$$

For the last condition, observe that when the firm's value grows to infinity, the possibility of default becomes meaningless, so that the value of the defaultable consol bond tends to the value c/r of the default-free consol bond. The general solution has the following form:

$$u^\infty(V) = \frac{c}{r} + K_1 V + K_2 V^{-\alpha},$$

where $\alpha = 2r/\sigma^2$ and K_1, K_2 are some constants, to be determined from boundary conditions. We find that $K_1 = 0$, and

$$K_2 = \begin{cases} \bar{v}^{\alpha+1} - (c/r)\bar{v}^\alpha, & \text{if } \bar{v} < c/r, \\ 0, & \text{if } \bar{v} \geq c/r. \end{cases}$$

Hence, if $\bar{v} < c/r$ then

$$u^\infty(V_t) = \frac{c}{r} + \left(\bar{v}^{\alpha+1} - \frac{c}{r}\bar{v}^\alpha \right) V_t^{-\alpha}$$

or, equivalently,

$$u^\infty(V_t) = \frac{c}{r}\left(1 - \left(\frac{\bar{v}}{V_t}\right)^\alpha \right) + \bar{v}\left(\frac{\bar{v}}{V_t}\right)^\alpha.$$

It is in the interest of the stockholders to select the bankruptcy level in such a way that the value of the debt, $D(V_t) = u^\infty(V_t)$, is minimized, and thus the value of firm's equity

$$E(V_t) = V_t - D(V_t) = V_t - \frac{c}{r}(1 - \bar{q}_t) - \bar{v}\bar{q}_t$$

is maximized. It is easy to check that the optimal level of the barrier does not depend on the current value of the firm, and it equals

$$v^* = \frac{c}{r}\frac{\alpha}{\alpha+1} = \frac{c}{r + \sigma^2/2}.$$

Given the optimal strategy of the stockholders, the price process of the firm's debt (i.e., of a consol bond) takes the form, on the set $\{\tau^* > t\}$,

$$D^*(V_t) = \frac{c}{r} - \frac{1}{\alpha V_t^\alpha}\left(\frac{c}{r + \sigma^2/2}\right)^{\alpha+1}$$

or, equivalently,

$$D^*(V_t) = \frac{c}{r}(1 - q_t^*) + v^* q_t^*,$$

where

$$q_t^* = \left(\frac{v^*}{V_t}\right)^\alpha = \frac{1}{V_t^\alpha}\left(\frac{c}{r + \sigma^2/2}\right)^\alpha.$$

Further Developments

We end this section by remarking that other important developments in the area of optimal capital structure were presented in the papers by Leland (1994), Leland and Toft (1996), Hilberink and Rogers (2002) and Christensen et al. (2002). It is probably worth noting that Hilberink and Rogers (2002) model the firm value process as a diffusion with jumps. The reason for this extension was to eliminate an undesirable feature of previously examined models, in which short spreads tend to zero when a bond approaches maturity date.

2.3 Stochastic Interest Rates

In this section, we assume that the underlying probability space $(\Omega, \mathcal{F}, \mathbb{P})$, endowed with the filtration $\mathbb{F} = (\mathcal{F}_t)_{t \geq 0}$, supports the short-term interest rate process r and the value process V. The dynamics under the martingale measure $\mathbb{P}^*$ of the firm's value and of the price of a default-free zero-coupon bond $B(t, T)$ are

$$dV_t = V_t\big((r_t - \kappa(t))\, dt + \sigma(t)\, dW_t^*\big)$$

and

$$dB(t, T) = B(t, T)\big(r_t\, dt + b(t, T)\, dW_t^*\big)$$

respectively, where W^* is a d-dimensional standard Brownian motion. Furthermore, $\kappa : [0, T] \to \mathbb{R}$, $\sigma : [0, T] \to \mathbb{R}^d$ and $b(\cdot, T) : [0, T] \to \mathbb{R}^d$ are assumed to be bounded functions. The *forward value* $F_V(t, T) = V_t/B(t, T)$ of the firm satisfies under the *forward martingale measure* $\mathbb{P}_T$ (see eg. Musiela and Rutkowski (1997), page 309, for definition of the forward martingale measure)

$$dF_V(t, T) = -\kappa(t) F_V(t, T)\, dt + F_V(t, T)\big(\sigma(t) - b(t, T)\big)\, dW_t^T$$

where the process $W_t^T = W_t^* - \int_0^t b(u, T)\, du$, $t \in [0, T]$, is a d-dimensional SBM under $\mathbb{P}_T$. For any $t \in [0, T]$, we set

$$F_V^\kappa(t, T) = F_V(t, T) e^{-\int_t^T \kappa(u)\, du}.$$

Then

$$dF_V^\kappa(t, T) = F_V^\kappa(t, T)\big(\sigma(t) - b(t, T)\big)\, dW_t^T.$$

Furthermore, it is apparent that $F_V^\kappa(T, T) = F_V(T, T) = V_T$. We consider the following modification of the Black and Cox approach:

$$X = L, \quad Z_t = \beta_2 V_t, \quad \tilde{X} = \beta_1 V_T, \quad \tau = \inf\{t \in [0, T] : V_t < v_t\},$$

where β_2, $\beta_1 \in [0, 1]$ are constants, and the barrier v is given by the formula

$$v_t = \begin{cases} K B(t, T) e^{\int_t^T \kappa(u)\, du} & \text{for } t < T, \\ L & \text{for } t = T, \end{cases}$$

with the constant K satisfying $0 < K \leq L$. Let us denote, for any $t \leq T$,

$$\kappa(t,T) = \int_t^T \kappa(u)\, du, \quad \sigma^2(t,T) = \int_t^T |\sigma(u) - b(u,T)|^2\, du$$

where $|\cdot|$ is the Euclidean norm in $\mathbb{R}^d$. For brevity, we write $F_t = F_V^\kappa(t,T)$, and we denote

$$\eta_+(t,T) = \kappa(t,T) + \tfrac{1}{2}\sigma^2(t,T), \quad \eta_-(t,T) = \kappa(t,T) - \tfrac{1}{2}\sigma^2(t,T).$$

The following result extends Black and Cox valuation formula for a corporate bond to the case of random interest rates.

Proposition 2.3. *For any $t < T$, the forward price of a defaultable bond $F_D(t,T) = D(t,T)/B(t,T)$ equals on the set $\{\tau > t\}$*

$$
\begin{aligned}
& L\big(N\big(\hat{h}_1(F_t,t,T)\big) - (F_t/K)e^{-\kappa(t,T)}N\big(\hat{h}_2(F_t,t,T)\big)\big) \\
& + \beta_1 F_t e^{-\kappa(t,T)}\big(N\big(\hat{h}_3(F_t,t,T)\big) - N\big(\hat{h}_4(F_t,t,T)\big)\big) \\
& + \beta_1 K\big(N\big(\hat{h}_5(F_t,t,T)\big) - N\big(\hat{h}_6(F_t,t,T)\big)\big) \\
& + \beta_2 K J_+(F_t,t,T) + \beta_2 F_t e^{-\kappa(t,T)} J_-(F_t,t,T),
\end{aligned}
$$

where

$$\hat{h}_1(F_t,t,T) = \frac{\ln(F_t/L) - \eta_+(t,T)}{\sigma(t,T)},$$

$$\hat{h}_2(F_t,T,t) = \frac{2\ln K - \ln(LF_t) + \eta_-(t,T)}{\sigma(t,T)},$$

$$\hat{h}_3(F_t,t,T) = \frac{\ln(L/F_t) + \eta_-(t,T)}{\sigma(t,T)},$$

$$\hat{h}_4(F_t,t,T) = \frac{\ln(K/F_t) + \eta_-(t,T)}{\sigma(t,T)},$$

$$\hat{h}_5(F_t,t,T) = \frac{2\ln K - \ln(LF_t) + \eta_+(t,T)}{\sigma(t,T)},$$

$$\hat{h}_6(F_t,t,T) = \frac{\ln(K/F_t) + \eta_+(t,T)}{\sigma(t,T)},$$

and for any fixed $0 \leq t < T$ and $F_t > 0$ we set

$$J_\pm(F_t,t,T) = \int_t^T e^{\kappa(u,T)}\, dN\left(\frac{\ln(K/F_t) + \kappa(t,T) \pm \tfrac{1}{2}\sigma^2(t,u)}{\sigma(t,u)}\right).$$

In the special case when $\kappa \equiv 0$, the formula of Proposition 2.3 covers as a special case the valuation result established by Briys and de Varenne (1997). In some other recent studies of first passage time models, in which the triggering barrier is assumed to be either a constant or an unspecified stochastic process, typically no closed-form solution for the value of a corporate debt is available, and thus a numerical approach is required (see, for instance, Kim et al. (1993), Longstaff and Schwartz (1995), Nielsen et al. (1993), or Saá-Requejo and Santa-Clara (1999)).

2.4 Credit Spreads: A Case Study

In the case of full information and Brownian filtration, the stopping time τ is predictable. This is no longer the case when we deal with incomplete information (as in Duffie and Lando (2001)), or when an additional source of randomness is present. We present here a formula for credit spreads arising in a special case of a totally inaccessible time of default. For a more detailed study we refer to Babbs and Bielecki (2003).

We postulate that the value process V is a geometric Brownian motion with a drift, that is, we set $V_t = e^{\Psi_t}$, where $\Psi_t = \mu t + \sigma W_t^*$. Let $v \in (0,1)$ denote a random default barrier. Specifically v is a random variable defined on $(\Omega, \mathcal{F}, \mathbb{P}^*)$ and independent of W^*. We define the default time as

$$\tau = \inf\{t \geq 0 : V_t \leq v\} = \inf\{t \geq 0 : \Psi_t \leq \psi\},$$

where $\psi = \ln v$. It is clear that we also have $\tau = \inf\{t \geq 0 : \Psi_t^* \leq \psi\}$, where Ψ^* is the *running minimum* of the process Ψ: $\Psi_t^* = \inf\{\Psi_s : 0 \leq s \leq t\}$. It is well known that (Ψ, Ψ^*) is a strong Markov process.

We choose the Brownian filtration as the reference filtration, i.e., we set $\mathbb{F} = \mathbb{F}^{W^*}$. This means that we assume that the value process V (hence also the process Ψ) is perfectly observed. In addition, we suppose that the bond investor can observe the occurrence of the default time. Thus, he can observe the process $H_t = \mathbb{1}_{\{\tau \leq t\}} = \mathbb{1}_{\{\Psi_t^* \leq \psi\}}$. We denote by $\mathbb{H}$ the natural filtration of the process H. The information available to the investor is represented by the (enlarged) filtration $\mathbb{G} = \mathbb{F} \vee \mathbb{H}$.

Let us denote by $G(z)$ the cumulative distribution function under $\mathbb{P}^*$ of the barrier ψ. We assume that $G(z) > 0$ for $z < 0$ and that G admits the density g with respect to the Lebesgue measure (note that $g(z) = 0$ for $z > 0$). In addition, we assume that the default time τ and interest rates are independent under $\mathbb{P}^*$. Then, it is possible to establish the following result (see Giesecke (2002) or Babbs and Bielecki (2003)). Note that the process Ψ^* is decreasing, so that the integral with respect to this process is a (pathwise) Stieltjes integral.

Proposition 2.4. *Under the assumptions stated above, and additionally assuming $L = 1$, $Z \equiv 0$ and $\tilde{X} = 0$, we have that for every $t < T$*

$$S(t,T) = -\mathbb{1}_{\{\tau > t\}} \frac{1}{T-t} \ln \mathbb{E}_{\mathbb{P}^*}\left(e^{\int_t^T \frac{g(\Psi_u^*)}{G(\Psi_u^*)} d\Psi_u^*} \,\Big|\, \mathcal{F}_t\right).$$

In the next chapter, we shall introduce the notion of a hazard process of a random time. For the default time τ defined above, the $\mathbb{F}$-hazard process Γ exists and is given by the formula

$$\Gamma_t = -\int_0^t \frac{g(\Psi_u^*)}{G(\Psi_u^*)}\, d\Psi_u^*.$$

This process is continuous, and thus the default time τ is a totally inaccessible stopping time with respect to the filtration $\mathbb{G}$.

2.5 Comments on Structural Models

We end this chapter by commenting on merits and drawbacks of the structural approach to credit risk.

Advantages

- An approach based on the volatility of the total value of the firm. The credit risk is thus measured in a standard way. The random time of default is defined in an intuitive way. The default event is linked to the notion of the firm's insolvency.
- Valuation and hedging of defaultable claims rely on similar techniques as the valuation and hedging of exotic options in the standard default-free Black-Scholes setup.
- The concept of the distance to default, which measures the obligor's leverage relative to the volatility of its assets value, may serve to reflect credit ratings.
- Dependent defaults are easy to handle through correlation of processes corresponding to different names.

Disadvantages

- A stringent assumption that the total value of the firm's assets can be easily observed. In practice, continuous-time observations of the value process V are not available. This issue was recently addressed by Crouhy et al. (1998) and Duffie and Lando (2001), who showed that a structural model with incomplete accounting data can be dealt with using the intensity-based methodology.
- An unrealistic postulate that the total value of the firm's assets is a tradeable security.
- This approach is known to generate low credit spreads for corporate bonds close to maturity. It requires a judicious specification of the default barrier in order to get a good fit to the observed spread curves.

Other issues

- A major problem with applying structural models is the difficulty with the estimation of the volatility of assets value. For the classical Merton's model, there exists a simple formula that relates this volatility to the volatility of the firm's equity, which in principle can be easily estimated. However, no such simple expression exists in case of first-passage-time models. Certain market-oriented technologies, such as CreditGrades, attempt to produce such a formula.
- Structural models discussed above were at most one-factor models, with the only factor being the short-term interest rate. Two- and three-factor structural models have also been developed and closed-form valuation formulae were derived in some special cases.

3 Intensity-Based Approach

A major motivation for the intensity-based approach (also known as the *reduced-form approach*) was to model a credit event as an unpredictable event, meaning that the date of its occurrence is a totally inaccessible stopping time with respect to an underlying filtration. The modeling of a default time is essentially reduced to the specification of the so-called hazard process with respect to some reference filtration. Under some circumstances, this is equivalent to the modeling of a default time in terms of its intensity process. The name reduced-form approach was probably well justified in the early stages of development of this approach when, typically, only exogenously given intensities were considered. However, it is possible to link a hazard process to economic fundamentals of a firm, such as the total asset value process, so that the hazard process can be specified endogenously.

The intensity-based approach to modeling of credit risk, was studied by, among others, Jarrow and Turnbull (1995), Jarrow et al. (1997), Duffie et al. (1996), Duffie (1998a), Lando (1998a), Duffie and Singleton (1999), Elliott et al. (2000), Schönbucher (2000a, 2000b), Bélanger et al. (2001), Jamshidian (2002), Collin-Dufresne et al. (2002), Brigo and Alfonsi (2003) and Chen and Filipović (2003a, 2003b).

3.1 Hazard Function

Before going deeper in the analysis of the reduced-form approach, we shall first examine a related technical question. Suppose we want to evaluate a conditional expectation $\mathbb{E}_{\mathbb{P}}(\mathbb{1}_{\{\tau > s\}} Y \,|\, \mathcal{G}_t)$, where τ is a stopping time on a probability space $(\Omega, \mathbb{G}, \mathbb{P})$, with respect to some filtration $\mathbb{G} = (\mathcal{G}_t)_{t \geq 0}$ and Y is an integrable, $\mathcal{G}_s$-measurable random variable for some $s \geq t$.

In financial applications, it is quite natural and convenient to model the filtration $\mathbb{G}$ as $\mathbb{G} = \mathbb{F} \vee \mathbb{H}$, where $\mathbb{H}$ is the filtration that carries full information about default events (that is, events such as $\{\tau \leq t\}$), whereas the *reference* filtration $\mathbb{F}$ carries information about other relevant financial and economic processes, but, typically, it does not carry full information about default event. The first question we address is how to compute the expectation $\mathbb{E}_{\mathbb{P}}(\mathbb{1}_{\{\tau > s\}} Y \mid \mathcal{G}_t)$ using the intensity of τ with respect to $\mathbb{F}$.

Hazard Function of a Random Time

In this section, we study the case where the reference filtration $\mathbb{F}$ is trivial, so that it does not carry any information whatsoever. Consequently, we have that $\mathbb{G} = \mathbb{H}$. Arguably, this is the simplest possible setup within the intensity-based approach; nevertheless, it is sometimes used in practical financial applications, as it leads to relatively easy calibration of the model. Most of the results obtained in this section can be considered as prototypes for the results presented in the remaining sections of this chapter, where the reference filtration is no longer assumed to be trivial.

We start by recalling the notion of a hazard function of a random time. Let τ be a finite, non-negative random variable on a probability space $(\Omega, \mathcal{G}, \mathbb{P})$, referred to as the *random time*. We assume that $\mathbb{P}\{\tau = 0\} = 0$ and τ is unbounded: $\mathbb{P}\{\tau > t\} > 0$ for every $t \in \mathbb{R}_+$.

The right-continuous cumulative distribution function F of τ satisfies $F(t) = \mathbb{P}\{\tau \le t\} < 1$ for every $t \in \mathbb{R}_+$. We also assume that $\mathbb{P}\{\tau < \infty\} = 1$, so that τ is a Markov time.

We introduce the right-continuous jump process $H_t = \mathbb{1}_{\{\tau \le t\}}$ and we write $\mathbb{H} = (\mathcal{H}_t)_{t \ge 0}$ to denote the (right-continuous and $\mathbb{P}$-completed) filtration generated by the process H. Of course, τ is an $\mathbb{H}$-stopping time.

We shall assume throughout that all random variables and processes that are used in what follows satisfy suitable integrability conditions. We begin with the following simple and important result.

Lemma 3.1. *For any $\mathcal{G}$-measurable (integrable) random variable Y we have*

$$\mathbb{E}_{\mathbb{P}}(Y \mid \mathcal{H}_t) = \mathbb{1}_{\{\tau \le t\}} \mathbb{E}_{\mathbb{P}}(Y \mid \tau) + \mathbb{1}_{\{\tau > t\}} \frac{\mathbb{E}_{\mathbb{P}}(\mathbb{1}_{\{\tau > t\}} Y)}{\mathbb{P}\{\tau > t\}}. \tag{2}$$

For any $\mathcal{H}_t$-measurable random variable Y we have

$$Y = \mathbb{1}_{\{\tau \le t\}} \mathbb{E}_{\mathbb{P}}(Y \mid \tau) + \mathbb{1}_{\{\tau > t\}} \frac{\mathbb{E}_{\mathbb{P}}(\mathbb{1}_{\{\tau > t\}} Y)}{\mathbb{P}\{\tau > t\}}, \tag{3}$$

that is, $Y = h(\tau)$ for a Borel measurable $h : \mathbb{R} \to \mathbb{R}$ which is constant on the interval $]t, \infty[$.

The hazard function is introduced through the following definition.

Definition 3.1. *The increasing right-continuous function $\Gamma : \mathbb{R}_+ \to \mathbb{R}_+$ given by the formula*

$$\Gamma(t) = -\ln(1 - F(t)), \quad \forall t \in \mathbb{R}_+,$$

is called the hazard function of a random time τ.

If the distribution function F is an absolutely continuous function, i.e., if we have

$$F(t) = \int_0^t f(u)\,du$$

for some function $f : \mathbb{R}_+ \to \mathbb{R}_+$, then we have

$$F(t) = 1 - e^{-\Gamma(t)} = 1 - e^{-\int_0^t \gamma(u)\,du}$$

where we set

$$\gamma(t) = \frac{f(t)}{1 - F(t)}.$$

It is clear that $\gamma : \mathbb{R}_+ \to \mathbb{R}$ is a non-negative function and it satisfies $\int_0^\infty \gamma(u)\,du = \infty$. The function γ is called the *hazard rate* or *intensity* of τ. Sometimes, in order to emphasize relevance of the measure $\mathbb{P}$ the terminology $\mathbb{P}$-*hazard rate* and $\mathbb{P}$-*intensity* is used.

The next two results follow from Lemma 3.1 and Definition 3.1.

Corollary 3.1. *For any $\mathcal{G}$-measurable random variable Y we have*

$$\mathbb{E}_\mathbb{P}(\mathbb{1}_{\{\tau > s\}} Y \mid \mathcal{H}_t) = \mathbb{1}_{\{\tau > t\}} e^{\Gamma(t)} \mathbb{E}_\mathbb{P}(\mathbb{1}_{\{\tau > s\}} Y).$$

Corollary 3.2. *Let Y be $\mathcal{H}_\infty$-measurable, so that $Y = h(\tau)$ for some function $h : \mathbb{R}_+ \to \mathbb{R}$. If the hazard function Γ is continuous then*

$$\mathbb{E}_\mathbb{P}(Y \mid \mathcal{H}_t) = \mathbb{1}_{\{\tau \le t\}} h(\tau) + \mathbb{1}_{\{\tau > t\}} \int_t^\infty h(u) e^{\Gamma(t) - \Gamma(u)}\,d\Gamma(u).$$

If, in addition, the random time τ admits the hazard rate function γ then we have

$$\mathbb{E}_\mathbb{P}(Y \mid \mathcal{H}_t) = \mathbb{1}_{\{\tau \le t\}} h(\tau) + \mathbb{1}_{\{\tau > t\}} \int_t^\infty h(u) \gamma(u) e^{-\int_t^u \gamma(v)\,dv}\,du.$$

In particular, for any $t \le s$ the last formula yields:

$$\mathbb{P}\{\tau > s \mid \mathcal{H}_t\} = \mathbb{1}_{\{\tau > t\}} e^{-\int_t^s \gamma(v)\,dv}$$

and

$$\mathbb{P}\{t < \tau < s \mid \mathcal{H}_t\} = \mathbb{1}_{\{\tau > t\}} \left(1 - e^{-\int_t^s \gamma(v)\,dv}\right).$$

Associated Martingales

Two particular processes associated with a random time τ are martingales with respect to the filtration $\mathbb{H}$. The first result is general, that is, it holds for an arbitrary (possibly discontinuous) hazard function Γ.

Lemma 3.2. *The process L given by the formula*

$$L_t = \frac{1 - H_t}{1 - F(t)} = \mathbb{1}_{\{\tau > t\}} e^{\Gamma(t)} = (1 - H_t) e^{\Gamma(t)}$$

is an $\mathbb{H}$-martingale. Equivalently,

$$\mathbb{E}_{\mathbb{P}}(H_s - H_t \mid \mathcal{H}_t) = \mathbb{1}_{\{\tau > t\}} \frac{F(s) - F(t)}{1 - F(t)}.$$

Notice that in the next result the hazard function Γ of τ is assumed to be continuous.

Lemma 3.3. *Assume that F (and thus also the hazard function Γ) is a continuous function. Then the process*

$$M_t = H_t - \Gamma(t \wedge \tau)$$

is an $\mathbb{H}$-martingale.

Change of a Probability Measure

Let $\mathbb{P}^*$ be any probability measure on $(\Omega, \mathcal{H}_\infty)$, which is equivalent to $\mathbb{P}$. Then there exists a Borel measurable function $h : \mathbb{R}_+ \to \mathbb{R}_+$ such that the Radon-Nikodým density of $\mathbb{P}^*$ with respect to $\mathbb{P}$ equals

$$\eta_\infty = \frac{d\mathbb{P}^*}{d\mathbb{P}} = h(\tau) > 0, \quad \mathbb{P}\text{-a.s.} \tag{4}$$

In particular, we have

$$\mathbb{E}_{\mathbb{P}}(h(\tau)) = \int_{]0,\infty[} h(u)\, dF(u) = 1.$$

Obviously, we have $\mathbb{P}^*\{\tau = 0\} = 0$ and $\mathbb{P}^*\{\tau > t\} > 0$ for every $t \in \mathbb{R}_+$. The cumulative distribution function F^* of τ under $\mathbb{P}^*$ is given by

$$F^*(t) = \mathbb{P}^*\{\tau \leq t\} = \mathbb{E}_{\mathbb{P}}(\mathbb{1}_{\{\tau \geq t\}} h(\tau)) = \int_{]0,t]} h(u)\, dF(u).$$

If the cumulative distribution function F (and thus F^*) is continuous, then from

$$d\Gamma^*(t) = \frac{dF^*(t)}{1 - F^*(t)}$$

one deduces that

$$d\Gamma^*(t) = \frac{h(t)dF(t)}{\int_t^\infty h(u)\, dF(u))} = \frac{h(t)\, dF(t)}{g(t)(1 - F(t))} = \frac{h(t)}{g(t)}\, d\Gamma(t),$$

where we set

$$g(t) = e^{\Gamma(t)} \, \mathbb{E}_{\mathbb{P}}\big(\mathbb{1}_{\{\tau > t\}} h(\tau)\big) = e^{\Gamma(t)} \int_t^\infty h(u) \, dF(u)$$

$$= e^{\Gamma(t)} \int_t^\infty h(u) e^{-\Gamma(u)} \, d\Gamma(u).$$

Proposition 3.1. *Let $\mathbb{P}^*$ and $\mathbb{P}$ be two equivalent probability measures on $(\Omega, \mathcal{H}_\infty)$. Assume that the hazard function Γ of τ under $\mathbb{P}$ is continuous. Then the hazard function Γ^* of τ under $\mathbb{P}^*$ is also continuous and*

$$d\Gamma^*(t) = (1 + \kappa(t)) \, d\Gamma(t),$$

where

$$\kappa(t) = \frac{h(t)}{g(t)} - 1 = -\frac{\int_t^\infty h(u) e^{-\Gamma(u)} \, d\Gamma(u) - h(t) e^{-\Gamma(t)}}{\int_t^\infty h(u) e^{-\Gamma(u)} \, d\Gamma(u)}.$$

In particular, Γ^* is absolutely continuous if and only if Γ is, and the intensity function of τ under $\mathbb{P}^*$ equals $\gamma^*(t) = (1 + \kappa(t))\gamma(t)$.

Radon-Nikodým Density Process

Let $\mathbb{P}$ and $\mathbb{P}^*$ be equivalent probability measures on $(\Omega, \mathcal{H}_\infty)$. We introduce the non-negative $\mathbb{P}$-martingale η by setting

$$\eta_t = \frac{d\mathbb{P}^*}{d\mathbb{P}}\bigg|_{\mathcal{H}_t} = \mathbb{E}_{\mathbb{P}}(\eta_\infty \,|\, \mathcal{H}_t) = \mathbb{E}_{\mathbb{P}}(h(\tau) \,|\, \mathcal{H}_t).$$

The process η is termed the Radon-Nikodým density process of $\mathbb{P}^*$ with respect to $\mathbb{P}$ (given the filtration $\mathbb{H}$). Notice that

$$\eta_t = \mathbb{1}_{\{\tau \le t\}} h(\tau) + \mathbb{1}_{\{\tau > t\}} \, e^{\Gamma(t)} \int_{]t, \infty[} h(u) \, dF(u),$$

so that $\eta_t = \mathbb{1}_{\{\tau \le t\}} h(\tau) + \mathbb{1}_{\{\tau > t\}} g(t)$. It is not difficult to check that

$$\eta_t = 1 + \int_{]0, t]} (h(u) - g(u)) \, dM_u. \tag{5}$$

Remarks. Representation (5) is a special case of a more general result. Consider the martingale $M_t^\chi = \mathbb{E}_{\mathbb{P}}(\chi(\tau) \,|\, \mathcal{H}_t)$ for some integrable function $\chi : \mathbb{R}_+ \to \mathbb{R}$, and denote

$$g^\chi(t) = e^{\Gamma(t)} \mathbb{E}_{\mathbb{P}}\big(\mathbb{1}_{\{\tau > t\}} \chi(\tau)\big).$$

Then, setting $\hat{\chi} = \chi - g^\chi$ and assuming that the cumulative distribution function F is continuous, we have

$$M_t^\chi = M_0^\chi + \int_{]0,t]} \hat{\chi}(u) \, dM_u. \tag{6}$$

Taking care of the jump of the process η at time τ, one can also prove that the process η solves the following SDE:

$$\eta_t = 1 + \int_{]0,t]} \eta_{u-} \kappa(u) \, dM_u. \tag{7}$$

It is not difficult to find an explicit solution to this equation, specifically,

$$\eta_t = \left(1 + 1\!\!1_{\{\tau \le t\}} \kappa(\tau)\right) \exp\left(-\int_0^{t \wedge \tau} \kappa(u) \, d\Gamma(u)\right). \tag{8}$$

Note that equation (7) is a special case of equation (9) that appears in the following version of a classical result.

Lemma 3.4. *Let Y be a process of finite variation. Consider the linear SDE*

$$Z_t = 1 + \int_{]0,t]} Z_{u-} \, dY_u. \tag{9}$$

The unique solution $Z_t = \mathcal{E}_t(Y)$, called the Doléans exponential of Y, equals

$$\mathcal{E}_t(Y) = e^{Y_t} \prod_{0 < u \le t} (1 + \Delta Y_u) e^{-\Delta Y_u}.$$

Equivalently, we have

$$\mathcal{E}_t(Y) = e^{Y_t^c} \prod_{0 < u \le t} (1 + \Delta Y_u) \tag{10}$$

where Y^c is the path-by-path continuous part of Y, that is,

$$Y_t^c = Y_t - \sum_{0 < u \le t} \Delta Y_u.$$

Since the process η satisfies (7), it is clear that it can be represented as follows:

$$\eta_t = \mathcal{E}_t\left(\int_{]0, \cdot]} \kappa(u) \, dM_u\right).$$

Thus, expression (8) for the random variable η_t can also be obtained from (10), upon setting $dY_u = \kappa(u) \, dM_u$. Notice also that equality (10) is merely a special case of the general formula for the Doléans exponential. Proposition 3.1 is a very special case of Girsanov's theorem (see, for instance, Elliott (1982), Protter (2003), or Revuz and Yor (1999)). Equality (6) is in turn a particular case of the predictable representation theorem (see Kusuoka (1999) or Blanchet-Scalliet and Jeanblanc (2003)).

Martingale Hazard Function

In view of the martingale property established in Lemma 3.3, the following definition is natural.

Definition 3.2. A function $\Lambda : \mathbb{R}_+ \to \mathbb{R}$ is called a *martingale hazard function* of a random time τ with respect to the filtration $\mathbb{H}$ if and only if the process $H_t - \Lambda(t \wedge \tau)$ is an $\mathbb{H}$-martingale.

Remarks. Since the bounded, increasing process[4] H is constant after time τ, its *compensator* is constant after τ as well. This explains why the function Λ has to be evaluated at time $t \wedge \tau$, rather than at time t.

It happens that the martingale hazard function can be found explicitly. In fact, we have the following

Proposition 3.2. *The unique martingale hazard function of τ with respect to the filtration $\mathbb{H}$ is the right-continuous increasing function Λ given by the formula*

$$\Lambda(t) = \int_{]0,t]} \frac{dF(u)}{1 - F(u-)} = \int_{]0,t]} \frac{d\mathbb{P}\{\tau \leq u\}}{1 - \mathbb{P}\{\tau < u\}}.$$

Observe that the martingale hazard function Λ is continuous if and only if F is continuous. In this case, we have $\Lambda(t) = -\ln(1 - F(t))$. We conclude that the martingale hazard function Λ coincides with the hazard function Γ if and only if F is a continuous function. In general, we have

$$e^{-\Gamma(t)} = e^{-\Lambda^c(t)} \prod_{0 \leq u \leq t} (1 - \Delta\Lambda(u)),$$

where $\Lambda^c(t) = \Lambda(t) - \sum_{0 \leq u \leq t} \Delta\Lambda(u)$ and $\Delta\Lambda(u) = \Lambda(u) - \Lambda(u-)$.

Defaultable Bonds: Deterministic Intensity

In order to value a defaultable claim, we need, of course, to specify the unit in which we would like to express all prices. Formally, this is done through a choice of a discount factor (a numeraire). For the sake of simplicity, we shall take the savings account

$$B_t = e^{\int_0^t r_v \, dv}, \quad \forall t \in [0, T^*], \tag{11}$$

as the numeraire, where r is the short-term interest rate process. We also postulate that some probability measure $\mathbb{Q}^*$ is a martingale measure relative to this numeraire. This assumption means, in particular, that the price of any contingent claim Y which settles at time T is given as the conditional expectation $B_t \, \mathbb{E}_{\mathbb{Q}^*}(B_T^{-1} Y \,|\, \mathcal{G}_t)$.

---

[4] The process H is thus a bounded $\mathbb{H}$-submartingale.

We shall now apply some results obtained earlier in this section, but using a martingale measure $\mathbb{Q}^*$, rather than an unspecified probability measure $\mathbb{P}$. In accordance with our assumption that the reference filtration is trivial, we also assume that:

(i) the default time τ admits the $\mathbb{Q}^*$-intensity function $\gamma(t)$,

(ii) the short-term interest rate $r(t)$ is a deterministic function of time.

In view of the latter assumption, the price at time t of a unit default-free zero-coupon bond of maturity T equals

$$B(t,T) = e^{-\int_t^T r(v)\,dv}.$$

In the market practice, the interest rate (more precisely, the yield curve) can be derived from the market price of the zero-coupon bond. In a similar way, the hazard rate can be deduced from the prices of the corporate zero-coupon bonds, or from market values of other actively traded credit derivatives.

In view of our notation for defaultable claims adopted in Chapter 1, for the corporate unit discount bond we have $C \equiv 0$ and $X = L = 1$. Recall that (since the reference filtration is assumed trivial) we have that $\mathbb{G} = \mathbb{H}$.

Zero Recovery. Consider first a corporate zero-coupon bond with unit face value, the maturity date T, and *zero recovery* at default (that is, $\tilde{X} = 0$ and $Z \equiv 0$). Formally, the bond can thus be identified with a claim of the form $\mathbb{1}_{\{\tau > T\}}$ which settles at T. It is clear that a corporate bond with zero recovery becomes worthless as soon as default occurs. Its time-t price is defined as

$$D^0(t,T) = B_t\, \mathbb{E}_{\mathbb{Q}^*}(B_T^{-1} \mathbb{1}_{\{\tau > T\}} | \mathcal{H}_t).$$

Consequently, in view of the results of Section 3.1, the price $D^0(t,T)$ can be represented as follows: $D^0(t,T) = \mathbb{1}_{\{\tau > t\}} \tilde{D}^0(t,T)$, where $\tilde{D}^0(t,T)$ is the bond's *pre-default value*, and is given by the formula

$$\tilde{D}^0(t,T) = e^{-\int_t^T (r(v)+\gamma(v))dv} = B(t,T)e^{-\int_t^T \gamma(v)dv}.$$

Fractional Recovery of Par Value (FRPV). According to this convention, we have $\tilde{X} = 0$ and the recovery process Z satisfy $Z_t = \delta$ for some constant *recovery rate* $\delta \in [0,1]$. This means that under FRPV the bond-holder receives at time of default a fixed fraction of the bond's par value.

Using Corollary 3.2, it is easy to check that the pre-default value $\tilde{D}^\delta(t,T)$ of a unit corporate zero-coupon bond with FRPV equals

$$\tilde{D}^\delta(t,T) = \delta \int_t^T e^{-\int_t^u \tilde{r}(v)dv} \gamma(u)\,du + e^{-\int_t^T \tilde{r}(v)dv},$$

where $\tilde{r} = r + \gamma$ is the default-risk-adjusted interest rate. Since the fraction of the par value is received at the time of default, in the case of full recovery, that is, for $\delta = 1$, we do not obtain the equality $\tilde{D}^\delta(t,T) = B(t,T)$, but rather the inequality $\tilde{D}^\delta(t,T) > B(t,T)$ (at least when the interest rate is strictly positive, so that $B(t,T) < 1$ for $t < T$).

Fractional Recovery of Treasury Value (FRTV). Assume now that $\tilde{X} = 0$ and that the recovery process equals $Z_t = \delta B(t,T)$. This means that the recovery payoff at the time of default τ represent a fraction of the price of the (equivalent) Treasury bond. The price of a corporate bond which is subject to this recovery scheme equals

$$S_t = B(t,T)\big(\delta\,\mathbb{Q}^*\{t < \tau \le T \,|\, \mathcal{H}_t\} + \mathbb{Q}^*\{\tau > T \,|\, \mathcal{H}_t\}\big).$$

Let us denote by $\hat{D}^\delta(t,T)$ the pre-default value of a unit corporate bond subject to the FRTV scheme. Then

$$\hat{D}^\delta(t,T) = \int_t^T \delta B(t,T) e^{-\int_t^u \gamma(v)dv} \gamma(u)\,du + e^{-\int_t^T \tilde{r}(v)dv}$$

or, equivalently,

$$\hat{D}^\delta(t,T) = B(t,T)\Big(\delta\Big(1 - e^{-\int_t^T \gamma(v)dv}\Big) + e^{-\int_t^T \gamma(v)dv}\Big).$$

In the case of full recovery, that is, for $\delta = 1$, we obtain $\hat{D}^\delta(t,T) = B(t,T)$, as expected.

Remarks. Let us stress that similar representations can be derived also in the case when the reference filtration $\mathbb{F}$ is not trivial, and under the assumption that *market risk* and *credit risk* are independent – that is:
(i) the default time admits the $\mathbb{F}$-intensity process γ,
(ii) the interest rate process r is independent of the filtration $\mathbb{F}$.

3.2 Hazard Processes

In the previous section, it was assumed that the reference filtration $\mathbb{F}$ carries no information. However, for practical purposes it is important to study the situation where the reference filtration is not trivial. This section presents some results to this effect.

We assume that a martingale measure $\mathbb{Q}^*$ is given, and we shall examine the valuation of defaultable contingent claims under this probability measure. Note that the defaultable market is incomplete if there are no defaultable assets traded in the market that are sensitive to the same default risk as the defaultable contingent claim we wish to price. Thus, the martingale measure may not be unique. Another important question is the relationship between the actual probability and a martingale measure, that is, the specification of market prices for risk (see Duffee (1999) or Jarrow et al. (2002)). We shall provide some discussion of the latter issue.

Hazard Process of a Random Time

Let $\tau : \Omega \to \mathbb{R}_+$ be a finite, non-negative random variable on a probability space $(\Omega, \mathcal{G}, \mathbb{P})$. Assume that $\mathcal{G}_t = \mathcal{F}_t \vee \mathcal{H}_t$ for some reference filtration $\mathbb{F}$, so that $\mathbb{G} = \mathbb{F} \vee \mathbb{H}$. We thus place ourselves here in a more general setting than in Section 3.1.

We start be extending some definitions and results to the present framework. We denote $F_t = \mathbb{P}\{\tau \le t \,|\, \mathcal{F}_t\}$, so that $G_t = 1 - F_t = \mathbb{P}\{\tau > t \,|\, \mathcal{F}_t\}$ is the *survival process* with respect to $\mathbb{F}$. It is easily seen that F is a bounded, non-negative, $\mathbb{F}$-submartingale. As a submartingale, this process admits a Doob-Meyer decomposition as $F_t = Z_t + A_t$ where A is an $\mathbb{F}$-predictable increasing process. Assume, in addition, that $F_t < 1$ for every $t \in \mathbb{R}_+$.

Definition 3.3. The $\mathbb{F}$-*hazard process* Γ of a random time τ is defined through the equality $1 - F_t = e^{-\Gamma_t}$, that is, $\Gamma_t = -\ln G_t$.

Notice that the existence of Γ implies that τ is not an $\mathbb{F}$-stopping time. Indeed, if the event $\{\tau > t\}$ belongs to the σ-field $\mathcal{F}_t$ for some $t > 0$ then $\mathbb{P}\{\tau > t \,|\, \mathcal{F}_t\} = \mathbb{1}_{\{\tau > t\}} > 0$ (P-almost surely), and thus $\tau = \infty$.

If the hazard process is absolutely continuous, so that $\Gamma_t = \int_0^t \gamma_u \, du$, for some process γ, then γ is called the $\mathbb{F}$-*intensity* of τ. Note that this is the case only if the process Γ is increasing (and thus γ is always non-negative). Note that if the reference filtration $\mathbb{F}$ is trivial, then the hazard process Γ is the same as the hazard function $\Gamma(\cdot)$. In this case, if Γ is absolutely continuous, then we have $\gamma_t = \gamma(t)$.

Terminal Payoff

The valuation of the terminal payoff $X^d(T)$ is based on the following generalization of Lemma 3.1, which first appeared in Dellacherie (1972) (see Page 122 therein). We return here to our original question: how to compute $\mathbb{E}_{\mathbb{P}}(\mathbb{1}_{\{\tau > s\}} Y \,|\, \mathcal{G}_t)$ for an $\mathcal{F}_s$-measurable random variable Y?

Lemma 3.5. *For any $\mathcal{G}$-measurable (integrable) random variable Y and arbitrary $s \ge t$ we have*

$$\mathbb{E}_{\mathbb{P}}(\mathbb{1}_{\{\tau > s\}} Y \,|\, \mathcal{G}_t) = \mathbb{1}_{\{\tau > t\}} \frac{\mathbb{E}_{\mathbb{P}}(\mathbb{1}_{\{\tau > s\}} Y \,|\, \mathcal{F}_t)}{\mathbb{P}\{\tau > t \,|\, \mathcal{F}_t\}}.$$

If, in addition, Y is $\mathcal{F}_s$-measurable then

$$\mathbb{E}_{\mathbb{P}}(\mathbb{1}_{\{\tau > s\}} Y \,|\, \mathcal{G}_t) = \mathbb{1}_{\{\tau > t\}} \mathbb{E}_{\mathbb{P}}(e^{\Gamma_t - \Gamma_s} Y \,|\, \mathcal{F}_t).$$

Assume that Y is $\mathcal{G}_t$-measurable. Then there exists an $\mathcal{F}_t$-measurable random variable $\tilde{Y}$ such that $\mathbb{1}_{\{\tau > t\}} Y = \mathbb{1}_{\{\tau > t\}} \tilde{Y}$.

The latter property can be extended to stochastic processes: for any $\mathbb{G}$-predictable process X there exists an $\mathbb{F}$-predictable process $\tilde{X}$ such that the equality $\mathbb{1}_{\{\tau > t\}} X_t = \mathbb{1}_{\{\tau > t\}} \tilde{X}_t$ is valid for every $t \in \mathbb{R}_+$, that is, both processes coincide on the random interval $[0, \tau[$.

Recovery Process

The following extension of Corollary 3.2 appears to be useful in the valuation of the recovery payoff Z_τ (recall that this payoff occurs at time τ).

Lemma 3.6. *Assume that the hazard process Γ is a continuous, increasing process, and let Z be a bounded, $\mathbb{F}$-predictable process. Then for any $t \leq s$ we have*

$$\mathbb{E}_\mathbb{P}(Z_\tau \mathbb{1}_{\{t < \tau \leq s\}} \mid \mathcal{G}_t) = \mathbb{1}_{\{\tau > t\}} \mathbb{E}_\mathbb{P}\left(\int_t^s Z_u e^{\Gamma_t - \Gamma_u} \, d\Gamma_u \,\Big|\, \mathcal{F}_t \right).$$

Promised Dividends

To value the promised dividends C that are paid prior to τ, it is convenient to make use of the following result.

Lemma 3.7. *Assume that the hazard process Γ is continuous. Let C be a bounded, $\mathbb{F}$-predictable process of finite variation. Then for every $t \leq s$*

$$\mathbb{E}_\mathbb{P}\left(\int_{]t,s]} (1 - H_u) \, dC_u \,\Big|\, \mathcal{G}_t \right) = \mathbb{1}_{\{\tau > t\}} \mathbb{E}_\mathbb{P}\left(\int_{]t,s]} e^{\Gamma_t - \Gamma_u} \, dC_u \,\Big|\, \mathcal{F}_t \right).$$

Valuation of Defaultable Claims

From now on, we assume that τ is given on a filtered probability space $(\Omega, \mathbb{G}, \mathbb{Q}^*)$, where $\mathbb{G} = \mathbb{F} \vee \mathbb{H}$ and $\mathbb{Q}^*\{\tau > t \mid \mathcal{F}_t\} > 0$ for every $t \in \mathbb{R}_+$ so that the $\mathbb{F}$-hazard process Γ of τ under $\mathbb{Q}^*$ is well defined. A default time τ is thus a $\mathbb{G}$-stopping time, but it is not an $\mathbb{F}$-stopping time.

The probability $\mathbb{Q}^*$ is assumed to be a martingale measure relative to a savings account process B, which is given by (11) for some $\mathbb{F}$-progressively measurable process r. In some sense, this probability, and thus also the $\mathbb{F}$-hazard process Γ of τ under $\mathbb{Q}^*$, are given by the market via calibration.

The ex-dividend price S_t of a defaultable claim $(X, C, \tilde{X}, Z, \tau)$ is given by Definition 2.2, with $\mathbb{P}^*$ substituted with $\mathbb{Q}^*$ and $\mathbb{F}$ replaced by $\mathbb{G}$. We postulate, in particular, that the processes Z and C are $\mathbb{F}$-predictable, and the random variables X and $\tilde{X}$ are $\mathcal{F}_T$-measurable and $\mathcal{G}_T$-measurable, respectively.

Using Lemmas 3.5–3.7 and the fact that the savings account process B is $\mathbb{F}$-adapted, it is easy to derive a convenient representation for the arbitrage price of a defaultable claim in terms of the $\mathbb{F}$-hazard process Γ.

Proposition 3.3. *The value process of a defaultable claim $(X, C, \tilde{X}, Z, \tau)$ admits the following representation for $t < T$*

$$S_t = \mathbb{1}_{\{\tau > t\}} G_t^{-1} B_t \, \mathbb{E}_{\mathbb{Q}^*}\left(\int_{]t,T]} B_u^{-1}(G_u \, dC_u - Z_u \, dG_u) \,\Big|\, \mathcal{F}_t \right) +$$

$$+ \mathbb{1}_{\{\tau > t\}} G_t^{-1} B_t \, \mathbb{E}_{\mathbb{Q}^*}\left(G_T B_T^{-1} X \mid \mathcal{F}_t \right) + B_t \, \mathbb{E}_{\mathbb{Q}^*}\left(B_T^{-1} \mathbb{1}_{\{\tau \leq T\}} \tilde{X} \mid \mathcal{G}_t \right).$$

If the hazard process Γ is an increasing, continuous process, then

$$S_t = \mathbb{1}_{\{\tau>t\}} B_t \, \mathbb{E}_{\mathbb{Q}^*} \left(\int_{]t,T]} B_u^{-1} e^{\Gamma_t - \Gamma_u} \left(dC_u + Z_u \, d\Gamma_u \right) \Big| \mathcal{F}_t \right)$$
$$+ \mathbb{1}_{\{\tau>t\}} B_t \, \mathbb{E}_{\mathbb{Q}^*} \left(B_T^{-1} e^{\Gamma_t - \Gamma_T} X \, \big| \, \mathcal{F}_t \right) + B_t \, \mathbb{E}_{\mathbb{Q}^*} \left(B_T^{-1} \mathbb{1}_{\{\tau \leq T\}} \tilde{X} \, \big| \, \mathcal{G}_t \right).$$

Remarks. Note that we have both conditioning with respect to $\mathcal{F}_t$ and $\mathcal{G}_t$ in the valuation formula. However, assuming that $\tilde{X}$ is $\mathcal{F}_T$-measurable, and that any $\mathbb{F}$-martingale is also a $\mathbb{G}$-martingale (the financial interpretation of this condition examined in some detail in Section 3.2), the value process of $(X, C, \tilde{X}, Z, \tau)$ can also be represented as follows, for $t < T$,

$$S_t = \mathbb{1}_{\{\tau>t\}} G_t^{-1} B_t \, \mathbb{E}_{\mathbb{Q}^*} \left(\int_{]t,T]} B_u^{-1} (G_u \, dC_u - Z_u \, dG_u) \Big| \mathcal{F}_t \right) +$$
$$+ \mathbb{1}_{\{\tau>t\}} G_t^{-1} B_t \, \mathbb{E}_{\mathbb{Q}^*} \left(G_T B_T^{-1} (X - \tilde{X}) \big| \mathcal{F}_t \right) + B_t \, \mathbb{E}_{\mathbb{Q}^*} \left(B_T^{-1} \tilde{X} \big| \mathcal{F}_t \right).$$

Under the present assumptions, the hazard process Γ is always increasing (see (H.3) in Section 3.2). If, in addition, Γ is continuous then we have

$$S_t = \mathbb{1}_{\{\tau>t\}} B_t \, \mathbb{E}_{\mathbb{Q}^*} \left(\int_{]t,T]} B_u^{-1} e^{\Gamma_t - \Gamma_u} \left(dC_u + Z_u \, d\Gamma_u \right) \Big| \mathcal{F}_t \right)$$
$$+ \mathbb{1}_{\{\tau>t\}} B_t \, \mathbb{E}_{\mathbb{Q}^*} \left(B_T^{-1} e^{\Gamma_t - \Gamma_T} (X - \tilde{X}) \big| \mathcal{F}_t \right) + B_t \, \mathbb{E}_{\mathbb{Q}^*} \left(B_T^{-1} \tilde{X} \big| \mathcal{F}_t \right).$$

The second formula in Proposition 3.3 yields the following result.

Corollary 3.3. *Assume that the $\mathbb{F}$-hazard process Γ is a continuous, increasing process. Then the value process of a defaultable contingent claim $(X, C, \tilde{X}, Z, \tau)$ coincides with the value process of a claim $(X, \hat{C}, \tilde{X}, 0, \tau)$, where we set $\hat{C}_t = C_t + \int_0^t Z_u \, d\Gamma_u$.*

Let us now consider the case when the default time admits a stochastic intensity γ with respect to $\mathbb{F}$. The valuation formula of Proposition 3.3 now takes the following form (we set $\tilde{X} = 0$ here)

$$S_t = \mathbb{1}_{\{\tau>t\}} \, \mathbb{E}_{\mathbb{Q}^*} \left(\int_{]t,T]} e^{-\int_t^u (r_v + \gamma_v) \, dv} \left(dC_u + \gamma_u Z_u \, du \right) \Big| \mathcal{F}_t \right)$$
$$+ \mathbb{1}_{\{\tau>t\}} \, \mathbb{E}_{\mathbb{Q}^*} \left(e^{-\int_t^T (r_v + \gamma_v) \, dv} X \, \big| \, \mathcal{F}_t \right).$$

To get a more concise representation for the last expression, we introduce the default-risk-adjusted interest rate $\tilde{r} = r + \gamma$ and the associated default-risk-adjusted savings account $\tilde{B}$, given by the formula

$$\tilde{B}_t = \exp \left(\int_0^t \tilde{r}_u \, du \right) = B_t e^{\Gamma_t}.$$

Although $\tilde{B}_t$ does not represent the price of a tradeable security, it has similar features as the savings account B; in particular, $\tilde{B}$ is an $\mathbb{F}$-adapted, continuous process of finite variation. In terms of the default-risk-adjusted savings account $\tilde{B}$, we obtain the following representation for the price S

$$S_t = \mathbb{1}_{\{\tau > t\}} \tilde{B}_t \, \mathbb{E}_{\mathbb{Q}^*} \left(\int_{]t,T]} \tilde{B}_u^{-1} \, dC_u + \int_t^T \tilde{B}_u^{-1} Z_u \gamma_u \, du + \tilde{B}_T^{-1} X \,\Big|\, \mathcal{F}_t \right).$$

It is noteworthy that the default time τ does not appear explicitly in the conditional expectation in the right-hand side of the last formula.

Alternative Recovery Rules

Now we will continue the study of a defaultable claim $(X, C, \tilde{X}, Z, \tau)$. Similarly as in the case of a corporate bond, the price of a generic defaultable claim depends on the choice of a recovery scheme. Formally, each recovery scheme corresponds to a specific choice of the recovery process Z.

Fractional Recovery of Par Value

We need to assume here that the par value (or the face value) of a defaultable claim is well defined. Denoting by L the constant representing the par value of the claim, and by δ the recovery rate of the claim, we set $Z_t = \delta L$. We thus deal here with a defaultable claim of the form $(X, C, \tilde{X}, \delta L, \tau)$. The ex-dividend price of this claim equals (for $\tilde{X} = 0$)

$$S_t = B_t \, \mathbb{E}_{\mathbb{Q}^*} \left(\int_{]t,T]} B_u^{-1} \big((1 - H_u) \, dC_u + \delta L \, dH_u \big) + B_T^{-1} X \mathbb{1}_{\{\tau > T\}} \,\Big|\, \mathcal{G}_t \right).$$

Consequently, by virtue of Proposition 3.3, it can be represented as follows

$$S_t = \mathbb{1}_{\{\tau > t\}} G_t^{-1} B_t \, \mathbb{E}_{\mathbb{Q}^*} \left(\int_{]t,T]} B_u^{-1} (G_u dC_u - \delta L \, dG_u) + G_T B_T^{-1} X \,\Big|\, \mathcal{F}_t \right),$$

where G is the survival process of τ with respect to the reference filtration $\mathbb{F}$. In the case of a continuous and increasing $\mathbb{F}$-hazard process Γ, the last formula yields the following expression for the pre-default value $\tilde{D}_t^\delta$

$$\tilde{D}_t^\delta = \mathbb{1}_{\{\tau > t\}} \tilde{B}_t \, \mathbb{E}_{\mathbb{Q}^*} \left(\int_{]t,T]} \tilde{B}_u^{-1} \, (dC_u + \delta L \, d\Gamma_u) + \tilde{B}_T^{-1} X \,\Big|\, \mathcal{F}_t \right),$$

where $\tilde{B}$ is the default-risk-adjusted savings account. As already observed in Corollary 3.3, if the $\mathbb{F}$-hazard process Γ is a continuous, increasing process, we may set $Z \equiv 0$ and substitute the promised dividends process C with the process $\hat{C}_t = C_t + \delta L \Gamma_t$. In the next result, the assumption that $\tilde{X} = 0$ is not needed.

Corollary 3.4. *The claims* $(X, C, \tilde{X}, \delta L, \tau)$ *and* $(X, C + \delta L\Gamma, \tilde{X}, 0, \tau)$ *are essentially equivalent if the* $\mathbb{F}$-*hazard process* Γ *is a continuous, increasing process.*

Assume now that the default time τ admits the $\mathbb{F}$-intensity process γ. Then we have

$$\tilde{D}_t^\delta = \mathbb{1}_{\{\tau > t\}} \tilde{B}_t \, \mathbb{E}_{\mathbb{Q}^*} \left(\int_{]t,T]} \tilde{B}_u^{-1} \, dC_u + \delta L \int_t^T \tilde{B}_u^{-1} \gamma_u \, du + \tilde{B}_T^{-1} X \, \Big| \, \mathcal{F}_t \right).$$

If, in addition, the sample paths of the process C are absolutely continuous functions, so that we have $C_t = \int_0^t c_u \, du$, then

$$\tilde{D}_t^\delta = \mathbb{1}_{\{\tau > t\}} \tilde{B}_t \, \mathbb{E}_{\mathbb{Q}^*} \left(\int_t^T \tilde{B}_u^{-1}(c_u + \delta L\gamma_u) \, du + \tilde{B}_T^{-1} X \, \Big| \, \mathcal{F}_t \right)$$

$$= \mathbb{1}_{\{\tau > t\}} \tilde{B}_t \, \mathbb{E}_{\mathbb{Q}^*} \left(\int_t^T \tilde{B}_u^{-1}(c_u \gamma_u^{-1} + \delta L)\gamma_u \, du + \tilde{B}_T^{-1} X \, \Big| \, \mathcal{F}_t \right),$$

where the last equality holds, provided that $\gamma > 0$. We may choose here, without loss of generality, $\mathbb{F}$-predictable versions of processes c and γ. In view of the considerations above, we are in a position to state the following corollary, which furnishes still another equivalent representation of a defaultable claim with fractional recovery of par value.

Corollary 3.5. *Assume that* $C_t = \int_0^t c_u \, du$ *and* $\Gamma_t = \int_0^t \gamma_u \, du$ *with* $\gamma > 0$. *Then a defaultable claim* $(X, C, \tilde{X}, \delta L, \tau)$ *is equivalent to a defaultable claim* $(X, 0, \tilde{X}, \tilde{Z}, \tau)$ *with the recovery process* $\tilde{Z}_t = \delta L + c_t \gamma_t^{-1}$.

Fractional Recovery of No-Default Value

In case of a general defaultable claim, the counterpart of the fractional recovery of Treasury value scheme is more aptly termed the *fractional recovery of no-default value*. In this scheme, it is assumed that the owner of a defaultable claim receives at time of default a fixed fraction of the market value of an equivalent non-defaultable security. By definition, the *no-default value* (also called the *Treasury value* by some authors) of a defaultable claim $(X, C, 0, Z, \tau)$ is equal to the expected discounted value of the promised dividends C and the promised contingent claim X, specifically:

$$U_t = B_t \, \mathbb{E}_{\mathbb{Q}^*} \left(\int_{]t,T]} B_u^{-1} \, dC_u + B_T^{-1} X \, \Big| \, \mathcal{G}_t \right).$$

Notice that U includes also the dividends paid at time t. When valuing a defaultable claim $(X, C, 0, Z, \tau)$ with the fractional recovery of no-default value, we set $Z_t = \delta U_t$. Put more explicitly, the ex-dividend price equals $S_t = I_t^1 + I_t^2$, where

$$I_t^1 = B_t \, \mathbb{E}_{\mathbb{Q}^*} \left(\int_{]t,T]} B_u^{-1} (1 - H_u) \, dC_u \, \Big| \, \mathcal{G}_t \right)$$

and

$$I_t^2 = B_t \, \mathbb{E}_{\mathbb{Q}^*} \left(\int_{]t,T]} B_u^{-1} \delta U_u \, dH_u + B_T^{-1} X \mathbb{1}_{\{\tau > T\}} \, \Big| \, \mathcal{G}_t \right).$$

The following result yields a convenient representation for S.

Proposition 3.4. *For any $t < T$, the ex-dividend price of a defaultable claim* $(X, C, 0, \delta U, \tau)$ *equals* $S_t = \mathbb{1}_{\{\tau > t\}} \left((1 - \delta) \tilde{D}_t^0 + \delta \tilde{U}_t \right)$, *where the process* $\tilde{D}_t^0$, *which equals*

$$\tilde{D}_t^0 = G_t^{-1} B_t \, \mathbb{E}_{\mathbb{Q}^*} \left(\int_{]t,T]} B_u^{-1} G_u \, dC_u + G_T B_T^{-1} X \, \Big| \, \mathcal{F}_t \right),$$

represents the pre-default value of a defaultable claim $(X, C, 0, 0, \tau)$ *with zero recovery, and the process* $\tilde{U}_t$ *is given by the following expression:*

$$\tilde{U}_t = B_t \, \mathbb{E}_{\mathbb{Q}^*} \left(\int_{]t,T]} B_u^{-1} \, dC_u + B_T^{-1} X \, \Big| \, \mathcal{F}_t \right).$$

Fractional Recovery of Pre-Default Value

Under the *fractional recovery of pre-default value* (also known as the *fractional recovery of market value*) scheme, the owner of a defaultable claim receives at time of default a fraction of its pre-default (market) value. Formally, this corresponds to an assumption that the recovery process satisfies $Z_t = \delta_t S_{t-}$, where δ is a given $\mathbb{F}$-predictable process taking values in $[0, 1]$, and S_t is the value of a claim at time t. We shall consider claims of the form $(X, 0, 0, Z, \tau)$, where $Z_t = \delta_t S_{t-}$. Using the definition of the ex-dividend price, we obtain the following equation for S

$$S_t = B_t \, \mathbb{E}_{\mathbb{Q}^*} \left(B_\tau^{-1} \delta_\tau S_{\tau-} \mathbb{1}_{\{t < \tau \leq T\}} + B_T^{-1} X \mathbb{1}_{\{\tau > T\}} \, \Big| \, \mathcal{G}_t \right), \qquad (12)$$

which can be interpreted as a backward stochastic differential equation (BSDE).

Assume that the BSDE (12) admits a unique solution, so that the price process S is well defined. Since $\mathbb{G} = \mathbb{F} \vee \mathbb{H}$, it is clear that we have $S_t = \mathbb{1}_{\{\tau > t\}} \tilde{S}_t$ for some $\mathbb{F}$-adapted process $\tilde{S}$. Obviously, we may thus substitute S_{t-} with $\tilde{S}_{t-}$ in (12), and thus we also have

$$S_t = \mathbb{1}_{\{\tau > t\}} \tilde{B}_t \, \mathbb{E}_{\mathbb{Q}^*} \left(\int_t^T \tilde{B}_u^{-1} \delta_u \tilde{S}_u \gamma_u \, du + \tilde{B}_T^{-1} X \, \Big| \, \mathcal{F}_t \right),$$

provided that τ has the $\mathbb{F}$-intensity γ. We conclude that the pre-default value process $\tilde{S}$ necessarily satisfies the following BSDE:

$$\tilde{S}_t = \tilde{B}_t \, \mathbb{E}_{\mathbb{Q}^*} \left(\int_t^T \tilde{B}_u^{-1} \delta_u \tilde{S}_u \gamma_u \, du + \tilde{B}_T^{-1} X \, \Big| \, \mathcal{F}_t \right). \qquad (13)$$

Proposition 3.5. *Assume that equations (12)-(13) admit unique solutions S and $\tilde{S}$. Then the process $\tilde{S}$ is given by the formula*

$$\tilde{S}_t = \hat{B}_t \, \mathbb{E}_{\mathbb{Q}^*}(\hat{B}_T^{-1} X \mid \mathcal{F}_t)$$

with

$$\hat{B}_t = \exp\left(\int_0^t (r_u + (1 - \delta_u)\gamma_u)\, du\right)$$

and the value of a defaultable claim, which is subject to the fractional recovery of pre-default value with $Z_t = \delta_t S_{t-}$, is equal to $S_t = \mathbb{1}_{\{\tau > t\}} \tilde{S}_t$.

Remarks. Existence and uniqueness of solutions to (12)-(13) may sometimes be deduced from general results on BSDEs (for instance when $\mathbb{F}$ is the Brownian filtration and τ is obtained through the canonical construction presented in Section 3.2). We do not go into details here.

General Recovery of Pre-Default Value

A more general recovery scheme is produced by postulating that the recovery process Z is defined through the equality $Z_t = p(t, S_{t-})$, where the recovery function $p(t, s)$ is jointly continuous with respect to the variables $(t, s) \in [0, T] \times \mathbb{R}$, and Lipschitz continuous with respect to s. Moreover, it is natural to postulate that $p(t, 0) = 0$ for every $t \in [0, T]$.

Assume that S is the unique solution to the BSDE:

$$S_t = B_t \, \mathbb{E}_{\mathbb{Q}^*}\left(B_\tau^{-1} p(\tau, S_{\tau-}) \mathbb{1}_{\{t < \tau \le T\}} + B_T^{-1} X \mathbb{1}_{\{\tau > T\}} \,\Big|\, \mathcal{G}_t\right), \qquad (14)$$

or equivalently, to the equation

$$S_t = \mathbb{E}_{\mathbb{Q}^*}\left(\int_t^T \left(p(u, S_u) h_u - r_u S_u\right) du + X \mathbb{1}_{\{\tau > T\}} \,\Big|\, \mathcal{G}_t\right).$$

Also, assume that $\tilde{S}$ is the unique solution to the BSDE

$$\tilde{S}_t = \tilde{B}_t \, \mathbb{E}_{\mathbb{Q}^*}\left(\int_t^T \tilde{B}_u^{-1} p(u, \tilde{S}_u) \gamma_u \, du + \tilde{B}_T^{-1} X \,\Big|\, \mathcal{F}_t\right) \qquad (15)$$

or, equivalently, to the equation

$$\tilde{S}_t = \mathbb{E}_{\mathbb{Q}^*}\left(\int_t^T \left(p(u, \tilde{S}_u)\gamma_u - (r_u + \gamma_u)\tilde{S}_u\right) du + X \,\Big|\, \mathcal{F}_t\right).$$

The following result shows that the solution $\tilde{S}$ to the BSDE represents the pre-default value of a defaultable claim, which is subject to a recovery at default with the recovery process of the feedback form $Z_t = p(t, S_{t-})$.

Proposition 3.6. *Let the default time τ have the $\mathbb{F}$-intensity γ. Assume that equations (14)-(15) admit unique solutions S and $\tilde{S}$. Then $S_t = \mathbb{1}_{\{\tau > t\}} \tilde{S}_t$.*

Remarks. We conclude this section by mentioning that Bélanger et al. (2001) derive several interesting relationships between values of defaultable claims that are subject to alternative recovery schemes.

Defaultable Bonds: Stochastic Intensity

Consider a defaultable zero-coupon bond with the par (face) value L and maturity date T. First, we shall re-examine the following recovery schemes: the fractional recovery of par value and the fractional recovery of Treasury value (recall that these schemes were already studied in Section 3.1 in the case of deterministic intensity). Subsequently, we shall deal with the fractional recovery of pre-default value. In this section, we assume that τ has the $\mathbb{F}$-intensity γ.

Fractional Recovery of Par Value

Under this scheme, a fixed fraction of the face value of the bond is paid to the bondholders at the time of default. Formally, we deal here with a defaultable claim $(X, 0, 0, Z, \tau)$, which settles at time T, with the promised payoff $X = L$, where L stands for the bond's face value, and with the recovery process $Z = \delta L$, where $\delta \in [0, 1]$ is a constant. The value at time $t < T$ of the bond is given by the expression

$$S_t = LB_t\, \mathbb{E}_{\mathbb{Q}^*}\big(\delta B_\tau^{-1}\mathbb{1}_{\{t < \tau \le T\}} + B_T^{-1}\mathbb{1}_{\{\tau > T\}} \,\big|\, \mathcal{G}_t\big).$$

If τ admits the $\mathbb{F}$-intensity γ, the pre-default value of the bond equals

$$\tilde{D}^\delta(t, T) = L\tilde{B}_t\, \mathbb{E}_{\mathbb{Q}^*}\Big(\delta \int_t^T \tilde{B}_u^{-1}\gamma_u\, du + \tilde{B}_T^{-1} \,\Big|\, \mathcal{F}_t\Big). \tag{16}$$

Remarks. The above setup is a special case of the fractional recovery of par value scheme with a general $\mathbb{F}$-predictable recovery process $Z_t = \delta_t$, where the process δ_t satisfies $\delta_t \in [0, 1]$ for every $t \in [0, T]$. A general version of formula (16) reads

$$\tilde{D}^\delta(t, T) = L\tilde{B}_t\, \mathbb{E}_{\mathbb{Q}^*}\Big(\int_t^T \tilde{B}_u^{-1}\delta_u\gamma_u\, du + \tilde{B}_T^{-1} \,\Big|\, \mathcal{F}_t\Big).$$

Fractional Recovery of Treasury Value

Here, in the case of default, the fixed fraction of the face value is paid to bondholders at maturity date T. A corporate zero-coupon bond is now represented by a defaultable claim $(X, 0, 0, Z, \tau)$ with the promised payoff $X = L$ and the recovery process $Z_t = \delta L B(t, T)$. As usual, $B(t, T)$ stands for the price at time t of a unit zero-coupon Treasury bond with maturity T. The corporate bond is now equivalent to a single contingent claim Y, which settles at time T and equals

$$Y = L\big(\mathbb{1}_{\{\tau > T\}} + \delta\mathbb{1}_{\{\tau \le T\}}\big).$$

The price of this claim at time $t < T$ equals

$$S_t = LB_t\, \mathbb{E}_{\mathbb{Q}^*}\big(B_T^{-1}(\delta\mathbb{1}_{\{\tau \le T\}} + \mathbb{1}_{\{\tau > T\}}) \,\big|\, \mathcal{G}_t\big),$$

or, equivalently,

$$S_t = L B_t \, \mathbb{E}_{\mathbb{Q}^*} \left(\delta B_\tau^{-1} B(\tau, T) \mathbb{1}_{\{t < \tau \leq T\}} + B_T^{-1} \mathbb{1}_{\{\tau > T\}} \,\big|\, \mathcal{G}_t \right).$$

The pre-default value $\hat{D}^\delta(t, T)$ of a defaultable bond with the fractional recovery of Treasury value equals

$$\hat{D}^\delta(t, T) = L \tilde{B}_t \, \mathbb{E}_{\mathbb{Q}^*} \left(\delta \int_t^T \tilde{B}_u^{-1} B(u, T) \gamma_u \, du + \tilde{B}_T^{-1} \,\Big|\, \mathcal{F}_t \right).$$

Again, the last formula is a special case of the general situation where $Z_t = \delta_t$, with some predictable recovery ratio process $\delta_t \in [0, 1)$.

Fractional Recovery of Pre-Default Value

Assume that δ_t is some predictable recovery ratio process $\delta_t \in [0, 1)$ and let us set $X = L$. By virtue of Proposition 3.5, we conclude that the pre-default value of the bond equals

$$D_M^\delta(t, T) = L \mathbb{E}_{\mathbb{Q}^*} \left(e^{-\int_t^T (r_u + (1 - \delta_t)\gamma_u) du} \,\Big|\, \mathcal{F}_t \right).$$

Choice of a Recovery Scheme

A challenging practical problem is the calibration of statistical properties of both the recovery process δ and the intensity process γ. According to a recent study by Guha (2003), the empirical evidence strongly suggests that the amount recovered at default is best modeled by the recovery of par value scheme. However, Bakshi et al. (2001) conclude that recovery concept that specifies the amount recovered as a fraction of appropriately discounted par value, that is, the fractional recovery of Treasury value, has broader empirical support.

Martingale Hazard Process

We now briefly discuss the relationship between the hazard process of a random time, and the so-called $\mathbb{F}$-*martingale hazard process* of a random time, which is of interest in the *martingale approach* to credit risk modeling. The next result is valid for any $\mathbb{F}$-hazard process Γ.

Lemma 3.8. *The process*

$$L_t = \mathbb{1}_{\{\tau > t\}} e^{\Gamma_t} = (1 - H_t) e^{\Gamma_t}, \quad \forall t \in \mathbb{R}_+,$$

is a $\mathbb{G}$-*martingale.*

If the $\mathbb{F}$-hazard process Γ of τ is continuous and of finite variation (hence, increasing), then it defines the compensator of the jump process H, as the following result shows.

Proposition 3.7. *Let the $\mathbb{F}$-hazard process Γ of τ be a continuous, increasing process. Then the process $M_t = H_t - \Gamma_{t \wedge \tau}$, $t \in \mathbb{R}_+$, is a $\mathbb{G}$-martingale.*

The last result suggests the following definition.

Definition 3.4. An $\mathbb{F}$-predictable, right-continuous and increasing process Λ (with $\Lambda_0 = 0$) is called an $\mathbb{F}$-*martingale hazard process* of a random time τ if and only if the process $H_t - \Lambda_{t \wedge \tau}$ is a $\mathbb{G}$-martingale.

In the *martingale approach*, the $\mathbb{F}$-martingale hazard process Λ is used, rather than the $\mathbb{F}$-hazard process Γ. An important issue thus arises: provide sufficient conditions for the equality $\Lambda = \Gamma$.

Properties of Λ

Condition (G). The process $F_t = \mathbb{Q}^* \{\tau \leq t \,|\, \mathcal{F}_t\}$ admits a modification with increasing sample paths (thus Γ is increasing as well).

Proposition 3.8. *Assume that* (G) *holds. If the process Λ*

$$\Lambda_t = \int_{]0,t]} \frac{dF_u}{1 - F_{u-}} = \int_{]0,t]} \frac{d\mathbb{Q}^*\{\tau \leq u \,|\, \mathcal{F}_u\}}{1 - \mathbb{Q}^*\{\tau < u \,|\, \mathcal{F}_u\}}$$

is $\mathbb{F}$-predictable, then Λ is the $\mathbb{F}$-martingale hazard process of the random time τ. If we additionally assume that the filtration $\mathbb{F}$ supports only continuous martingales, and if the $\mathbb{F}$-martingale hazard process Λ is continuous, then the $\mathbb{F}$-hazard process Γ is also continuous, and we have $\Gamma = \Lambda$.

If (G) fails to hold, we have the following more general result.

Proposition 3.9. *The $\mathbb{F}$-martingale hazard process of τ equals*

$$\Lambda_t = \int_{]0,t]} \frac{d\hat{F}_u}{1 - F_{u-}}$$

where $\hat{F}$ denotes the $\mathbb{F}$-compensator of the $\mathbb{F}$-submartingale F; that is, the unique $\mathbb{F}$-predictable, increasing process, such that $F - \hat{F}$ is an $\mathbb{F}$-martingale.

A counter-example given in Elliott et al. (2000) shows that if condition (G) is not assumed, the continuity of processes Γ and Λ is not sufficient for the equality $\Gamma = \Lambda$ to hold.

Martingale Hypothesis

The valuation results for defaultable claims, established in Sections 3.2–3.2, are valid when $\mathbb{G} = \mathbb{F} \vee \mathbb{H}$ for some reference filtration $\mathbb{F}$. Note that the choice of $\mathbb{F}$ is basically unrestricted. The only assumption which is always imposed is that $G_t = \mathbb{Q}^*\{\tau > t \mid \mathcal{F}_t\} > 0$ for every $t \in \mathbb{R}_+$. As already mentioned, this condition implies that τ is not an $\mathbb{F}$-stopping time.

Remarks. Note that if we consider only defaultable claims that settle prior to or at some fixed date T, it is enough to assume the strict positivity of the survival process G_t for $t \in [0, T]$. Under this assumption, τ is not an $\mathbb{F}$-stopping time, unless the inequality $\tau \geq T$ holds $\mathbb{Q}^*$-almost surely.

Typically, a reference filtration $\mathbb{F}$ represents the information flow in a model of a financial market without default. It is thus essential to ensure that the Brownian motion (or, more generally, an $\mathbb{F}$-martingale) with respect to $\mathbb{F}$ remains a Brownian motion with respect to $\mathbb{G}$ (a $\mathbb{G}$- martingale). We shall now introduce suitable assumptions related to the conditional independence of the two filtrations $\mathbb{F}$ and $\mathbb{H}$.

(H.1) For any $t \in \mathbb{R}_+$, the σ-fields $\mathcal{F}_\infty$ and $\mathcal{G}_t$ are conditionally independent given $\mathcal{F}_t$. Equivalently, for any $t \in \mathbb{R}_+$ and any bounded, $\mathcal{F}_\infty$-measurable random variable ξ we have: $\mathbb{E}_{\mathbb{Q}^*}(\xi \mid \mathcal{G}_t) = \mathbb{E}_{\mathbb{Q}^*}(\xi \mid \mathcal{F}_t)$.

(H.2) For any $t \in \mathbb{R}_+$, the σ-fields $\mathcal{F}_\infty$ and $\mathcal{H}_t$ are conditionally independent given $\mathcal{F}_t$.

(H.3) For any $t \in \mathbb{R}_+$ and any $u \leq t$, we have

$$\mathbb{Q}^*\{\tau \leq u \mid \mathcal{F}_t\} = \mathbb{Q}^*\{\tau \leq u \mid \mathcal{F}_\infty\}.$$

It is known that condition (H.2) (as well as (H.3)) is equivalent to (H.1). We shall thus refer to them collectively as hypotheses (H).

The next definition describes a specific link between the two filtrations, $\mathbb{F}$ and $\mathbb{G}$, under a given probability measure $\mathbb{Q}^*$. It should be stressed that the martingale invariance property introduced in this definition is not preserved, in general, under an equivalent change of a probability measure (for a counterexample, see Section 3.2).

Definition 3.5. We say that a filtration $\mathbb{F}$ has the *martingale invariance property* with respect to a filtration $\mathbb{G}$ if every $\mathbb{F}$-martingale is also a $\mathbb{G}$-martingale.

Hypothesis (H.1) (or, equivalently, (H.2) or (H.3)) is equivalent to the martingale invariance property of $\mathbb{F}$ with respect to $\mathbb{G}$.

Lemma 3.9. *A filtration $\mathbb{F}$ has the martingale invariance property with respect to a filtration $\mathbb{G}$ if and only if condition (H.1) is satisfied.*

Financial Interpretation

We stated above that the martingale invariance property is *essential*. We shall argue that there are good reasons for this claim:

- First, the martingale invariance property ensures the compatibility of default-free and defaultable market models. Let us call the financial market that encompasses both default-free and defaultable securities – the *full market*. The relevant information for this market is carried by the *full filtration* $\mathbb{G}$. Suppose that for the default-free sub-market, the one that only admits non-defaultable securities, the relevant information is carried by the reference filtration $\mathbb{F}$. Thus, properly discounted prices of default-free securities should naturally be $\mathbb{F}$-martingales. However, these prices should also be relevant for the full market, and so they should be $\mathbb{G}$-martingales as well. Thus, from the arbitrage pricing point of view, the martingale invariance hypothesis is indeed a very natural requirement.
- Second, there is an obvious computational advantage of models in which the hypotheses (H) hold. This comes from the fact that conditioning with respect to the larger σ-fields can be replaced with conditioning with respect to smaller σ-fields. Frequently, the (smaller) filtration $\mathbb{F}$ is a Markovian filtration (i.e., it is generated by some Markov process), which allows for utilizing Markovian properties.

Remarks. (i) In the next section, we shall present the so-called *canonical construction* of a full market, for which the martingale hypothesis is satisfied. It should be acknowledged that the construction presented is the next section is only the most basic example of a construction of a default time. More sophisticated variants and non-trivial extensions of this construction will be presented later, specifically, in Section 4.2 (conditionally independent defaults), Section 4.3 (copula-based approach), Section 4.4 (dependent intensities), and Section 4.8 (conditionally Markov credit migrations). To conclude, in most existing reduced-form models of credit risk a certain form of hypotheses (H) is satisfied, at least for a judicious choice of a probability measure and a reference filtration.

(ii) It should be made clear that in some models the reference filtration $\mathbb{F}$ carries full information about credit events, including default events. These, typically, are models where credit events are given in terms of some factor processes. For example, a structural model with a constant default barrier is such a model. In these models, we manifestly have that $\mathbb{G} = \mathbb{F}$ in which case the hypotheses (H) are trivially satisfied (but τ is typically an $\mathbb{F}$-stopping time, and thus the $\mathbb{F}$-hazard process is not well defined).

Canonical Construction

A random time obtained through the *canonical construction* has certain specific features that are not necessarily shared by all random times with a given $\mathbb{F}$-hazard process Γ.

Assume that we are given an $\mathbb{F}$-adapted, right-continuous, increasing process Γ defined on a probability space $(\tilde{\Omega}, \mathbb{F}, \mathbb{P}^*)$ such that $\Gamma_0 = 0$ and $\Gamma_\infty = +\infty$. To construct a random time τ such that Γ is the $\mathbb{F}$-hazard process of τ, we enlarge the underlying probability space $\tilde{\Omega}$. This means that Γ will be the $\mathbb{F}$-hazard process of τ under a certain extension $\mathbb{Q}^*$ of $\mathbb{P}^*$.

Let ξ be a random variable, defined on some probability space $(\hat{\Omega}, \hat{\mathcal{F}}, \hat{\mathbb{Q}})$, and uniformly distributed on the interval $[0,1]$ under $\hat{\mathbb{Q}}$. We consider the product space $\Omega = \tilde{\Omega} \times \hat{\Omega}$ with the σ-field $\mathcal{G} = \mathcal{F}_\infty \otimes \hat{\mathcal{F}}$ and the probability measure $\mathbb{Q}^* = \mathbb{P}^* \otimes \hat{\mathbb{Q}}$. The latter equality means that for any events $A \in \mathcal{F}_\infty$ and $B \in \hat{\mathcal{F}}$ we have $\mathbb{Q}^*(A \times B) = \mathbb{P}^*(A)\hat{\mathbb{Q}}(B)$.

Define the random time $\tau : \Omega \to \mathbb{R}_+$ by setting

$$\tau(\tilde{\omega}, \hat{\omega}) = \inf\{t \in \mathbb{R}_+ : e^{-\Gamma_t(\tilde{\omega})} \le \xi(\hat{\omega})\} = \inf\{t \in \mathbb{R}_+ : \Gamma_t(\tilde{\omega}) \ge \theta(\hat{\omega})\},$$

where the random variable $\theta = -\ln \xi$ has a unit exponential law under $\mathbb{Q}^*$. Let us find the conditional survival process

$$G_t = \mathbb{Q}^*\{\tau > t \mid \mathcal{F}_t\}.$$

Since $\{\tau > t\} = \{\xi < e^{-\Gamma_t}\}$ and Γ_t is $\mathcal{F}_\infty$-measurable, we obtain

$$\mathbb{Q}^*\{\tau > t \mid \mathcal{F}_\infty\} = \mathbb{Q}^*\{\xi < e^{-\Gamma_t} \mid \mathcal{F}_\infty\} = \hat{\mathbb{Q}}\{\xi < e^x\}_{x=\Gamma_t} = e^{-\Gamma_t}.$$

Consequently, we have

$$G_t = \mathbb{Q}^*\{\tau > t \mid \mathcal{F}_t\} = \mathbb{E}_{\mathbb{Q}^*}\{\mathbb{Q}^*\{\tau > t \mid \mathcal{F}_\infty\} \mid \mathcal{F}_t\} = e^{-\Gamma_t},$$

and thus G is an $\mathbb{F}$-adapted, right-continuous, and decreasing process. Thus Γ is the $\mathbb{F}$-hazard process of τ under $\mathbb{Q}^*$. In addition, we obtain the following property of the canonical construction:

$$\mathbb{Q}^*\{\tau \le t \mid \mathcal{F}_\infty\} = \mathbb{Q}^*\{\tau \le t \mid \mathcal{F}_t\}.$$

Consequently, for any two dates $0 \le u \le t$

$$\mathbb{Q}^*\{\tau \le u \mid \mathcal{F}_\infty\} = \mathbb{Q}^*\{\tau \le u \mid \mathcal{F}_t\} = \mathbb{Q}^*\{\tau \le u \mid \mathcal{F}_u\} = e^{-\Gamma_u}.$$

The latter equality shows the conditional independence under $\mathbb{Q}^*$ of the σ-fields $\mathcal{H}_t$ and $\mathcal{F}_t$, given $\mathcal{F}_\infty$. It is clear that (H.3), and thus also (H.1)-(H.2), hold. This concludes our analysis of the canonical construction of τ.

Let us observe that in fact working under (H.1) is essentially equivalent to the canonical construction. Recall that (H.1) is equivalent to (H.3), which reads: for any $t \in \mathbb{R}_+$ and any $u \le t$ we have

$$\mathbb{Q}^*\{\tau \le u \mid \mathcal{F}_t\} = \mathbb{Q}^*\{\tau \le u \mid \mathcal{F}_\infty\}. \tag{17}$$

Suppose that we are given a filtered probability space $(\Omega, \mathbb{G}, \mathbb{Q}^*)$, and τ is an $\mathbb{G}$-stopping time such that for a certain subfiltration $\mathbb{F} \subset \mathbb{G}$ we have

$$\mathbb{Q}^*\{\tau > t \mid \mathcal{F}_\infty\} = e^{-\Gamma_t}, \quad \forall t \in \mathbb{R}_+, \tag{18}$$

for some $\mathbb{F}$-adapted, continuous, strictly increasing process Γ. Then there exists a random variable $\theta : \Omega \to \mathbb{R}_+$, with exponential law under $\mathbb{Q}^*$, which is independent of the σ-field $\mathcal{F}_\infty$ and such that

$$\tau(\omega) = \inf \{ t \in \mathbb{R}_+ : \Gamma_t(\omega) \geq \theta(\omega) \}.$$

To establish this property, it is enough to set $\theta = \Gamma_\tau$. Indeed, we then have

$$\{\theta > t\} = \{\Gamma_\tau > t\} = \{\tau > A_t\},$$

where $A = \Gamma^{-1}$ is the inverse of Γ, so that $\Gamma_{A_t} = t$ for every $t \in \mathbb{R}_+$. In view of (18) and the fact that A_t is $\mathcal{F}_\infty$-measurable, we obtain for every $t \in \mathbb{R}_+$

$$\mathbb{Q}^*\{\theta > t \,|\, \mathcal{F}_\infty\} = \mathbb{Q}^*\{\tau > A_t \,|\, \mathcal{F}_\infty\} = e^{-\Gamma_{A_t}} = e^{-t},$$

as expected. Finally, it is clear that

$$\tau = \inf \{ t \in \mathbb{R}_+ : \Gamma_t \geq \Gamma_\tau \} = \inf \{ t \in \mathbb{R}_+ : \Gamma_t \geq \theta \}.$$

Remarks. It is worthwhile to observe that $\mathbb{Q}^*\{\tau > t \,|\, \mathcal{F}_\infty\}$ is deterministic if and only if the default time is independent of $\mathcal{F}_\infty$. In this case, we have

$$\mathbb{Q}^*\{\tau > t \,|\, \mathcal{F}_\infty\} = \mathbb{Q}^*\{\tau > t \,|\, \mathcal{F}_t\} = \mathbb{Q}^*\{\tau > t\}.$$

Kusuoka's Counter-Example

We shall now present a counter-example, due to Kusuoka (1999), which shows that the martingale invariance property introduced in Definition 3.5 is not necessarily preserved, in general, under an equivalent change of a probability measure.

First, we postulate that under the original probability measure $\mathbb{Q}$ the random times τ_1, τ_2 are independent random variables, with exponential laws with parameters λ_1 and λ_2, respectively. Next, for a fixed $T > 0$, we introduce an equivalent probability measure $\mathbb{Q}^*$ on $(\Omega, \mathcal{G})$ by setting

$$\frac{d\mathbb{Q}^*}{d\mathbb{Q}} = \eta_T \quad \mathbb{Q}\text{-a.s.}$$

where η_t, $t \in [0, T]$, satisfies

$$\eta_t = 1 + \sum_{i=1}^{2} \int_{]0,t]} \eta_{u-} \kappa_u^i \, dM_u^i,$$

with

$$M_t^i = H_t^i - \int_0^{t \wedge \tau_i} \lambda_i \, du,$$

where $H_t^i = \mathbb{1}_{\{\tau_i \leq t\}}$, and the auxiliary processes κ^1 and κ^2 satisfy

$$\kappa_t^1 = \mathbb{1}_{\{\tau_2 < t\}}\left(\frac{\alpha_1}{\lambda_1} - 1\right), \quad \kappa_t^2 = \mathbb{1}_{\{\tau_1 < t\}}\left(\frac{\alpha_2}{\lambda_2} - 1\right).$$

Note that the process κ^1 (κ^2, resp.) is $\mathbb{H}^2$-predictable ($\mathbb{H}^1$-predictable, resp.), where $\mathbb{H}^i$ is the filtration generated by H^i. It is easily seen that

$$H_t^i - \int_0^{t \wedge \tau_i} \lambda_u^{i*}\, du, \ i = 1, 2,$$

are $\mathbb{H}^1 \vee \mathbb{H}^2$-martingales, where

$$\lambda_t^{*1} = \lambda_1(1 - H_t^2) + \alpha_1 H_t^2 = \lambda_1 \mathbb{1}_{\{\tau_2 > t\}} + \alpha_1 \mathbb{1}_{\{\tau_2 \leq t\}},$$

and

$$\lambda_t^{*2} = \lambda_2(1 - H_t^1) + \alpha_2 H_t^1 = \lambda_2 \mathbb{1}_{\{\tau_1 > t\}} + \alpha_2 \mathbb{1}_{\{\tau_1 \leq t\}}.$$

This means that the $\mathbb{H}^2$-martingale intensity λ_1^* of default time τ_1 under $\mathbb{Q}^*$ jumps from λ_1 to α_1 after τ_2. The second default time has an analogous property. It can be checked by straightforward calculations that we have

$$\mathbb{Q}^*\{\tau_1 > s \,|\, \mathcal{H}_t^1 \vee \mathcal{H}_t^2\} \neq \mathbb{1}_{\{\tau_1 > t\}} \mathbb{E}_{\mathbb{Q}^*}(e^{\Lambda_t^{1*} - \Lambda_s^{1*}} \,|\, \mathcal{H}_t^2).$$

Let $\mathbb{G} = \mathbb{H}^1 \vee \mathbb{H}^2$ and let $\mathbb{F} = \mathbb{H}^2$ play the role of the reference filtration. The martingale invariance property of $\mathbb{F}$ with respect to $\mathbb{G}$ under $\mathbb{Q}^*$ is equivalent to the following statement: for any bounded, $\mathcal{H}_\infty^2$-measurable random variable ξ, the equality $\mathbb{E}_{\mathbb{Q}^*}(\xi \,|\, \mathcal{H}_t^1 \vee \mathcal{H}_t^2) = \mathbb{E}_{\mathbb{Q}^*}(\xi \,|\, \mathcal{H}_t^2)$ is satisfied for every $t \in \mathbb{R}_+$ (see condition (H.1)).

It is possible to check by direct calculations, that the last condition fails to hold in Kusuoka's example under $\mathbb{Q}^*$ (although, obviously, this condition is satisfied under $\mathbb{Q}$ in view of the assumed independence of τ_1 and τ_2 under $\mathbb{Q}$). This example is closely related to the valuation of *basket credit derivatives*, e.g., the *first-to-default claims* (see Chapter 4).

Change of a Probability

Until now, we have worked under a martingale measure $\mathbb{Q}^*$, which was associated with the choice of the savings account as a numeraire. We shall now relate $\mathbb{Q}^*$ to the *statistical* (or *real-world*) probability that will be denoted by $\mathbb{Q}$ in what follows. Since at this stage we do not impose any specific restrictions on the choice of $\mathbb{Q}$, we adopt the following definition.

Definition 3.6. For a fixed horizon date $T^* > 0$, the *statistical* probability $\mathbb{Q}$ is an arbitrary probability measure on $(\Omega, \mathcal{G}_{T^*})$ equivalent to $\mathbb{Q}^*$.

We denote by $\hat{\eta}$ the Radon-Nikodým density process of $\mathbb{Q}$ with respect to $\mathbb{Q}^*$ and the filtration $\mathbb{G}$, so that

$$\hat{\eta}_t = \frac{d\mathbb{Q}}{d\mathbb{Q}^*}\Big|_{\mathcal{G}_t} = \mathbb{E}_{\mathbb{Q}^*}(\hat{\eta}_{T^*}\,|\,\mathcal{G}_t), \quad \forall\, t \in [0, T^*]. \tag{19}$$

Of course, the process $\eta = 1/\hat{\eta}$ represents the density of $\mathbb{Q}^*$ with respect to $\mathbb{Q}$, that is,

$$\eta_t = \frac{d\mathbb{Q}^*}{d\mathbb{Q}}\Big|_{\mathcal{G}_t} = \frac{1}{\hat{\eta}_t}, \quad \forall\, t \in [0, T^*].$$

Let us first consider the following general question: how does the hazard process behave under an equivalent change of a probability measure?

To examine this question, let us consider an arbitrary equivalent change of a probability measure on $(\Omega, \mathcal{G}_{T^*})$. To be more specific, we postulate that a probability measure $\tilde{\mathbb{Q}}$ is given by

$$\tilde{\eta}_t = \frac{d\tilde{\mathbb{Q}}}{d\mathbb{Q}^*}\Big|_{\mathcal{G}_t} = \mathbb{E}_{\mathbb{Q}^*}(\tilde{\eta}_{T^*}\,|\,\mathcal{G}_t), \quad \forall\, t \in [0, T^*].$$

Let us define

$$\tilde{G}_t = 1 - \tilde{F}_t = 1 - \tilde{\mathbb{Q}}\{\tau \le t\,|\,\mathcal{F}_t\} = \tilde{\mathbb{Q}}\{\tau > t\,|\,\mathcal{F}_t\}$$

and let us set $\tilde{\Gamma}_t = -\ln \tilde{G}_t$ (provided, of course, that $\tilde{G}_t > 0$). Notice that the Bayes rule combined with the tower rule for conditional expectations yields (we use the property $\mathcal{F}_t \subset \mathcal{G}_t$)

$$\tilde{G}_t = \tilde{\mathbb{Q}}\{\tau > t\,|\,\mathcal{F}_t\} = \frac{\mathbb{E}_{\mathbb{Q}^*}(\tilde{\eta}_{T^*}\,\mathbb{1}_{\{\tau>t\}}\,|\,\mathcal{F}_t)}{\mathbb{E}_{\mathbb{Q}^*}(\tilde{\eta}_{T^*}\,|\,\mathcal{F}_t)} = \frac{\mathbb{E}_{\mathbb{Q}^*}(\tilde{\eta}_t\,\mathbb{1}_{\{\tau>t\}}\,|\,\mathcal{F}_t)}{\mathbb{E}_{\mathbb{Q}^*}(\tilde{\eta}_t\,|\,\mathcal{F}_t)}.$$

Lemma 3.10. *If the density process $\tilde{\eta}$ is $\mathbb{F}$-adapted then we have $\tilde{\Gamma}_t = \Gamma_t$. In general, the density process $\tilde{\eta}$ is $\mathbb{G}$-adapted and we have*

$$\tilde{G}_t = \frac{\mathbb{E}_{\mathbb{Q}^*}(\tilde{\eta}_t\,\mathbb{1}_{\{\tau>t\}}\,|\,\mathcal{F}_t)}{\mathbb{E}_{\mathbb{Q}^*}(\tilde{\eta}_t\,|\,\mathcal{F}_t)}. \tag{20}$$

The $\mathbb{F}$-hazard process $\tilde{\Gamma}$ of τ under $\mathbb{Q}$ exists if and only if $\mathbb{F}$-hazard process Γ of τ under $\mathbb{Q}^*$ is well defined. In other words, $\tilde{G}_t > 0$ for every $t \in \mathbb{R}_+$ if and only if $G_t > 0$ for every $t \in \mathbb{R}_+$.

To check the last statement, observe that, by virtue of Lemma 3.5, we have $\tilde{\eta}_t \mathbb{1}_{\{\tau>t\}} = \tilde{\xi}_t \mathbb{1}_{\{\tau>t\}}$ for some $\mathcal{F}_t$-measurable random variable $\tilde{\xi}_t$. Since $\tilde{\eta}_t$ is strictly positive, it is clear that $\tilde{\xi}_t$ has this property as well. Thus

$$\tilde{G}_t = \frac{\mathbb{E}_{\mathbb{Q}^*}(\tilde{\xi}_t \mathbb{1}_{\{\tau>t\}}\,|\,\mathcal{F}_t)}{\mathbb{E}_{\mathbb{Q}^*}(\tilde{\eta}_t\,|\,\mathcal{F}_t)} = \frac{\tilde{\xi}_t G_t}{\mathbb{E}_{\mathbb{Q}^*}(\tilde{\eta}_t\,|\,\mathcal{F}_t)} > 0.$$

Although Lemma 3.10 is too general to yield a sufficiently explicit expressions for practical purposes, it will nevertheless prove useful in what follows.

Statistical Probability

To get more explicit representations for the hazard process under a statistical probability (or, more generally, under an arbitrary equivalent probability measure), which generalize some results of Section 3.1, we shall work from now on under the following assumptions:

(M.1) Reference filtration. We have $\mathbb{G} = \mathbb{F} \vee \mathbb{H}$, where the reference filtration $\mathbb{F}$ is generated by some Brownian motion W^*.

(M.2) Martingale invariance. A Brownian motion W^* is also a martingale (and thus a Brownian motion) with respect to $\mathbb{G}$ under $\mathbb{Q}^*$.

(M.3) Regularity of the hazard process. We assume that the hazard process Γ of τ under $\mathbb{Q}^*$ is a continuous, increasing process.

The subsequent auxiliary result is an immediate consequence of a suitable version of the predictable representation theorem (see, for instance, Corollary 5.2.4 in Bielecki and Rutkowski (2002)).

Proposition 3.10. *Assume that (M.1)-(M.3) hold. Then the density process $\hat{\eta}$ of the statistical probability $\mathbb{Q}$ with respect to $\mathbb{Q}^*$ admits the following integral representation*

$$\hat{\eta}_t = 1 + \int_0^t \xi_u \, dW_u^* + \int_0^t \zeta_u \, dM_u,$$

where $M_t = H_t - \Gamma_{t \wedge \tau}$ is a (purely discontinuous) $\mathbb{G}$-martingale under $\mathbb{Q}^*$.

Since the density process $\hat{\eta}$ defined by (19) is strictly positive, its dynamics can also be put in the form of a SDE:

$$\hat{\eta}_t = 1 - \int_0^t \hat{\eta}_{u-}\beta_u \, dW_u^* + \int_0^t \hat{\eta}_{u-}\kappa_u \, dM_u, \tag{21}$$

where β and $\kappa > -1$ are some $\mathbb{G}$-predictable processes. It is well known that SDE (21) has a unique solution $\hat{\eta}$, which is given by the explicit formula (cf. Section 3.1)

$$\hat{\eta}_t = \mathcal{E}_t\left(-\int_0^{\cdot} \beta_u \, dW_u^*\right)\mathcal{E}_t\left(\int_{]0,\cdot]} \kappa_u \, dM_u\right),$$

where the first term is the classic Doléans exponential

$$\mathcal{E}_t\left(-\int_0^{\cdot} \beta_u \, dW_u^*\right) = \exp\left(-\int_0^t \beta_u \, dW_u^* - \frac{1}{2}\int_0^t |\beta_u|^2 \, du\right)$$

and the second term is given by Lemma 3.4. We are in the position to state a version of Girsanov's theorem.

Proposition 3.11. *Let the statistical probability $\mathbb{Q}$ be a probability measure on $(\Omega, \mathcal{G}_{T^*})$ equivalent to $\mathbb{Q}^*$. Assume that the Radon-Nikodým density $\hat{\eta}$ of $\mathbb{Q}$ with respect to $\mathbb{Q}^*$ satisfies SDE (21). Then the process*

$$W_t = W_t^* + \int_0^t \beta_u \, du, \quad \forall t \in [0, T^*],\tag{22}$$

is a Brownian motion with respect to $\mathbb{G}$ under $\mathbb{Q}$, and the process $\hat{M}$ given by the formula

$$\hat{M}_t = H_t - \int_0^{t \wedge \tau} (1 + \kappa_u) \, d\Gamma_u, \quad \forall t \in [0, T^*],\tag{23}$$

is a $\mathbb{G}$-martingale under $\mathbb{Q}$.

We assume from now on that the two processes intervening in the Radon-Nikodým density, β and κ, are $\mathbb{F}$-predictable. Let us emphasize that this assumption is essential for the next result to be valid.

Corollary 3.6. *The process $\hat{\Gamma}_t = \int_0^t (1 + \kappa_u) \, d\Gamma_u$ is the $\mathbb{F}$-hazard process of τ under the statistical probability $\mathbb{Q}$. In particular, if τ admits the $\mathbb{F}$-intensity γ under $\mathbb{Q}^*$ then the $\mathbb{F}$-intensity of τ under $\mathbb{Q}$ equals $\hat{\gamma}_t = (1 + \kappa_t)\gamma_t$.*

Assume, for instance, that the process β is $\mathbb{F}$-predictable and $\kappa = 0$. Then, of course, we have $\hat{\Gamma} = \Gamma$ so that the hazard processes of τ under $\mathbb{Q}$ and under $\mathbb{Q}^*$ coincide. Under these assumptions, the solution to (21) is manifestly $\mathbb{F}$-adapted, and thus the equality $\hat{\Gamma} = \Gamma$ is also an immediate consequence of Lemma 3.10.

Financial Interpretation

The financial interpretation of the processes β and κ is not straightforward. Since the default intensity is frequently modeled as an Itô process with respect to the filtration $\mathbb{F}$ generated by some Brownian motion, the impact of the equivalent change of the underlying probability measure on the intensity process γ is twofold. First, γ can be modified through a multiplication by some strictly positive process $1 + \kappa$. Second, even for the special case when $\kappa = 0$, we still observe an important effect of changing the drift in dynamics of γ. Indeed, if under $\mathbb{Q}^*$ the intensity γ satisfies

$$d\gamma_t = \mu_t \, dt + \sigma_t \, dW_t^*$$

then under $\mathbb{Q}$ we have

$$d\gamma_t = (\mu_t + \beta_t \sigma_t) \, dt + \sigma_t \, dW_t.$$

Therefore, the choice of the process β does really matter not only for the real-world dynamics of default-free securities, but also, in general, for the real-world dynamics of the default intensity process γ. Put another way, for $\kappa = 0$ the real-world probability of default depends on the risk-neutral default intensity and on the choice of the process β. Indeed, the equality $\tilde{\Gamma} = \Gamma$ does not mean that the default probabilities are the same under $\mathbb{Q}$ and $\mathbb{Q}^*$. We only have for $s \geq t$ (see Lemma 3.5)

$$\mathbb{Q}\{\tau > s \mid \mathcal{G}_t\} = \mathbb{1}_{\{\tau > t\}} \, \mathbb{E}_{\mathbb{Q}}(e^{\Gamma_t - \Gamma_s} \mid \mathcal{F}_t)$$

and

$$\mathbb{Q}^*\{\tau > s \mid \mathcal{G}_t\} = \mathbb{1}_{\{\tau > t\}} \, \mathbb{E}_{\mathbb{Q}^*}(e^{\Gamma_t - \Gamma_s} \mid \mathcal{F}_t).$$

It thus seems justified to adopt the following terminological convention:

- The process β is called the *premium for the market risk*.
- The process κ is called the *premium for the event risk*.

The term *market risk* encompasses, for instance, the interest rate risk of Treasury bonds, as well as the part of default risk associated with the variations of Treasury yields. The *event risk* represents that portion of the default risk, which is associated with other factors than the overall market risk. A practically interesting and theoretically challenging issue is whether the equality $\kappa = 0$ is compatible with the observed yields on corporate bonds (for econometric studies, see Duffee (1999)), and whether it can be explained by postulating some form of diversification of default risk (see Jarrow et al. (2002) in this regard). In a recent work by Brigo and Alfonsi (2003), the authors analyze the calibration procedure of a particular example of an intensity-based model. Since we are not in the position to present these papers here, the interested reader is referred to original works for further results and a thorough discussion of the above-mentioned issues.

Change of a Numeraire

A judicious choice of a numeraire asset is well known to be a very efficient tool in arbitrage pricing of derivative securities of various kinds (see, for instance, Geman et al. (1995)). Until now, we have invariably used the savings account, with the price process B, as a numeraire. As one might guess, from the practical viewpoint this choice is rarely the most convenient one, and thus it is essential to examine the impact of the change of a numeraire on default probabilities and valuation formulae. In the present setup, one may use either a default-free or a defaultable security as a new numeraire. The only requirement for an asset to be eligible as a universal numeraire is to have a strictly positive price process. We thus adopt the following definition, in which we write α to denote the price process of a generic, either a default-free or a defaultable, security, which pays no dividends.

Definition 3.7. A $\mathbb{G}$-adapted process α is called a *numeraire* if it is strictly positive and it represents the arbitrage price of some asset. A probability measure $\mathbb{Q}^\alpha$, equivalent to $\mathbb{Q}^*$ on $(\Omega, \mathcal{G}_{T^*})$, is termed a *martingale measure associated with* α if the relative price S/α of any (non-dividend paying) security S is a $\mathbb{Q}^\alpha$-martingale.

For instance, let us choose as a numeraire the process $A_t = B_t \hat{\eta}_t^{-1}$, where the density process $\hat{\eta}$ is given by (19). Notice that

$$B_t \, \mathbb{E}_{\mathbb{Q}^*}(B_{T^*}^{-1} A_{T^*} \,|\, \mathcal{G}_t) = B_t \, \mathbb{E}_{\mathbb{Q}^*}(\hat{\eta}_{T^*}^{-1} \,|\, \mathcal{G}_t) = B_t \hat{\eta}_t^{-1} = A_t,$$

so that A represent the price process of some contingent claim, and $A_0 = 1$.

The statistical probability $\mathbb{Q}$ is a martingale measure associated with A, and thus it becomes a pricing measure, in the sense that the price of any contingent claim will be given by the formula

$$S_t = A_t \, \mathbb{E}_{\mathbb{Q}}\left(\int_{]t,T]} A_u^{-1} \, dD_u \,\Big|\, \mathcal{G}_t \right). \tag{24}$$

If denote $\xi_t = A_t^{-1} = B_t^{-1} \hat{\eta}_t$, the last formula becomes

$$S_t = \xi_t^{-1} \, \mathbb{E}_{\mathbb{Q}}\left(\int_{]t,T]} \xi_u \, dD_u \,\Big|\, \mathcal{G}_t \right). \tag{25}$$

One recognizes here the classic idea of a *state-price density* (also known as a *deflator*). Indeed, the price at time 0 of a contingent claim which settles at time T and equals $\mathbb{1}_E$ for some event E in $\mathcal{G}_T$, equals $\int_E \xi_T \, d\mathbb{Q}$. Let us notice that formula (25) was adopted by Jamshidian (2002) as the definition of the price. Of course, it is formally equivalent to ours.

Let us now consider an arbitrary numeraire α. Since α pays no dividends, the discounted price process α/B is a strictly positive $\mathbb{Q}^*$-martingale. Then the associated martingale measure $\mathbb{Q}^\alpha$, equivalent to $\mathbb{Q}^*$ on $(\Omega, \mathcal{G}_{T^*})$, is given by

$$\frac{d\mathbb{Q}^\alpha}{d\mathbb{Q}^*} = \frac{B_0 \alpha_{T^*}}{\alpha_0 B_{T^*}} = \eta_{T^*}^\alpha, \quad \mathbb{Q}^*\text{-a.s.} \tag{26}$$

Consequently, the Radon-Nikodým density process of $\mathbb{Q}^\alpha$ with respect to $\mathbb{Q}^*$ equals, for every $t \in [0, T^*]$,

$$\eta_t^\alpha = \frac{d\mathbb{Q}^\alpha}{d\mathbb{Q}^*}\Big|_{\mathcal{G}_t} = \mathbb{E}_{\mathbb{Q}^*}(\eta_{T^*}^\alpha \,|\, \mathcal{G}_t) = \frac{B_0 \alpha_t}{\alpha_0 B_t}. \tag{27}$$

The fact that the probability $\mathbb{Q}^\alpha$ given by (26) is a martingale measure associated with the numeraire α follows from the next lemma, which is a straightforward consequence of the Bayes formula.

Lemma 3.11. *Let the process S_t/B_t, $t \in [0, T^*]$, be a $\mathbb{Q}^*$-martingale. Then the process S_t/α_t, $t \in [0, T^*]$, is a $\mathbb{Q}^\alpha$-martingale, where the probability measure $\mathbb{Q}^\alpha$ is given by (26).*

More generally, if α and β are numeraires, then the corresponding martingale measures $\mathbb{Q}^\alpha$ and $\mathbb{Q}^\beta$ satisfy on $(\Omega, \mathcal{G}_{T^*})$

$$\frac{d\mathbb{Q}^\alpha}{d\mathbb{Q}^\beta} = \frac{B_0 \alpha_{T^*}}{\alpha_0 \beta_{T^*}}, \quad \mathbb{Q}^\beta\text{-a.s.}$$

More generally, we have, for every $t \in [0, T^*]$,

$$\eta_t^{\alpha, \beta} = \frac{d\mathbb{Q}^\alpha}{d\mathbb{Q}^\beta}\Big|_{\mathcal{G}_t} = \mathbb{E}_{\mathbb{Q}^*}(\eta_{T^*}^{\alpha, \beta} | \mathcal{G}_t) = \frac{\beta_0 \alpha_t}{\alpha_0 \beta_t}. \tag{28}$$

Default-Free Numeraires

By convention, a numeraire α is termed a *default-free numeraire* if α is an $\mathbb{F}$-adapted process. Since the Radon-Nikodým density process η^α, given by formula (26), is manifestly $\mathbb{F}$-adapted, by invoking Lemma 3.10, we conclude that $\Gamma^\alpha = \Gamma$, where Γ^α is the $\mathbb{F}$-hazard process of τ under $\mathbb{Q}^\alpha$. It is thus an easy exercise to rewrite the valuation formulae of Sections 3.2–3.2 in terms of α and $\mathbb{Q}^\alpha$. Let us consider a particular example. A quite common choice of a default-free numeraire is a T-maturity default-free zero-coupon bond, with the strictly positive, $\mathbb{F}$-adapted price $\alpha_t = B(t, T)$. The corresponding martingale measure, denoted as $\mathbb{Q}_T$, is known as the *T-forward measure*. The probability $\mathbb{Q}_T$ is defined on $(\Omega, \mathcal{F}_T)$ (i.e., the maturity date T is now the horizon date), and we have

$$\frac{d\mathbb{Q}_T}{d\mathbb{Q}^*}\Big|_{\mathcal{G}_t} = \frac{B_0 B(t, T)}{B_t B(0, T)}, \quad \forall t \in [0, T].$$

Again, it is rather straightforward to express the valuation formulae for defaultable claims in terms of the forward measure $\mathbb{Q}_T$. For instance, the second formula in Proposition 3.3 becomes

$$S_t = \mathbb{1}_{\{\tau > t\}} B(t, T) \mathbb{E}_{\mathbb{Q}_T}\left(\int_{]t, T]} B^{-1}(u, T) e^{\Gamma_t - \Gamma_u} (dC_u + Z_u \, d\Gamma_u) \Big| \mathcal{F}_t\right)$$
$$+ \mathbb{1}_{\{\tau > t\}} B(t, T) \mathbb{E}_{\mathbb{Q}_T}(X e^{\Gamma_t - \Gamma_T} | \mathcal{F}_t).$$

Although the $\mathbb{F}$-hazard process under $\mathbb{Q}_T$ is the same as under $\mathbb{Q}^*$, the computations of expected values under $\mathbb{Q}^*$ and under $\mathbb{Q}_T$ are not identical. If under $\mathbb{Q}^*$ we have

$$d\gamma_t = \mu_t \, dt + \sigma_t \, dW_t^*$$

then under $\mathbb{Q}_T$ the dynamics of γ are

$$d\gamma_t = (\mu_t + b(t, T)\sigma_t) \, dt + \sigma_t \, dW_t^T,$$

where $b(t, T)$ is the volatility of the bond price, and the process W^T is a Brownian motion under $\mathbb{Q}_T$.

Defaultable Numeraires

A numeraire α is called a *defaultable numeraire* if it is not $\mathbb{F}$-adapted (of course, α is a $\mathbb{G}$-adapted process). As we know from Lemma 3.10, the $\mathbb{F}$-hazard process Γ^α of τ under $\mathbb{Q}^\alpha$ is always well defined, but it no longer coincides with the $\mathbb{F}$-hazard process Γ of τ under $\mathbb{Q}^*$.

To examine this issue, let us observe that, using formula (20) of Lemma 3.10 and equality (28), we obtain the following expression for the conditional survival process $G_t^\alpha = \mathbb{Q}^\alpha \{\tau > t \mid \mathcal{F}_t\}$ of τ under $\mathbb{Q}^\alpha$

$$G_t^\alpha = \frac{\mathbb{E}_{\mathbb{Q}^\beta}(\alpha_t \beta_t^{-1} \mathbb{1}_{\{\tau > t\}} \mid \mathcal{F}_t)}{\mathbb{E}_{\mathbb{Q}^\beta}(\alpha_t \beta_t^{-1} \mid \mathcal{F}_t)}.$$

It is convenient to introduce, following Jamshidian (2002), the notion of $\mathbb{F}$-coadapted numeraires. Numeraires α and β are said to be $\mathbb{F}$-*coadapted* (with respect to a reference filtration $\mathbb{F}$) if the process $\alpha \beta^{-1}$ are $\mathbb{F}$-adapted. Of course, if α and β are $\mathbb{F}$-coadapted numeraires then $G^\alpha = G^\beta$.

An arbitrary numeraire α is not necessarily $\mathbb{F}$-coadapted with the savings account B, and thus G^α may differ from G. Consequently, it may happen that $\Gamma^\alpha \neq \Gamma$, in general.

Preprice of a Defaultable Claim

We shall now examine more closely an application of a change of numeraire to the valuation of defaultable (or default-free) claims. Let α be an arbitrary numeraire. We restrict our attention to a generic claim, which settles at time T, and thus is represented by a $\mathcal{G}_T$-measurable random variable Y_T. First, it should be checked that the price is invariant with respect to a choice of numeraire. To this end, observe that the arbitrage price Y_t at time $t \leq T$ satisfies

$$Y_t := B_t \, \mathbb{E}_{\mathbb{Q}^*}(B_T^{-1} Y_T \mid \mathcal{G}_t) = \alpha_t \, \mathbb{E}_{\mathbb{Q}^\alpha}(\alpha_T^{-1} Y_T \mid \mathcal{G}_t) =: Y_t^\alpha,$$

where the second equality is an easy consequence of (27) of the Bayes formula. Thus, as expected, the price of Y_T is independent of the choice of a numeraire. Specifically, for an arbitrary choice of a numeraire α we have $Y_t^\alpha = Y_t$ for every $t \in [0, T]$. Following Jamshidian (2002), we introduce the concept of a preprice of a claim Y_T relative to α and $\mathbb{F}$.

Definition 3.8. For any date $t \leq T$, the $(\mathbb{F}, \alpha)$-*preprice* $\hat{Y}_t^\alpha$ of a claim Y_T is defined by the formula $\hat{Y}_t^\alpha = \alpha_t \, \mathbb{E}_{\mathbb{Q}^\alpha}(\alpha_T^{-1} Y_T \mid \mathcal{F}_t)$.

Let us examine briefly the properties of the preprice. It is easily seen that the equality $\hat{Y}_t^\alpha = Y_t$ is satisfied for every $t \leq T$ if and only if the relative price Y/α follows an $\mathbb{F}$-adapted process, that is, processes Y and α are $\mathbb{F}$-coadapted. The definition of the preprice is fairly general, but the last property can be given a nice financial interpretation in the case of defaultable assets. Namely, it says that if Y and α are two defaultable claims (with a common default time) that exhibit the same pattern of behavior at default, then the valuation of Y in terms of α requires only the knowledge of the reference filtration $\mathbb{F}$.

More generally, if α and β are two $\mathbb{F}$-coadapted numeraires then for any contingent claim Y_T we have $\hat{Y}_t^\alpha = \hat{Y}_t^\beta$. This follows, for instance, from the fact that the Radon-Nikodým density process $\eta^{\alpha,\beta}$ is $\mathbb{F}$-adapted (see 28), and thus

$$\alpha_t \, \mathbb{E}_{\mathbb{Q}^\alpha}(\alpha_T^{-1} Y_T \mid \mathcal{F}_t) = \alpha_t \, \frac{\mathbb{E}_{\mathbb{Q}^\beta}(\alpha_T^{-1} \alpha_T \beta_T^{-1} Y_T \mid \mathcal{F}_t)}{\mathbb{E}_{\mathbb{Q}^\beta}(\alpha_T \beta_T^{-1} \mid \mathcal{F}_t)}$$

$$= \alpha_t \, \frac{\mathbb{E}_{\mathbb{Q}^\beta}(\beta_T^{-1} Y_T \mid \mathcal{F}_t)}{\alpha_t \beta_t^{-1}} = \beta_t \, \mathbb{E}_{\mathbb{Q}^\beta}(\beta_T^{-1} Y_T \mid \mathcal{F}_t).$$

The first equality above yields easily the following general result.

Lemma 3.12. *If α and β are numeraires then we have $\hat{Y}_t^\alpha = \alpha_t (\hat{\alpha}_t^\beta)^{-1} \hat{Y}_t^\beta$, where $\hat{\alpha}_t^\beta = \beta_t \, \mathbb{E}_{\mathbb{Q}^\alpha}(\beta_T^{-1} \alpha_T \mid \mathcal{F}_t)$ is the $(\mathbb{F}, \beta)$-preprice of the claim α_T.*

It is also clear that the notion of the $(\mathbb{F}, \alpha)$-*preprice* is related to the concept of the pre-default price, at least in the special case of the form $Y_T = X \mathbb{1}_{\{\tau > T\}}$. Any claim of this form will be called a *survival claim* in what follows.

For any survival claim Y_T, by virtue of Lemma 3.5, we have

$$Y_t = \mathbb{1}_{\{\tau > t\}} B_t e^{\Gamma_t} \, \mathbb{E}_{\mathbb{Q}^*}(B_T^{-1} Y_T \mid \mathcal{F}_t) = \mathbb{1}_{\{\tau > t\}} \tilde{Y}_t,$$

where $\tilde{Y}_t$ is the pre-default price at time t of Y_T. By an application of Lemma 3.5, we also obtain

$$Y_t = Y_t^\alpha = \alpha_t \, \mathbb{E}_{\mathbb{Q}^\alpha}(\alpha_T^{-1} Y_T \mid \mathcal{G}_t) = \mathbb{1}_{\{\tau > t\}} \alpha_t e^{\Gamma_t^\alpha} \, \mathbb{E}_{\mathbb{Q}^\alpha}(\alpha_T^{-1} Y_T \mid \mathcal{F}_t).$$

We thus obtain the following result, due to Jamshidian (2002).

Proposition 3.12. *Let $Y_T = X \mathbb{1}_{\{\tau > T\}}$, where X is an $\mathcal{F}_T$-measurable random variable, and let α be a numeraire. Then the price Y_t of Y_T equals*

$$Y_t = \mathbb{1}_{\{\tau > t\}} e^{\Gamma_t^\alpha} \hat{Y}_t^\alpha, \quad \forall t \in [0, T],$$

where $\hat{Y}_t^\alpha$ is the $(\mathbb{F}, \alpha)$-preprice of Y_T.

In view of the last result, the pre-default price $\tilde{Y}_t$ of a survival claim Y_T satisfies $\tilde{Y}_t = e^{\Gamma_t^\alpha} \hat{Y}_t^\alpha$ for every $t \in [0, T]$ and an arbitrary choice of a numeraire α. Let us stress that the concept of a preprice corresponds to the general notion of the cum-dividend price of a contingent claim, while the pre-default price is associated with the idea of the ex-dividend price. The last two notions clearly coincide in the case of a survival claim, but not in general. It seems likely that the preprice will prove to be a more convenient tool than the pre-default price if one wishes to analyze simultaneously several default-free and defaultable claims (possibly with different default times, maturity dates, etc.).

Jamshidian (2002) correctly argues that the concept of the preprice can be used in the analysis of an option to exchange two survival claims. Let A_T and B_T be two survival claims that settle at T. Consider a claim $C_T = (A_T - B_T)^+$

(note that C_T is also a survival claim). We wish to evaluate the price C_t for $t \leq T$ using the process α as a numeraire. In view of Proposition 3.12, we have

$$C_t = 1\!\!1_{\{\tau > t\}} e^{\Gamma_t^\alpha} \hat{C}_t^\alpha = 1\!\!1_{\{\tau > t\}} \alpha_t e^{\Gamma_t^\alpha} \, \mathbb{E}_{\mathbb{Q}^\alpha} (\alpha_T^{-1} (A_T - B_T)^+ \mid \mathcal{F}_t).$$

Proposition 3.13. *We shall show that last formula can also be represented as follows*

$$C_t = 1\!\!1_{\{\tau > t\}} \alpha_t e^{\Gamma_t^\alpha} \, \mathbb{E}_{\mathbb{Q}^\alpha} (\alpha_T^{-1} (\hat{A}_T^\alpha - \hat{B}_T^\alpha)^+ \mid \mathcal{F}_t).$$

Furthermore, we have

$$C_T = 1\!\!1_{\{\tau > T\}} e^{\Gamma_T^\alpha} (\hat{A}_T^\alpha - \hat{B}_T^\alpha)^+.$$

To establish the first formula in Proposition 3.13, it suffices to apply Proposition 3.12 for $t = T$ to survival claims A_T and B_T. We obtain

$$\mathbb{E}_{\mathbb{Q}^\alpha} (\alpha_T^{-1} (A_T - B_T)^+ \mid \mathcal{F}_t) = \mathbb{E}_{\mathbb{Q}^\alpha} \left(\alpha_T^{-1} 1\!\!1_{\{\tau > T\}} e^{\Gamma_T^\alpha} (\hat{A}_T^\alpha - \hat{B}_T^\alpha)^+ \, \middle| \, \mathcal{F}_t \right)$$

$$= \mathbb{E}_{\mathbb{Q}^\alpha} \left(1\!\!1_{\{\tau > T\}} e^{\Gamma_T^\alpha} (\mathbb{E}_{\mathbb{Q}^\alpha} (\alpha_T^{-1} Y_T \mid \mathcal{F}_T) - \mathbb{E}_{\mathbb{Q}^\alpha} (\alpha_T^{-1} Y_T \mid \mathcal{F}_T))^+ \, \middle| \, \mathcal{F}_t \right)$$

$$= \mathbb{E}_{\mathbb{Q}^\alpha} \left((\mathbb{E}_{\mathbb{Q}^\alpha} (\alpha_T^{-1} Y_T \mid \mathcal{F}_T) - \mathbb{E}_{\mathbb{Q}^\alpha} (\alpha_T^{-1} Y_T \mid \mathcal{F}_T))^+ \, \middle| \, \mathcal{F}_t \right)$$

$$= \mathbb{E}_{\mathbb{Q}^\alpha} (\alpha_T^{-1} (\hat{A}_T^\alpha - \hat{B}_T^\alpha)^+ \mid \mathcal{F}_t).$$

Since $\alpha_T^{-1} (\hat{A}_T^\alpha - \hat{B}_T^\alpha)^+$ is $\mathcal{F}_T$-measurable, the second equality follows from the first formula. Let us finally observe that

$$C_t = 1\!\!1_{\{\tau > t\}} \alpha_t e^{\Gamma_t^\alpha} \, \mathbb{E}_{\mathbb{Q}^\alpha} (\alpha_T^{-1} e^{-\Gamma_T^\alpha} (\tilde{A}_T - \tilde{B}_T)^+ \mid \mathcal{F}_t).$$

We thus conclude that in the case of survival claims working with preprices is equivalent to working with pre-default prices. For this reason, in our presentation of Jamshidian (2002) approach to survival swaptions, we shall make use of pre-default prices, rather than preprices. It should be observed, however, that the concept of the preprice is more general, and it is more suitable for an analysis of general default-free and defaultable claims and numeraires.

Credit Default Swaption

As a non-trivial application of the hazard process approach to the valuation of defaultable claims, we shall now briefly describe a challenging issue of valuation of a credit default swaption. By definition, the option expires worthless if there was default of the reference entity prior to the start date, denoted by T_0 in the sequel. Essentially, we follow here Jamshidian (2002), although our presentation is slightly different and definitely less detailed than the original one.

In particular, we do not make an explicit use of the general concept of the $(\mathbb{F}, \alpha)$-pre-default price if a contingent claim, but we instead refer directly to the idea of the pre-default price of a defaultable claim. As we already mentioned, both concepts are essentially equivalent in the case of survival claims.

Valuation of Survival Swaptions

Consider a generic contingent claim Y_T which settles at time T (that is, Y_T is a $\mathcal{G}_T$-measurable random variable), and assume that Y_T is a survival claim. This means that $Y_T = 0$ on the set $\{\tau \le T\}$ or, equivalently, that $Y_T = \mathbb{1}_{\{\tau > T\}} Y_T$. By applying Lemma 3.5, we obtain

$$Y_T = \mathbb{E}_{\mathbb{Q}^*}(Y_T \,|\, \mathcal{G}_T) = \mathbb{E}_{\mathbb{Q}^*}(\mathbb{1}_{\{\tau > T\}} Y_T \,|\, \mathcal{G}_T) = \mathbb{1}_{\{\tau > T\}} \frac{\mathbb{E}_{\mathbb{Q}^*}(\mathbb{1}_{\{\tau > T\}} Y_T \,|\, \mathcal{F}_T)}{\mathbb{Q}^*\{\tau > T \,|\, \mathcal{F}_T\}}.$$

In view of the last equality, we may and do assume, without loss of generality, that $Y_T = \mathbb{1}_{\{\tau > T\}} \tilde{Y}_T$, where $\tilde{Y}_T$ is an $\mathcal{F}_T$-measurable random variable. We thus deal here with a defaultable claim that is subject to zero recovery, of the form $(\tilde{Y}_T, 0, 0, 0, \tau)$.

Remarks. Recall that the property used here is quite general, namely, for any $\mathcal{G}_t$-measurable random variable Y, and for any $t \in \mathbb{R}_+$ there exists an $\mathcal{F}_t$-measurable random variable $\tilde{Y}_t$ such that $\mathbb{1}_{\{\tau > t\}} Y = \mathbb{1}_{\{\tau > t\}} \tilde{Y}_t$.

Assume, in addition, that the terminal payoff Y_T can be represented as follows: $Y_T = (V_T^1 - \kappa V_T^2)^+$ for some $\mathcal{G}_T$-measurable random variables V_T^1 and V_T^2, and some constant $\kappa > 0$. Such a claim will be termed a *survival swaption* in what follows. We have

$$Y_T = \mathbb{1}_{\{\tau > T\}} (V_T^1 - \kappa V_T^2)^+ = (\mathbb{1}_{\{\tau > T\}} V_T^1 - \kappa \mathbb{1}_{\{\tau > T\}} V_T^2)^+$$
$$= (\mathbb{1}_{\{\tau > T\}} \tilde{V}_T^1 - \kappa \mathbb{1}_{\{\tau > T\}} \tilde{V}_T^2)^+ = \mathbb{1}_{\{\tau > T\}} (\tilde{V}_T^1 - \kappa \tilde{V}_T^2)^+,$$

where the random variables $\tilde{V}_T^i$, $i = 1, 2$ are $\mathcal{F}_T$-measurable, specifically,

$$\tilde{V}_T^i = \frac{\mathbb{E}_{\mathbb{Q}^*}(\mathbb{1}_{\{\tau > T\}} V_T^i \,|\, \mathcal{F}_T)}{\mathbb{Q}^*\{\tau > T \,|\, \mathcal{F}_T\}}.$$

Lemma 3.13. *The ex-dividend price of Y_T at time $t < T$ equals*

$$S_t = \mathbb{1}_{\{\tau > t\}} \tilde{B}_t \, \mathbb{E}_{\mathbb{Q}^*}\big(\tilde{B}_T^{-1}(\tilde{V}_T^1 - \kappa \tilde{V}_T^2)^+ \,\big|\, \mathcal{F}_t\big),$$

where $\tilde{B}_t = B_t e^{\Gamma_t}$.

Suppose, in addition, that the random variable $\tilde{V}_T^2$ is strictly positive, so that the process $\tilde{V}_t^2$ representing the pre-default price of V_T^2 is also strictly positive. Note that we have

$$\tilde{V}_t^2 = \tilde{B}_t \, \mathbb{E}_{\mathbb{Q}^*}\big(\tilde{B}_T^{-1} \tilde{V}_T^2 \,\big|\, \mathcal{F}_t\big).$$

We define a probability measure $\tilde{\mathbb{Q}}$, equivalent to $\mathbb{Q}^*$ on $(\Omega, \mathcal{G}_T)$, by setting for every $t \in [0, T]$

$$\tilde{\eta}_t = \frac{d\tilde{\mathbb{Q}}}{d\mathbb{Q}^*}\bigg|_{\mathcal{G}_t} = \frac{d\tilde{\mathbb{Q}}}{d\mathbb{Q}^*}\bigg|_{\mathcal{F}_t} = \frac{\tilde{B}_0 \tilde{V}_t^2}{\tilde{B}_t \tilde{V}_0^2}.$$

Using the Bayes formula, it is easy to verify that

$$S_t = 1\!\!1_{\{\tau > t\}} \tilde{V}_t^2 \, \mathbb{E}_{\tilde{\mathbb{Q}}}\big((\tilde{\kappa}_T - \kappa)^+ \,\big|\, \mathcal{F}_t\big), \tag{29}$$

where $\tilde{\kappa}_T = \tilde{V}_T^1 / \tilde{V}_T^2$. In addition, the process $\tilde{\kappa}_t = \tilde{V}_t^1 / \tilde{V}_t^2$, which represents an abstract *swap rate*, is a martingale with respect to $\mathbb{F}$ under $\tilde{\mathbb{Q}}$.

Suppose now that the reference filtration $\mathbb{F}$ is generated by some $\tilde{\mathbb{Q}}$-Brownian motion $\tilde{W}$. Assuming, in addition, that the process $\tilde{\kappa}$ is strictly positive, we conclude that the dynamics of this process under $\tilde{\mathbb{Q}}$ are

$$d\tilde{\kappa}(t) = \tilde{\kappa}(t)\tilde{\nu}(t)\, d\tilde{W}_t \tag{30}$$

for some volatility process $\tilde{\nu}$. The following result is a consequence of (29).

Proposition 3.14. *Suppose that $\tilde{\nu}$ is a deterministic function. The price S_t of a survival option equals*

$$S_t = 1\!\!1_{\{\tau > t\}} \tilde{V}_t^2 \Big(\tilde{\kappa}(t) N\big(d_+(t, T)\big) - \kappa N\big(d_-(t, T)\big) \Big), \tag{31}$$

where

$$d_\pm(t, T_j) = \frac{\ln(\tilde{\kappa}(t)/\kappa) \pm \frac{1}{2}\tilde{v}^2(t, T)}{\tilde{v}(t, T)}$$

and $\tilde{v}^2(t, T) = \int_t^T |\tilde{\nu}(u)|^2 \, du$.

Jamshidian's Model

Before proceeding to credit default swaptions, we shall describe briefly the lognormal model of co-terminal (non-defaultable) swap rates (see Jamshidian (1997)). We consider here a family of forward swap rate associated with interest rate swaps in which a fixed rate is periodically exchanged for a floating (LIBOR) rate. For a given collection of dates $0 < T_0 < T_1 < \cdots < T_n$, referred to as the *tenor structure*, the corresponding swap rate processes are given by the formula (we denote $\alpha_j = T_j - T_{j-1}$)

$$\kappa(t, T_j, n - j) = \frac{B(t, T_j) - B(t, T_n)}{\alpha_{j+1} B(t, T_{j+1}) + \cdots + \alpha_n B(t, T_n)}.$$

According to Jamshidian's model, these processes are lognormal martingales. More precisely, for any $j = 1, \ldots, n$ we have

$$d\kappa(t, T_j, n - j) = \kappa(t, T_j, n - j)\nu(t, T_j)\, d\tilde{W}_t^{T_{j+1}},$$

where $\tilde{W}^{T_{j+1}}$, $t \in [0, T_{j+1}]$ is a d-dimensional standard Brownian motion under the corresponding *swap measure* $\tilde{\mathbb{P}}_{T_{j+1}}$.

It follows that for $j = 1, \ldots, n$ the price of the jth swaption in Jamshidian's model is given by the Black formula for swaptions:

$$S_t^j = \sum_{k=j+1}^{n} \alpha_k B(t, T_k) \Big(\kappa(t, T_j, n-j) N\big(d_+(t, T_j)\big) - \kappa N\big(d_-(t, T_j)\big) \Big),$$

where

$$d_\pm(t, T_j) = \frac{\ln(\kappa(t, T_j, n-j)/\kappa) \pm \frac{1}{2} v^2(t, T_j)}{v(t, T_j)}$$

and $v^2(t, T_j) = \int_t^{T_j} |\nu(u, T_j)|^2 \, du$.

Credit Default Swaption

We assume that the tenor structure $0 < T_0 < T_1 < \cdots < T_n$ is given.

Spread premium leg. By a *survival annuity stream* we mean the sequence of cash flows

$$\sum_{i=1}^{n} \kappa_i \mathbb{1}_{\{\tau > T_i\}} \mathbb{1}_{\{t=T_i\}},$$

where κ_i is an $\mathcal{F}_{T_i}$-measurable random variable. Each component represents a payoff κ_i, which is due to be paid at T_i, provided that default has not occurred prior to or at this date. For instance, $\kappa_i = \kappa$ may represent the credit spread over LIBOR of a defaultable bond.

Default protection leg. A generic *default protection stream* is defined as the following sequence of cash flows

$$\sum_{i=1}^{n} L_i \mathbb{1}_{\{T_{i-1} < \tau \le T_i\}} \mathbb{1}_{\{t=T_i\}},$$

where L_i is an $\mathcal{F}_{T_{i-1}}$-measurable random variable. Each component represents a protection payoff L_i, which is received at T_i if default has occurred between T_{i-1} and T_i. In practice, L_i may equal, for instance, a constant rate plus LIBOR $L(T_{i-1})$ times the daycount fraction α_i.

Valuation. Observe that if default has occurred prior to or at T_i for $i = 1, \ldots, n$, then clearly both legs of the contract become worthless. Note that this remark also applies to the start date T_0. We may thus expect that Lemma 3.13 is applicable, at least in principle, to the valuation of a credit default swaption with the expiration date $T \le T_0$. Using Proposition 3.14, combined with a suitable approximation of the volatility process in (30), Jamshidian (2002) derives an approximate valuation formula (31) for a credit default swaption (a similar result was obtained in a different setup by Schönbucher (2000b)).

A Practical Example

Although, theoretically speaking, the features of general intensity-based approach are well understood, practical implementations of reduced - form methodology are usually done under several simplifying assumptions. We present here a brief summary of the StepCredit model for pricing credit instruments, developed by a major U.S. bank. The bank employs a reduced-form methodology in order to manage risks associated with trading credit default swaps. Basic modeling assumptions are:

- The default intensity and the interest rates are independent.
- The recovery scheme is consistent with the fractional recovery of par value, with the stochastic recovery rate δ_t.
- The recovery rate process δ_t is stationary, with $\mathbb{E}_{\mathbb{Q}^*}(\delta_t) = \delta$.
- The recovery rate process is independent of both the interest rates and the default intensity.

The reference leg of a CDS: corporate coupon bond. This leg of the credit default swap is modeled at present time $t = 0$ as (it is implicitly assumed that the reference σ-field is trivial at time 0)

$$\sum_{M_k > 0} c_k B(0, M_k) \mathbb{E}_{\mathbb{Q}^*} \left(e^{-\int_0^{M_k} \gamma_u du} \right)$$

$$+ \delta \int_0^T c(u) B(0, u) \mathbb{E}_{\mathbb{Q}^*} \left(\gamma_u e^{-\int_0^u \gamma_v dv} \right) du,$$

where T is the maturity of the bond, the M_ks are coupon payment dates, and the quantities c_k and $c(s)$ are related to coupon rates and amortized outstanding principal.

The premium/protection payment leg. This represents the PV of the CDS contract and is evaluated at time $t = 0$ as

$$\sum_{T_k > 0} P_k B(0, T_k) \mathbb{E}_{\mathbb{Q}^*} \left(e^{-\int_0^{T_k} \gamma_u du} \right)$$

$$- \int_0^T P^\delta(u) B(0, u) \mathbb{E}_{\mathbb{Q}^*} \left(\gamma_u e^{-\int_0^u \gamma_v dv} \right) du,$$

where the T_ks are swap premium payment dates, and the quantities P_k and $P^\delta(u)$ are related to swap premia and the bond's insurance payment, respectively. StepCredit postulates that the stochastic intensity γ follows a sufficiently regular process, so that for any future date $t > 0$ we have

$$\mathbb{E}_{\mathbb{Q}^*} \left(e^{-\int_0^t \gamma_u du} \right) = e^{-\int_0^t s(u) du}$$

for some right-continuous, piecewise constant function s. Thus, for example, the present value of the CDS can be written as

$$\sum_{T_k > 0} P_k B(0, T_k) e^{-\int_0^{T_k} s(u) du} - \int_0^T P^\delta(u) B(0, u) s(u) e^{-\int_0^u s(v) dv} du.$$

The model is then calibrated at time 0 for the intensity function s, and thus the stochastic intensity γ becomes spurious. To conclude, under the present set of assumptions there is no real advantage of introducing the concept of a stochastic intensity if we are only interested in the valuation of the CDS (or other contingent claims of similar features) at time 0.

Suppose, however, that we wish to find the price at some future date $t > 0$ of the CDS within the present setup. By reasoning in a similar manner as above, we see that it is now convenient to introduce a random field $s(t, u)$, which is implicitly defined through the following relationship

$$\mathbb{E}_{\mathbb{Q}^*}\left(e^{-\int_t^T \gamma_u du} \,\Big|\, \mathcal{F}_t\right) = e^{-\int_t^T s(t,u)du}.$$

Notice that $s(u) = s(0, u)$ for every $u \in \mathbb{R}_+$. The random field $s(t, u)$ appears to be useful in the valuation of contingent claims at future dates. For instance, the value of the CDS at time $t > 0$ can be represented in terms of this random field

$$\sum_{T_k > t} P_k B(t, T_k) e^{-\int_t^{T_k} s(t,u)du} - \int_t^T P^\delta(u) B(t, u) s(t, u) e^{-\int_t^u s(t,v)dv} \, du.$$

Of course, the knowledge of the intensity function $s(u) = s(0, u)$ is insufficient for the calculation of the price of the CDS at time $t > 0$.

Remarks. It is interesting to notice that the field $s(t, u)$ plays a role analogous to the instantaneous forward rate in the representation of price process of a discount bond. Under mild assumptions on default-free and defaultable term structures, it can be shown that $s(t, u)$ represents the so-called *forward credit spread*, that is, the difference between the instantaneous forward rates implied by prices of defaultable bonds with zero recovery and default-free bonds. To be more specific, we have

$$\tilde{D}^0(t, T) = B(t, T) e^{-\int_0^t s(t,u) \, du}.$$

For similar representations in a more general framework, see Section 4.8.

3.3 Martingale Approach

The term *martingale approach* refers to a specific version of the intensity-based approach, in which we work directly with the martingale hazard process of a default time. In the preceding section, we have assumed that the information was made of two parts, the reference filtration (corresponding to market risk) and the filtration of the default time (associated with default risk). It may be difficult to separate the two kinds of available information coming from the market data. For this reason, it seems useful to develop an approach in which we deal only with a filtration $\mathbb{G}$, such that τ is a $\mathbb{G}$-stopping time. However, as we shall see in that follows, the results obtained within this setup are less explicit than results of Section 3.2. In addition, some crucial assumptions appearing in valuation results are difficult to verify.[5] Therefore, in our opinion this attempt to develop a more general methodology than the standard approach based on the notion of an $\mathbb{F}$-hazard process can not be judged at present as fully successful. Nevertheless, we decided to present it here for the sake of completeness.

Standing Assumptions

We work under the following standing assumptions (see Duffie et al. (1996)).

(A.1) Martingale measure. We are given a probability space $(\Omega, \mathbb{G}, \mathbb{Q}^*)$, with $\mathbb{Q}^*$ interpreted as a martingale measure. A $\mathbb{G}$-adapted process r represents the short-term interest rate, and the process $B_t = \exp\left(\int_0^t r_u \, du\right)$ models the savings account.

(A.2) Promised claim. A $\mathcal{G}_T$-measurable random variable X represents the *promised claim*, that is, the amount of cash which the owner of a defaultable claim is entitled to receive at time T, provided that the default has not occurred prior to T.

(A.3) Recovery process. A $\mathbb{G}$-predictable process Z, called *recovery process*, models the payoff which is actually received by the owner of a defaultable claim in case the default occurs prior to the maturity T.

(A.4) Default time. We assume that the *default time* τ is a $\mathbb{G}$-stopping time. In addition, we postulate that there exists a $\mathbb{G}$-predictable process λ, such that the process $M_t = H_t - \int_0^{t \wedge \tau} \lambda_u \, du$ is a $\mathbb{G}$-martingale under $\mathbb{Q}^*$. The process λ is a $\mathbb{G}$-martingale intensity of τ under $\mathbb{Q}^*$. Consequently, the process $\Lambda_t = \int_0^t \lambda_u \, du$ is a $\mathbb{G}$-martingale hazard process of τ under $\mathbb{Q}^*$.

Remarks. In the context of (A.4), one needs to realize that when working with $\mathbb{G}$-adapted intensities, we face the situation where the intensity process is not unique. In fact, if $\lambda_t, t \geq 0$ is an intensity, then any process $\hat{\lambda}$ such that $\hat{\lambda}_t \mathbb{1}_{\{\tau > t\}} = \lambda_t \mathbb{1}_{\{\tau > t\}}$ is an intensity (for example, the process $\lambda_t \mathbb{1}_{\{\tau > t\}} + a \mathbb{1}_{\{\tau \geq t\}}$ is an intensity for any choice of the constant a). This suggests that the concept of the $\mathbb{G}$-martingale intensity is more ambiguous that the notion of the $\mathbb{F}$-intensity for some reference filtration $\mathbb{F}$.

Valuation of Defaultable Claims

According to (A.2)-(A.3), we have $C \equiv 0$ (the promised dividends are zero) and $\tilde{X} = 0$ (the recovery payoff at T equals 0), so that a generic *defaultable claim* is now represented by a triplet (X, Z, τ). As in Section 2.1, we postulate

---

[5] Unless, for instance, we work within the classic framework of Cox processes. In this case, however, the martingale approach is not required at all, since the direct approach easily yields all desired results.

that the ex-dividend price S_t at time $t < T$ of a defaultable claim (X, Z, τ) equals

$$S_t = B_t \, \mathbb{E}_{\mathbb{Q}^*} \left(\int_{]t,T]} B_u^{-1} \, dD_u \, \Big| \, \mathcal{G}_t \right),$$

where

$$D_t = X \mathbb{1}_{\{\tau>T\}} \mathbb{1}_{\{t=T\}} + Z_\tau \mathbb{1}_{\{\tau \le T\}} \mathbb{1}_{\{t=\tau\}}$$

is the dividend process. More explicitly, for any $t \in [0, T[$,

$$S_t = B_t \, \mathbb{E}_{\mathbb{Q}^*} \left(B_\tau^{-1} Z_\tau \mathbb{1}_{\{t < \tau \le T\}} + B_T^{-1} X \mathbb{1}_{\{\tau > T\}} \, \Big| \, \mathcal{G}_t \right).$$

In addition, we set $S_T = X^d(T) = X \mathbb{1}_{\{\tau>T\}}$. Notice that the ex-dividend price always vanishes on the set $\{\tau \le t\}$. Therefore, we have $S_t = \mathbb{1}_{\{\tau>t\}} \tilde{S}_t$, where $\tilde{S}$ is termed the *pre-default value* of a claim.

Our goal is to provide an alternative representation for the process S with an explicit use of a $\mathbb{G}$-martingale intensity λ. To this end, let us introduce an auxiliary process $h_t = \lambda_t \mathbb{1}_{\{t \le \tau\}}$. Since

$$\hat{M}_t = H_t - \int_0^t h_u du$$

is a $\mathbb{G}$-martingale, the process $A_t = \int_0^t h_u \, du = \Lambda_{t \wedge \tau}$ is the compensator of the bounded $\mathbb{G}$-submartingale H. The following preliminary result provides an alternative representation for the price S in terms of h.

Lemma 3.14. *The value process S satisfies*

$$S_t = B_t \, \mathbb{E}_{\mathbb{Q}^*} \left(\int_t^T B_u^{-1} Z_u h_u \, du + B_T^{-1} X \mathbb{1}_{\{T < \tau\}} \, \Big| \, \mathcal{G}_t \right)$$

or, equivalently,

$$S_t = \mathbb{E}_{\mathbb{Q}^*} \left(\int_t^T (Z_u h_u - r_u S_u) \, du + X \mathbb{1}_{\{\tau>T\}} \, \Big| \, \mathcal{G}_t \right).$$

The representation of S furnished by the last result can be further improved. Note the formulae of Lemma 3.14 correspond to the assumption that the intensity process λ vanishes after τ, but other choices are also possible. It seems that the choice of $\lambda = h$ is probably the worst possible choice from the viewpoint of valuation of defaultable claims.

In general, for any choice of the intensity λ, we introduce an auxiliary process V by setting

$$V_t = \tilde{B}_t \, \mathbb{E}_{\mathbb{Q}^*} \left(\int_t^T \tilde{B}_u^{-1} Z_u \lambda_u \, du + \tilde{B}_T^{-1} X \, \Big| \, \mathcal{G}_t \right) \qquad (32)$$

where $\tilde{B}$ is the *default-risk-adjusted savings account*, which corresponds to the increased interest rate $\tilde{r}_t = r_t + \lambda_t$. More explicitly, we define

$$\tilde{B}_t = \exp\left(\int_0^t (r_u + \lambda_u)\, du\right) = B_t e^{\Lambda_t}.$$

Thus (32) can also be rewritten as follows:

$$V_t = B_t\, \mathbb{E}_{\mathbb{Q}^*}\left(\int_t^T B_u^{-1} e^{\Lambda_t - \Lambda_u} Z_u\, d\Lambda_u + B_T^{-1} X e^{\Lambda_t - \Lambda_T}\,\Big|\, \mathcal{G}_t\right) \qquad (33)$$

Note that the last formula mimics the second equality in Proposition 3.3. However, the desired equality $S_t = 1_{\{\tau > t\}} V_t$ does not always hold. In other words, the auxiliary process V represents the pre-default value of a defaultable claim (X, Z, τ) only in some specific circumstances. In general, we have the following weaker result.

Proposition 3.15. *Assume that (A.1)-(A.4) hold. Then the value process S satisfies*

$$S_t = 1_{\{\tau > t\}}\left\{V_t - B_t\, \mathbb{E}_{\mathbb{Q}^*}\left(B_\tau^{-1} 1_{\{t < \tau \leq T\}} \Delta V_\tau\,\Big|\, \mathcal{G}_t\right)\right\}.$$

If

$$\mathbb{E}_{\mathbb{Q}^*}\left(B_\tau^{-1} 1_{\{t < \tau \leq T\}} \Delta V_\tau\,\Big|\, \mathcal{G}_t\right) = 0 \qquad (34)$$

then

$$S_t = 1_{\{\tau > t\}} \tilde{B}_t\, \mathbb{E}_{\mathbb{Q}^*}\left(\int_t^T \tilde{B}_u^{-1} Z_u \lambda_u\, du + \tilde{B}_T^{-1} X\,\Big|\, \mathcal{G}_t\right).$$

Remarks. It is crucial to observe that condition (34) is difficult to verify. In general, it depends on the specification of a defaultable claim, as well as on the choice of the intensity. If condition (34) happens to hold, we get a formula similar to the one derived in the previous section, up to the conditioning. In the previous section, we have worked with a specific choice of the intensity: the $\mathbb{F}$-adapted one, and the conditioning was with respect to $\mathcal{F}_t$. One can ask whether the two kinds of conditioning happen to coincide in the present setup; we shall examine this important issue in the next subsection.

Martingale Approach under (H.1)

In order to be able to apply condition (H.1) within the setup (A.1)-(A.4), we assume that $\mathbb{G} = \mathbb{F} \vee \mathbb{H}$, where $\mathbb{F}$ is some reference filtration. Also, we postulate that the $\mathbb{G}$-martingale intensity λ is adapted to the reference filtration $\mathbb{F}$ (so that λ is also the $\mathbb{F}$-martingale intensity of τ under $\mathbb{Q}^*$). In addition, we modify (A.2)-(A.3), by assuming that the promised claim X is $\mathcal{F}_T$-measurable, and the recovery process Z is $\mathbb{F}$-predictable.

Under the present assumptions, in view of (H.1), the conditioning with respect to the σ-field $\mathcal{G}_t$ in (32) can be replaced by conditioning with respect to the σ-field $\mathcal{F}_t$. We thus obtain

$$V_t = \tilde{B}_t \, \mathbb{E}_{\mathbb{Q}^*} \Big(\int_t^T \tilde{B}_u^{-1} Z_u \lambda_u \, du + \tilde{B}_T^{-1} X \, \Big| \, \mathcal{F}_t \Big). \tag{35}$$

The next result is valid under the set of assumptions introduced above. Note that for $\mathbb{F} = \mathbb{G}$ all assumptions of this section hold, and thus Proposition 3.15 is a direct extension of Proposition 3.16.

Proposition 3.16. *Suppose that for the process V given by (35) condition (34) is satisfied. Then*

$$S_t = \mathbb{1}_{\{\tau > t\}} \tilde{B}_t \, \mathbb{E}_{\mathbb{Q}^*} \Big(\int_t^T \tilde{B}_u^{-1} Z_u \lambda_u \, du + \tilde{B}_T^{-1} X \, \Big| \, \mathcal{F}_t \Big). \tag{36}$$

Note that when $\mathbb{F}$ is the Brownian filtration, the continuity of the process V is rather clear. In this case $\Delta V_\tau = 0$ and thus (34) holds.

As already mentioned in Section 3.2, working under (H.1) is essentially equivalent to working with a default time given by the canonical construction. Thus, the martingale approach under an additional assumption (H.1) does not appear to be more general than the standard approach.

3.4 Further Developments

In this section, we shall present some recent developments in the area of intensity-based modeling of default risk.

Default-Adjusted Martingale Measure

In a recent paper by Collin-Dufresne et al. (2003), the authors furnish an alternative representation for the value process S within the framework of the martingale approach. Their main goal was to get rid of the no-jump condition (34), and thus, to simplify considerably the pricing procedure, even in these models in which this condition is known to be violated.

To this end, they first define a probability measure $\mathbb{Q}'$, absolutely continuous with respect to $\mathbb{Q}^*$, by setting, for a fixed $T > 0$ (a similar approach was proposed by Schönbucher (2000b); his goal was different, however)

$$\frac{d\mathbb{Q}'}{d\mathbb{Q}^*} \Big|_{\mathcal{G}_T} = \mathbb{1}_{\{\tau > T\}} e^{\Lambda_T}.$$

Using this new probability, which can be termed the *default-adjusted martingale measure*, they arrive at the following result.

Proposition 3.17. *For every* $t \leq T$ *we have* $S_t = \mathbb{1}_{\{\tau > t\}} V'_t$, *where*

$$V'_t = \tilde{B}_t \, \mathbb{E}_{\mathbb{Q}'} \left(\int_t^T \tilde{B}_u^{-1} Z_u \lambda_u \, du + \tilde{B}_T^{-1} X \, \Big| \, \mathcal{G}'_t \right),$$

and where the filtration $\mathbb{G}'$ *is the completion of* $\mathbb{G}$ *under* $\mathbb{Q}'$.

In view of the last result, it is clear that $\mathbb{Q}'$ is a suitable tool for the valuation of a particular survival claim with a given default time τ. However, the fact that $\mathbb{Q}'$ is merely absolutely continuous with respect to original pricing measure $\mathbb{Q}^*$ suggests that this probability is unlikely to be a universal tool for the valuation of all defaultable and default-free claims present in a market model.

Remarks. From the technical perspective, the above construction is in the spirit of Kusuoka's (1999) example, discussed earlier in this chapter. Also, we believe that there is a strong link between the above result and some important issues arising in the theory of initial enlargement of a filtration.

Hybrid Models

Some authors attempt to partially reconcile the structural and reduced-form approaches by postulating that the default intensity is directly linked to the firm's value. For instance, Madan and Unal (1998) consider the discounted equity value (including reinvested dividends) process $E_t^* = E_t / B_t$ as the unique Markovian state variable in their intensity-based model. The dynamics of E^* under the martingale measure $\mathbb{P}^*$ are:

$$dE_t^* = \sigma E_t^* \, dW_t^*, \quad E_0^* > 0,$$

for some constant volatility coefficient σ. They postulate that the intensity of default satisfies $\lambda_t = \lambda(E_t^*)$ for some function $\lambda : \mathbb{R}_+ \to \mathbb{R}_+$. The default time τ is specified through the canonical construction, so that it is defined on an enlarged probability space $(\Omega, \mathbb{G}, \mathbb{Q}^*)$, where a martingale measure $\mathbb{Q}^*$ is an extension of $\mathbb{P}^*$. Madan and Unal (1998) propose to take the function $\lambda(x) = c \, (\ln(x/\bar{v}))^{-2}$, where c and $\bar{v}$ are strictly positive constants. It is interesting to notice that the stochastic intensity $\lambda_t = \lambda(E_t^*)$ tends to infinity, when the discounted equity value E_t^* approaches, either from above or from below, the critical level $\bar{v}$. To avoid making a particular choice of a default-free term structure model, they focus on the futures price of a corporate bond. It is well known (see Duffie and Stanton (1992) or Section 15.2 in Musiela and Rutkowski (1997)) that the futures price $\pi^f(X)$ of a contingent claim X, for the settlement date T, is given by the conditional expectation under the martingale measure $\pi_t^f(X) = \mathbb{E}_{\mathbb{Q}^*}(X \,|\, \mathcal{G}_t)$. In particular, the futures price $D^f(t, T)$ of a defaultable bond with zero recovery is given by the formula $D^f(t, T) = \mathbb{Q}^* \{\tau > T \,|\, \mathcal{G}_t\}$. More explicitly

$$D^f(t,T) = \mathbb{1}_{\{\tau>t\}}\, \mathbb{E}_{\mathbb{P}^*}\left(e^{-\int_t^T \lambda(E_u^*,u)\,du}\,\Big|\,\mathcal{F}_t\right) = \mathbb{1}_{\{\tau>t\}} v(E_t^*, t)$$

for some function $v : \mathbb{R}_+ \to \mathbb{R}_+$. By virtue of the Feynman-Kac theorem, the function v satisfies, under mild technical assumptions, the following pricing PDE

$$v_t(x,t) + \tfrac{1}{2}\sigma^2(x,t)v_{xx}(x,t) - \lambda(x,t)v(x,t) = 0$$

subject to the terminal condition $v(x,T) = 1$. For the sake of notational simplicity, we have assumed here that the process W^* is one-dimensional.

Madan and Unal (1998) show that under these assumptions the futures price of a corporate bond equals $G(h(E_t^*, T - t))$, where the function h is explicitly known, and G satisfies a certain second-order ODE.

Unified Approach

Let us finally mention that Bélanger et al. (2001) make an attempt to unify the structural and reduced-form approaches. They work under the standard assumption that the reference filtration $\mathbb{F}$ is the augmented Brownian filtration. They construct time τ by means of an extension of the canonical construction (see Section 3.2)

$$\tau = \inf\{t \in [0, T^*] \,:\, \Psi_t \geq \theta\},$$

where Ψ is a càdlàg process adapted to a reference filtration $\mathbb{F}$, and $\eta > 0$ is a random variable independent of $\mathbb{F}$ (not necessarily exponentially distributed). Since this is a slight generalization of the canonical construction, condition (H.1) is manifestly satisfied in this setup. Since their framework is slightly more general than the one presented above, process Λ may not be absolutely continuous, and it takes the form (cf. Proposition 3.8)

$$\Lambda_t = \int_{]0,t]} \frac{dF_u}{1 - F_{u-}},$$

where, as usual, $F_t = \mathbb{Q}^*\{\tau \leq t \,|\, \mathcal{F}_t\}$. However, in view of their construction, the process Λ is in fact the $\mathbb{F}$-hazard process of τ, that is, $\Lambda = \Gamma$. Unfortunately, in view of space limits, it is not possible to present the unified approach in more detail here.

3.5 Comments on Intensity-Based Models

We end this chapter by giving few comments on the reduced-form (that is, the intensity-based) approach to the modeling of credit risk. It should be emphasized that the advantages and disadvantages listed below are mainly relative to the alternative structural approach, which was presented in the preceding chapter. Note that at least some of the disadvantages listed below disappear in a hybrid approach to credit risk modeling.

Advantages

- The specifications of the value-of-the-firm process and the default - triggering barrier are not needed.
- The level of the credit risk is reflected in a single quantity: the risk - neutral default intensity.
- The random time of default is an unpredictable stopping time, and thus the default event comes as an almost total surprise.
- The valuation of defaultable claims is rather straightforward. It resembles the valuation of default-free contingent claims in term structure models, through well understood techniques.
- Credit spreads are much easier to quantify and manipulate than in structural models of credit risk. Consequently, the credit spreads are more realistic and risk premia are easier to handle.

Disadvantages

- Typically, current data regarding the level of the firm's assets and the firm's leverage are not taken into account.
- Specific features related to safety covenants and debt's seniority are not easy to handle.
- All (important) issues related to the capital structure of a firm are beyond the scope of this approach.
- Most practical approaches to portfolio's credit risk are linked to the value-of-the-firm approach.

In the next chapter, we shall present more elaborated versions of the reduced-form approach, which aim to cover the case of multi-name credit derivatives (i.e., basket credit derivatives), as well as derivative products related to credit migrations (i.e., changes of credit ratings).

4 Dependent Defaults and Credit Migrations

Let us start by providing a tentative, and definitely non-exhaustive, classification of issues and techniques that arise in the context of modeling dependent defaults and credit ratings.

Valuation of basket credit derivatives covers, in particular:

- Default swaps of type F (Duffie 1998b, Kijima and Muromachi 2000) – a protection against the first default in a basket of defaultable claims.
- Default swaps of type D (Kijima and Muromachi 2000) – a protection against the first two defaults in a basket of defaultable claims.
- The ith-to-default claims (Bielecki and Rutkowski 2003) – a protection against the first i defaults in a basket of defaultable claims.

Technical issues arising in the context of dependent defaults include:

- Conditional independence of default times (Kijima and Muromachi 2000).
- Simulation of correlated defaults (Duffie and Singleton 1998).
- Modeling of infectious defaults (Davis and Lo 1999).
- Asymmetric default intensities (Jarrow and Yu 2001).
- Copulas (Schönbucher and Schubert 2001, Laurent and Gregory 2001).
- Dependent credit ratings (Lando 1998b, Bielecki and Rutkowski 2003).
- Simulation of dependent credit migrations (Kijima et al. 2002, Bielecki 2002).

4.1 Basket Credit Derivatives

Basket credit derivatives are credit derivatives deriving their cash flows values (and thus their values) from credit risks of several reference entities (or prespecified credit events).

Standing assumptions. We assume that:

- We are given a collection of default times $\tau_1, \ldots, \tau_n$ defined on a common probability space $(\Omega, \mathcal{G}, \mathbb{Q}^*)$.
- $\mathbb{Q}^*\{\tau_i = 0\} = 0$ and $\mathbb{Q}^*\{\tau_i > t\} > 0$ for every i and t.
- $\mathbb{Q}^*\{\tau_i = \tau_j\} = 0$ for arbitrary $i \neq j$ (in a continuous time setup).

We associate with the collection $\tau_1, \ldots, \tau_n$ of default times the ordered sequence $\tau_{(1)} < \tau_{(2)} < \cdots < \tau_{(n)}$, where $\tau_{(i)}$ stands for the random time of the ith default. Formally,

$$\tau_{(1)} = \min \{\tau_1, \tau_2, \ldots, \tau_n\}$$

and for $i = 2, \ldots, n$

$$\tau_{(i)} = \min \{\tau_k : k = 1, \ldots, n, \ \tau_k > \tau_{(i-1)}\}.$$

In particular,

$$\tau_{(n)} = \max \{\tau_1, \tau_2, \ldots, \tau_n\}.$$

The ith-to-Default Contingent Claims

We set $H_t^i = \mathbb{1}_{\{\tau_i \leq t\}}$ and we denote by $\mathbb{H}^i$ the filtration generated by the process H^i, that is, by the observations of the default time τ_i. In addition, we are given a reference filtration $\mathbb{F}$ on the space $(\Omega, \mathcal{G}, \mathbb{Q}^*)$. The filtration $\mathbb{F}$ is related to some other market risks, for instance, to the interest rate risk. Finally, we introduce the enlarged filtration $\mathbb{G}$ by setting

$$\mathbb{G} = \mathbb{F} \vee \mathbb{H}^1 \vee \mathbb{H}^2 \vee \cdots \vee \mathbb{H}^n.$$

The σ-field $\mathcal{G}_t$ models the information available at time t.

A general i^{th}-to-default contingent claim which matures at time T is specified by the following covenants:

- If $\tau_{(i)} = \tau_k \leq T$ for some $k = 1, \ldots, n$ it pays at time $\tau_{(i)}$ the amount $Z^k_{\tau_{(i)}}$ where Z^k is an $\mathbb{F}$-predictable recovery process.
- If $\tau_{(i)} > T$ it pays at time T an $\mathcal{F}_T$-measurable promised amount X.

Case of Two Entities

For the sake of notational simplicity, we shall frequently consider the case of two reference credit risks.

Cash flows of the first-to-default contract (FDC):

- If $\tau_{(1)} = \min\{\tau_1, \tau_2\} = \tau_i \leq T$ for $i = 1, 2$, the claim pays at time τ_i the amount $Z^i_{\tau_i}$.
- If $\min\{\tau_1, \tau_2\} > T$, it pays at time T the amount X.

Cash flows of the last-to-default contract (LDC):

- If $\tau_{(2)} = \max\{\tau_1, \tau_2\} = \tau_i \leq T$ for $i = 1, 2$, the claim pays at time τ_i the amount $Z^i_{\tau_i}$.
- If $\max\{\tau_1, \tau_2\} > T$, it pays at time T the amount X.

We recall that throughout these lectures the savings account B equals

$$B_t = \exp\left(\int_0^t r_u \, du\right),$$

and $\mathbb{Q}^*$ stands for the martingale measure for our model of the financial market (including defaultable securities, such as: corporate bonds and credit derivatives). Consequently, the price $B(t,T)$ of a zero-coupon default-free bond equals

$$B(t,T) = B_t \, \mathbb{E}_{\mathbb{Q}^*}\left(B_T^{-1} \,|\, \mathcal{G}_t\right) = B_t \, \mathbb{E}_{\mathbb{Q}^*}\left(B_T^{-1} \,|\, \mathcal{F}_t\right).$$

Values of FDC and LDC

In general, the value at time t of a defaultable claim (X, Z, τ) is given by the *risk-neutral valuation formula*

$$S_t = B_t \, \mathbb{E}_{\mathbb{Q}^*}\left(\int_{]t,T]} B_u^{-1} \, dD_u \,\Big|\, \mathcal{G}_t\right)$$

where D is the *dividend process*, which describes all the cash flows of the claim. Consequently, the value at time t of the FDC equals:

$$S_t^{(1)} = B_t \, \mathbb{E}_{\mathbb{Q}^*}\left(B_{\tau_1}^{-1} Z_{\tau_1}^1 \mathbb{1}_{\{\tau_1 < \tau_2,\, t < \tau_1 \leq T\}} \,\Big|\, \mathcal{G}_t\right)$$

$$+ B_t \, \mathbb{E}_{\mathbb{Q}^*}\left(B_{\tau_2}^{-1} Z_{\tau_2}^2 \mathbb{1}_{\{\tau_2 < \tau_1,\, t < \tau_2 \leq T\}} \,\Big|\, \mathcal{G}_t\right)$$

$$+ B_t \, \mathbb{E}_{\mathbb{Q}^*}\left(B_T^{-1} X \mathbb{1}_{\{T < \tau_{(1)}\}} \,\Big|\, \mathcal{G}_t\right).$$

The value at time t of the LDC equals:

$$S_t^{(2)} = B_t \, \mathbb{E}_{\mathbb{Q}^*} \left(B_{\tau_1}^{-1} Z_{\tau_1}^1 \, \mathbb{1}_{\{\tau_2 < \tau_1, \, t < \tau_1 \leq T\}} \, \middle| \, \mathcal{G}_t \right)$$

$$+ B_t \, \mathbb{E}_{\mathbb{Q}^*} \left(B_{\tau_2}^{-1} Z_{\tau_2}^2 \, \mathbb{1}_{\{\tau_1 < \tau_2, \, t < \tau_2 \leq T\}} \, \middle| \, \mathcal{G}_t \right)$$

$$+ B_t \, \mathbb{E}_{\mathbb{Q}^*} \left(B_T^{-1} X \, \mathbb{1}_{\{T < \tau_{(2)}\}} \, \middle| \, \mathcal{G}_t \right).$$

Both expressions above are merely special cases of a general formula. The goal is to derive more explicit representations under various assumptions about τ_1 and τ_2, or to provide ways of efficient calculation of involved expected values by means of simulation (using perhaps another probability measure).

4.2 Conditionally Independent Defaults

Relatively simple representations for prices of basket credit derivatives can be obtained under the assumption of conditional independence of default times.

Definition 4.1. The random times τ_i, $i = 1, \ldots, n$ are said to be *conditionally independent* with respect to $\mathbb{F}$ under $\mathbb{Q}^*$ if for any $T > 0$ and any $t_1, \ldots, t_n \in [0, T]$ we have:

$$\mathbb{Q}^* \{ \tau_1 > t_1, \ldots, \tau_n > t_n \mid \mathcal{F}_T \} = \prod_{i=1}^{n} \mathbb{Q}^* \{ \tau_i > t_i \mid \mathcal{F}_T \}.$$

Let us comment briefly on Definition 4.1.

- Conditional independence has the following intuitive interpretation: the reference credits (credit names) are subject to common risk factors that may trigger credit (default) events. In addition, each credit name is subject to idiosyncratic risks that are specific for this name.
- Conditional independence of default times means that once the common risk factors are fixed then the idiosyncratic risk factors are independent of each other.
- The property of conditional independence is not invariant with respect to an equivalent change of a probability measure.
- Conditional independence fits into static and dynamic theories of default times.

Canonical Construction

Let Γ^i, $i = 1, \ldots, n$ be a given family of $\mathbb{F}$-adapted, increasing, continuous processes, defined on a probability space $(\tilde{\Omega}, \mathbb{F}, \mathbb{P}^*)$. We assume that $\Gamma_0^i = 0$ and $\Gamma_\infty^i = \infty$. Let $(\hat{\Omega}, \hat{\mathcal{F}}, \hat{\mathbb{P}})$ be an auxiliary probability space with a sequence ξ_i, $i = 1, \ldots, n$ of mutually independent random variables uniformly distributed on $[0, 1]$. We set

$$\tau_i(\tilde{\omega}, \hat{\omega}) = \inf \left\{ t \in \mathbb{R}_+ \; : \; \Gamma_t^i(\tilde{\omega}) \geq -\ln \xi_i(\hat{\omega}) \right\}$$

on the product probability space $(\Omega, \mathcal{G}, \mathbb{Q}^*) = (\tilde{\Omega} \times \hat{\Omega}, \mathcal{F}_\infty \otimes \hat{\mathcal{F}}, \mathbb{P}^* \otimes \hat{\mathbb{P}})$. We endow the space $(\Omega, \mathcal{G}, \mathbb{Q}^*)$ with the filtration $\mathbb{G} = \mathbb{F} \vee \mathbb{H}^1 \vee \cdots \vee \mathbb{H}^n$.

Proposition 4.1. *The process Γ^i is the $\mathbb{F}$-hazard process of τ_i:*

$$\mathbb{Q}^*\{\tau_i > s \,|\, \mathcal{F}_t \vee \mathcal{H}_t^i\} = \mathbb{1}_{\{\tau_i > t\}} \, \mathbb{E}_{\mathbb{Q}^*}\left(e^{\Gamma_t^i - \Gamma_s^i} \,|\, \mathcal{F}_t\right).$$

We have $\mathbb{Q}^*\{\tau_i = \tau_j\} = 0$ for every $i \neq j$. Moreover, default times $\tau_1, \ldots, \tau_n$ are conditionally independent with respect to $\mathbb{F}$ under $\mathbb{Q}^*$.

Recall that if $\Gamma_t^i = \int_0^t \gamma_u^i \, du$ then γ^i is the $\mathbb{F}$-intensity of τ_i. Intuitively

$$\mathbb{Q}^*\{\tau_i \in [t, t+dt] \,|\, \mathcal{F}_t \vee \mathcal{H}_t^i\} \approx \mathbb{1}_{\{\tau_i > t\}} \gamma_t^i \, dt.$$

Independent Default Times

We shall first examine the case of default times $\tau_1, \ldots, \tau_n$ that are mutually independent under $\mathbb{Q}^*$. Suppose that for every $k = 1, \ldots, n$ we know the cumulative distribution function $F_k(t) = \mathbb{Q}^*\{\tau_k \leq t\}$ of the default time of the kth reference entity. The cumulative distribution functions of $\tau_{(1)}$ and $\tau_{(n)}$ are:

$$F_{(1)}(t) = \mathbb{Q}^*\{\tau_{(1)} \leq t\} = 1 - \prod_{k=1}^n (1 - F_k(t))$$

and

$$F_{(n)}(t) = \mathbb{Q}^*\{\tau_{(n)} \leq t\} = \prod_{k=1}^n F_k(t).$$

More generally, for any $i = 1, \ldots, n$ we have

$$F_{(i)}(t) = \mathbb{Q}^*\{\tau_{(i)} \leq t\} = \sum_{m=i} \sum_{\pi \in \Pi^m} \prod_{j \in \pi} F_{k_j}(t) \prod_{l \notin \pi} (1 - F_{k_l}(t))$$

where Π^m denote the family of all subsets of $\{1, \ldots, n\}$ consisting of m elements.

Suppose, in addition, that the default times $\tau_1, \ldots, \tau_n$ admit intensity functions $\gamma_1(t), \ldots, \gamma_n(t)$. It is easily seen that the default time $\tau_{(1)}$ has the intensity function

$$\gamma_{(1)}(t) = \gamma_1(t) + \cdots + \gamma_n(t)$$

and for any $t \in \mathbb{R}_+$

$$\mathbb{Q}^*\{\tau_{(1)} > t\} = e^{-\int_0^t \gamma_{(1)}(v) \, dv}.$$

By direct calculations, it is also possible to find the intensity function of the ith default time. We do not necessarily need to assume that the reference filtration $\mathbb{F}$ is trivial, so that the case of random interest rates is also covered.

Example 4.1. We shall consider a digital default put of basket type. To be more specific, we postulate that a contract pays a fixed amount (e.g., one unit of cash) at the ith default time $\tau_{(i)}$ provided that $\tau_{(i)} \leq T$. Assume that the interest rates are non-random. Then the value at time 0 of the contract equals

$$S_0 = \mathbb{E}_{\mathbb{Q}^*}\left(B_{\tau_{(i)}}^{-1}\mathbb{1}_{\{\tau_{(i)}\leq T\}}\right) = \int_{]0,T]} B_u^{-1}\,dF_{(i)}(u).$$

If $\tau_1,\ldots,\tau_n$ admit intensities then

$$S_0 = \int_0^T B_u^{-1}\,dF_{(i)}(u) = \int_0^T B_u^{-1}\gamma_{(i)}(u)e^{-\int_0^u \gamma_{(i)}(v)\,dv}\,du.$$

Signed Intensities

Some authors (e.g., Kijima and Muromachi (2000)) examine credit risk models in which the negative values of intensities are not precluded. Negative values of the intensity process clearly contradict the interpretation of the intensity as the conditional probability of survival over an infinitesimal time interval.

Nevertheless, the canonical construction of conditionally independent random times also works in this case. For a given collection Γ^i, $i = 1,\ldots,n$ of $\mathbb{F}$-adapted continuous stochastic processes, with $\Gamma_0^i = 0$, defined on $(\hat{\Omega}, \mathbb{F}, \hat{\mathbb{P}})$. We define τ_i, $i = 1,\ldots,n$, on the enlarged probability space $(\Omega, \mathcal{G}, \mathbb{Q}^*)$:

$$\tau_i = \inf\left\{\,t \in \mathbb{R}_+ : \Gamma_t^i(\hat{\omega}) \geq -\ln\xi_i(\hat{\omega})\,\right\}.$$

Let us denote $\hat{\Gamma}_t^i = \max_{u\leq t}\Gamma_t^i$. Observe that if the process Γ^i is absolutely continuous, than so it the process $\hat{\Gamma}^i$; in this case the intensity of τ_i is obtained as the derivative of $\hat{\Gamma}^i$ with respect to the time variable.

The following result examines the case of signed intensities.

Lemma 4.1. *Random times τ_i, $i = 1,\ldots,n$ are conditionally independent with respect to $\mathbb{F}$ under $\mathbb{Q}^*$. In particular, for every $t_1,\ldots,t_n \leq T$,*

$$\mathbb{Q}^*\{\tau_1 > t_1,\ldots,\tau_n > t_n \,|\, \mathcal{F}_T\} = \prod_{i=1}^n e^{-\hat{\Gamma}_{t_i}^i} = e^{-\sum_{i=1}^n \hat{\Gamma}_{t_i}^i}.$$

Valuation of FDC and LDC

Valuation of the first-to-default or last-to-default contingent claim is relatively straightforward under the assumption of conditional independence of default times. We have the following result in which, for notational simplicity, we consider only the case of two entities. As usual, we do not state explicitly integrability conditions that should be imposed on recovery processes Z^j and the terminal payoff X.

Proposition 4.2. *Let the default times τ_j, $j = 1, 2$ be $\mathbb{F}$-conditionally independent with $\mathbb{F}$-intensities γ^j. Assume that Z^j are $\mathbb{F}$-predictable processes, and that the terminal payoff X is $\mathcal{F}_T$-measurable. Then the price at time $t = 0$ of the first-to-default claim equals*

$$S_0^{(1)} = \sum_{i,j=1,\, i\neq j}^{2} \mathbb{E}_{\mathbb{Q}^*}\left(\int_0^T B_u^{-1} Z_u^j e^{-\Gamma_u^i} \gamma_u^j e^{-\Gamma_u^j}\, du \right) + \mathbb{E}_{\mathbb{Q}^*}\left(B_T^{-1} X G \right),$$

where we denote

$$G = e^{-(\Gamma_T^1 + \Gamma_T^2)} = \mathbb{Q}^*\{\tau_1 > T,\, \tau_2 > T \,|\, \mathcal{F}_T\}.$$

The price at time $t = 0$ of the last-to-default claim equals

$$S_0^{(2)} = \sum_{i,j=1,\, i\neq j}^{2} \mathbb{E}_{\mathbb{Q}^*}\left(\int_0^T B_u^{-1} Z_u^j (1 - e^{-\Gamma_u^i}) \gamma_u^j e^{-\Gamma_u^j}\, du \right) + \mathbb{E}_{\mathbb{Q}^*}\left(B_T^{-1} X H \right),$$

where we denote

$$H = \left(1 - (1 - e^{-\Gamma_T^2})(1 - e^{-\Gamma_T^1}) \right) = 1 - \mathbb{Q}^*\{\tau_1 \leq T,\, \tau_2 \leq T \,|\, \mathcal{F}_T\}.$$

General Valuation Formula

We shall examine the case of a generic i^{th}-to-default contingent claims. Recall that we have introduced the notation

$$\tau_{(1)} < \tau_{(2)} < \cdots < \tau_{(n)}$$

for the ordered sequence of default times.

Recall that according to our notational convention:

- If the ith default occurs before or at the maturity date and $\tau_{(i)} = \tau_k$ for some $k \in \{1, \ldots, n\}$, then an immediate recovery cash flow $Z_{\tau_{(i)}}^k = Z_{\tau_k}^k$ is received at time of the ith default.
- The terminal promised payment occurs at the maturity date if the ith default does happen not prior to or at T.

We assume that $\tau_1, \ldots, \tau_n$ are $\mathbb{F}$-conditionally independent with stochastic intensities $\gamma^1, \ldots, \gamma^n$. Then we have the following result (recall that, by convention, $B_0 = 0$).

Proposition 4.3. *The price at time $t = 0$ of the ith-to-default claim equals*

$$S_0^{(i)} = \sum_{j=1}^{n} \mathbb{E}_{\mathbb{Q}^*}\left(\int_0^T B_u^{-1} Z_u^j g_{ij}(u) \gamma_u^j e^{-\Gamma_u^j}\, du \right)$$

$$+ \sum_{j=1}^{n} \mathbb{E}_{\mathbb{Q}^*}\left(B_T^{-1} X \int_T^\infty g_{ij}(u) \gamma_u^j e^{-\Gamma_u^j}\, du \right),$$

where for every $u \in \mathbb{R}_+$

$$g_{ij}(u) = \sum_{\pi \in \Pi^{(i,j)}} e^{-\sum_{l \in \pi_+} \Gamma_u^l} \prod_{k \in \pi_-} \left(1 - e^{-\Gamma_u^k}\right),$$

and where by $\Pi^{i,j}$ we denote the collection of specific partitions of the set $\{1, \ldots, n\}$. Specifically, if $\pi \in \Pi^{(i,j)}$ then $\pi = \{\pi_-, \{j\}, \pi_+\}$, where

$$\pi_- = \{k_1, k_2, \ldots, k_{i-1}\}, \quad \pi_+ = \{k_{i+1}, k_{i+2}, \ldots, k_n\},$$

and: $j \notin \pi_-$, $j \notin \pi_+$, $\pi_- \cap \pi_+ = \emptyset$ and

$$\pi_- \cup \pi_+ \cup \{j\} = \{1, \ldots, n\}.$$

Consider, for instance, $n = 2$ credit entities. For $i = 1$ (i.e., in the case of the first-to-default claim) and $j = 1, 2$ we have

$$\Pi^{(1,1)} = \left\{\{\emptyset, \{1\}, \{2\}\}\right\}, \quad \Pi^{(1,2)} = \left\{\{\emptyset, \{2\}, \{1\}\}\right\}.$$

Likewise, in the case of the second-to-default claim, we have

$$\Pi^{(2,1)} = \left\{\{\{2\}, \{1\}, \emptyset\}\right\}, \quad \Pi^{(2,2)} = \left\{\{\{1\}, \{2\}, \emptyset\}\right\}.$$

In this example, each set $\Pi^{(i,j)}$ contains only one partition; for example, the only element of $\Pi^{(1,1)}$ is the partition $\pi = \{\emptyset, \{1\}, \{2\}\}$.

Default Swap of Basket Type

Let us consider a portfolio of n corporate bonds. The kth bond has the face value L_k and maturity T_k. Its price process is denoted by $D_k(t, T_k)$, $k = 1, \ldots, n$. By τ_k we denote the default time of the kth bond, and, as usual, $\tau_{(i)}$ stands for the random time of the ith default. We shall examine a default swap, which matures at some future date $T < \min\{T_1, \ldots, T_k\}$ and whose covenants are described as follows. If $\tau_{(i)} \leq T$, the contract holder (i.e., the protection buyer) receives at time $\tau_{(i)}$ the recovery payment

$$\sum_{k=1}^n \left(L_k - D_k(\tau_{(i)}, T_k)\right) \mathbb{1}_{\{\tau_{(i)} = \tau_k\}}.$$

This means that if the ith defaulting bond was issued by the kth reference entity, the recovery payment is based on the value of the kth bond only. A default swap premium in the amount κ is paid by the contract holder at each of prespecified time instants $t_p \leq T$, $p = 1, 2, \ldots, m$ prior to the ith default time or to the maturity T, whichever comes first.

Default Swap Premium

We assume that all corporate bonds are subject to the fractional recovery of par value scheme. Specifically, δ_j is the constant recovery rate of jth bond. We also assume that each default time τ_j, $j = 1, \ldots, n$, admits the $\mathbb{F}$-intensity process γ^j so that $\Gamma_t^j = \int_0^t \gamma_u^j \, du$. Then, the following result gives the value κ of the default swap premium,

Proposition 4.4. *The default swap premium $\kappa = J_1 / J_2$ where*

$$
J_1 = \sum_{j=1}^n \mathbb{E}_{\mathbb{Q}^*} \left\{ L_j (1 - \delta_j) \int_0^T B_u^{-1} \left(\sum_{\pi \in \Pi^{(i,j)}} \left[\prod_{k \in \pi_-} \left(1 - e^{-\Gamma_u^k} \right) \right] \right. \right.
$$
$$
\left. \left. \times \left[e^{-\sum_{l \in \pi_+} \Gamma_u^l} \right] \right) \gamma_u^j \, e^{-\Gamma_u^j} \, du \right\}
$$

$$
J_2 = \sum_{p=1}^m \mathbb{E}_{\mathbb{Q}^*} \left\{ B_{t_p}^{-1} \sum_{j=1}^n \left(\int_{t_p}^T \left(\sum_{\pi \in \Pi^{(i,j)}} \left[\prod_{k \in \pi_-} \left(1 - e^{-\Gamma_u^k} \right) \right] \right. \right. \right.
$$
$$
\left. \left. \left. \times \left[e^{-\sum_{l \in \pi_+} \Gamma_u^l} \right] \right) \gamma_u^j \, e^{-\Gamma_u^j} \, du \right) \right\}
$$
$$
+ \sum_{p=1}^m \mathbb{E}_{\mathbb{Q}^*} \left\{ B_{t_p}^{-1} \sum_{j=1}^n \left(\int_T^\infty \left(\sum_{\pi \in \Pi^{(i,j)}} \left[\prod_{k \in \pi_-} \left(1 - e^{-\Gamma_u^k} \right) \right] \right. \right. \right.
$$
$$
\left. \left. \left. \times \left[e^{-\sum_{l \in \pi_+} \Gamma_u^l} \right] \right) \gamma_u^j \, e^{-\Gamma_u^j} \, du \right) \right\}.
$$

4.3 Copula-Based Approaches

The concept of a *copula function* allows to produce various multidimensional probability distributions with prespecified univariate marginal laws.

Definition 4.2. A function $C : [0,1]^n \to [0,1]$ is called a *copula* if the following conditions are satisfied:
(i) $C(1, \ldots, 1, v_i, 1, \ldots, 1) = v_i$ for any i and any $v_i \in [0,1]$,
(ii) C is an n-dimensional cumulative distribution function (c.d.f.).

Let us give few examples of copulas:
- Product copula: $\Pi(v_1, \ldots, v_n) = \Pi_{i=1}^n v_i$,
- Gumbel copula: for $\theta \in [1, \infty)$ we set

$$
C(v_1, \ldots, v_n) = \exp \left(- \left[\sum_{i=1}^n (- \ln v_i)^\theta \right]^{1/\theta} \right),
$$

- Gaussian copula:

$$C(v_1, \ldots, v_n) = N_{\Sigma}^n \left(N^{-1}(v_1), \ldots, N^{-1}(v_n) \right),$$

where N_{Σ}^n is the c.d.f for the n-variate central normal distribution with the linear correlation matrix Σ, and N^{-1} is the inverse of the c.d.f. for the univariate standard normal distribution.

- t-copula:

$$C(v_1, \ldots, v_n) = \Theta_{\nu, \Sigma}^n \left(t_{\nu}^{-1}(v_1), \ldots, t_{\nu}^{-1}(v_n) \right),$$

where $\Theta_{\nu, \Sigma}^n$ is the c.d.f for the n-variate t-distribution with ν degrees of freedom and with the linear correlation matrix Σ, and t_{ν}^{-1} is the inverse of the c.d.f. for the univariate t-distribution with ν degrees of freedom.

The following theorem is the fundamental result underpinning the theory of copulas.

Theorem 4.1. (Sklar) *For any cumulative distribution function F on $\mathbb{R}^n$ there exists a copula function C such that*

$$F(x_1, \ldots, x_n) = C(F_1(x_1), \ldots, F_n(x_n))$$

where F_i is the ith marginal cumulative distribution function. If, in addition, F is continuous then C is unique.

Direct Application

In a direct application, we first postulate a (univariate marginal) probability distribution for each random variable τ_i. Let us denote it by F_i for $i = 1, 2, \ldots, n$. Then, a suitable copula function C is chosen in order to introduce an appropriate dependence structure of the random vector $(\tau_1, \tau_2, \ldots, \tau_n)$. Finally, the joint distribution of the random vector $(\tau_1, \tau_2, \ldots, \tau_n)$ is derived, specifically,

$$\mathbb{Q}^* \{ \tau_i \leq t_i, \ i = 1, 2, \ldots, n \} = C \left(F_1(t_1), \ldots, F_n(t_n) \right).$$

In the finance industry, the most commonly used are elliptical copulas (such as the Gaussian copula and the t-copula). The direct approach has an apparent drawback. It is essentially a static approach; it makes no account of changes in credit ratings, and no conditioning on the flow of information is present. Let us mention, however, an interesting theoretical issue, namely, the study of the effect of a change of probability measures on the copula structure.

Indirect Application

A less straightforward application of copulas is based on an extension of the canonical construction of conditionally independent default times. This can be considered as the first step towards a dynamic theory, since the techniques of copulas is merged with the flow of available information, in particular, the information regarding the observations of defaults.

Assume that the cumulative distribution function of $(\xi_1, \ldots, \xi_n)$ in the canonical construction (cf. Section 4.2) is given by an n-dimensional copula C, and that the univariate marginal laws are uniform on $[0, 1]$. Similarly as in Section 4.2, we postulate that $(\xi_1, \ldots, \xi_n)$ are independent of $\mathbb{F}$, and we set

$$\tau_i(\tilde{\omega}, \hat{\omega}) = \inf \{ t \in \mathbb{R}_+ \; : \; \Gamma_t^i(\tilde{\omega}) \geq -\ln \xi_i(\hat{\omega}) \}.$$

Then:

- The case of default times conditionally independent with respect to $\mathbb{F}$ corresponds to the choice of the product copula Π. In this case, for $t_1, \ldots, t_n \leq T$ we have

$$\mathbb{Q}^* \{\tau_1 > t_1, \ldots, \tau_n > t_n \,|\, \mathcal{F}_T\} = \Pi(Z_{t_1}^1, \ldots, Z_{t_n}^n),$$

  where we set $Z_t^i = e^{-\Gamma_t^i}$.
- In general, for $t_1, \ldots, t_n \leq T$ we obtain

$$\mathbb{Q}^* \{\tau_1 > t_1, \ldots, \tau_n > t_n \,|\, \mathcal{F}_T\} = C(Z_{t_1}^1, \ldots, Z_{t_n}^n),$$

  where C is the copula used in the construction of $\tau_1, \ldots, \tau_n$.

Survival Intensities

Schönbucher and Schubert (2001) show that for arbitrary $s \leq t$ on the set $\{\tau_1 > s, \ldots, \tau_n > s\}$ we have

$$\mathbb{Q}^* \{\tau_i > t \,|\, \mathcal{G}_s\} = \mathbb{E}_{\mathbb{Q}^*} \left(\frac{C(Z_s^1, \ldots, Z_t^i, \ldots, Z_s^n)}{C(Z_s^1, \ldots, Z_s^n)} \,\middle|\, \mathcal{F}_s \right).$$

Consequently, assuming that the derivatives $\gamma_t^i = \frac{d\Gamma_t^i}{dt}$ exist, the ith intensity of survival equals, on the set $\{\tau_1 > t, \ldots, \tau_n > t\}$,

$$\lambda_t^i = \gamma_t^i \, Z_t^i \, \frac{\frac{\partial}{\partial v_i} C(Z_t^1, \ldots, Z_t^n)}{C(Z_t^1, \ldots, Z_t^n)} = \gamma_t^i \, Z_t^i \, \frac{\partial}{\partial v_i} \ln C(Z_t^1, \ldots, Z_t^n),$$

where λ_t^i is understood as the limit:

$$\lambda_t^i = \lim_{h \downarrow 0} h^{-1} \mathbb{Q}^* \{t < \tau_i \leq t + h \,|\, \mathcal{F}_t, \tau_1 > t, \ldots, \tau_n > t\}.$$

It appears that, in general, the ith intensity of survival jumps at time t, if the jth entity defaults at time t for some $j \neq i$. In fact, it holds that

$$\lambda_t^{i,j} = \gamma_t^i \, Z_t^i \, \frac{\frac{\partial^2}{\partial v_i \partial v_j} C(Z_t^1, \ldots, Z_t^n)}{\frac{\partial}{\partial v_j} C(Z_t^1, \ldots, Z_t^n)},$$

where

$$\lambda_t^{i,j} = \lim_{h \downarrow 0} h^{-1} \, \mathbb{Q}^* \{ t < \tau_i \le t + h \, | \, \mathcal{F}_t, \tau_k > t, k \ne j, \tau_j = t \}.$$

Schönbucher and Schubert (2001) also examine the intensities of survival after the default times of some entities. Let us fix s, and let $t_i \le s$ for $i = 1, 2, \ldots, k < n$, and $T_i \ge s$ for $i = k+1, k+2, \ldots, n$. Then,

$$\mathbb{Q}^* \{ \tau_i > T_i, \, i = k+1, k+2, \ldots, n \, | \, \mathcal{F}_s, \, \tau_j = t_j, \, j = 1, 2, \ldots, k,$$

$$\tau_i > s, \, i = k+1, k+2, \ldots, n \}$$

$$= \frac{\mathbb{E}_{\mathbb{Q}^*} \left(\frac{\partial^k}{\partial v_1 \ldots \partial v_k} \, C(Z_{t_1}^1, \ldots, Z_{t_k}^k, Z_{T_{k+1}}^{k+1}, \ldots, Z_{T_n}^n) \, \Big| \, \mathcal{F}_s \right)}{\frac{\partial^k}{\partial v_1 \ldots \partial v_k} \, C(Z_{t_1}^1, \ldots, Z_{t_k}^k, Z_s^{k+1}, \ldots, Z_s^n)}. \tag{37}$$

Remarks. Jumps of intensities cannot be efficiently controlled, except for the choice of C. In the approach described above, the dependence between the default times is implicitly introduced through Γ^is, and explicitly introduced by the choice of a copula C.

Simplified Version

Laurent and Gregory (2002) examine a simplified version of the framework of Schönbu–cher and Schubert (2001). Namely, they assume that the reference filtration is trivial – that is, $\mathcal{F}_t = \{ \Omega, \emptyset \}$ for every $t \in \mathbb{R}_+$. This implies, in particular, that the default intensities γ^i are deterministic functions, and

$$\mathbb{Q}^* \{ \tau_i > t \} = 1 - F_i(t) = e^{- \int_0^t \gamma_u^i \, du}.$$

They obtain closed-form expressions for certain conditional intensities of default, by making specific assumptions regarding the choice of a copula C.

Example 4.2. This example describes the use of one-factor Gaussian copula (Bank of International Settlements (BIS) standard). Let

$$X_i = \rho_i V + \sqrt{1 - \rho_i^2} \, \bar{V}_i,$$

where $V, \bar{V}_i, \, i = 1, 2, \ldots, n$, are independent, standard Gaussian variables under the probability measure $\mathbb{Q}^*$. Define the copula function C as

$$C(v_1, \ldots, v_n) = \mathbb{Q}^* \{ X_i < N^{-1}(v_i), \, i = 1, 2, \ldots, n \}.$$

Then, a special case of formula (37) takes the form (for $i > 1$)

$$\mathbb{Q}^* \{ \tau_i \ge T_i \, | \, \tau_1 = s, \, \tau_j \ge s, \, j = 1, 2, \ldots, n \}$$

$$= \frac{\int_{-\infty}^{\infty} \prod_{j=2}^{n} N\left(\frac{\rho_j\sqrt{1-\rho_1^2}u+\rho_j\rho_1 x_1-x_j}{\sqrt{1-\rho_j^2}}\right) n(u)\,du}{\int_{-\infty}^{\infty} \prod_{j=2}^{n} N\left(\frac{\rho_j\sqrt{1-\rho_1^2}u+\rho_j\rho_1 y_1-y_j}{\sqrt{1-\rho_j^2}}\right) n(u)\,du}$$

with $x_j = y_j = N^{-1}(F_j(s))$ for $j \neq i$ and

$$x_i = N^{-1}(F_i(T_i)), \quad y_i = N^{-1}(F_i(s)),$$

where n is the univariate standard normal density function.

4.4 Jarrow and Yu Model

Jarrow and Yu (2001) approach can be considered as another step towards a dynamic theory of dependence between default times. For a given finite family of reference credit names, Jarrow and Yu (2001) propose to make a distinction between the *primary firms* and the *secondary firms*.
At the intuitive level:

- The class of primary firms encompasses those entities whose probabilities of default are influenced by macroeconomic conditions, but not by the credit risk of counterparties. The pricing of bonds issued by primary firms can be done through the standard intensity-based methodology.
- It suffices to focus on securities issued by secondary firms, that is, firms for which the intensity of default depends on the status of some other firms.

Formally, the construction is based on the assumption of asymmetric information. Unilateral dependence is not possible in the case of complete (i.e., symmetric) information.

Construction and Properties of the Model

Let $\{1, \ldots, n\}$ represent the set of all firms, and let $\mathbb{F}$ be the reference filtration. We postulate that:

- For any firm from the set $\{1, \ldots, k\}$ of primary firms, the 'default intensity' depends only on $\mathbb{F}$.
- The 'default intensity' of each firm belonging to the set $\{k+1, \ldots, n\}$ of secondary firms may depend not only on the filtration $\mathbb{F}$, but also on the status (default or no-default) of the primary firms.

Construction of Default Times $\tau_1, \ldots, \tau_n$

First step. We first model default times of primary firms. To this end, we assume that we are given a family of $\mathbb{F}$-adapted 'intensity processes' $\lambda^1, \ldots, \lambda^k$ and we produce a collection $\tau_1, \ldots, \tau_k$ of $\mathbb{F}$-conditionally independent random times through the canonical method:

$$\tau_i = \inf \left\{ t \in \mathbb{R}_+ \ : \ \int_0^t \lambda_u^i \, du \geq -\ln \xi_i \right\}$$

where ξ_i, $i = 1, \ldots, k$ are mutually independent identically distributed random variables with uniform law on $[0,1]$ under the martingale measure $\mathbb{Q}^*$.

Second step. We now construct default times of secondary firms. We assume that:

- The probability space $(\Omega, \mathcal{G}, \mathbb{Q}^*)$ is large enough to support a family ξ_i, $i = k+1, \ldots, n$ of mutually independent random variables, with uniform law on $[0,1]$.
- These random variables are independent not only of the filtration $\mathbb{F}$, but also of the default times $\tau_1, \ldots, \tau_k$ of primary firms already constructed in the first step.

The default times τ_i, $i = k+1, \ldots, n$ are also defined by means of the standard formula:

$$\tau_i = \inf \left\{ t \in \mathbb{R}_+ \ : \ \int_0^t \lambda_u^i \, du \geq -\ln \xi_i \right\}.$$

However, the 'intensity processes' λ^i for $i = k+1, \ldots, n$ are now given by the following expression:

$$\lambda_t^i = \mu_t^i + \sum_{l=1}^k \nu_t^{i,l} \mathbb{1}_{\{\tau_l \leq t\}},$$

where μ^i and $\nu^{i,l}$ are $\mathbb{F}$-adapted stochastic processes. If the default of the jth primary firm does not affect the default intensity of the ith secondary firm, we set $\nu^{i,j} \equiv 0$.

Main Features

Let $\mathbb{G} = \mathbb{F} \vee \mathbb{H}^1 \vee \cdots \vee \mathbb{H}^n$ stand for the enlarged filtration and let $\hat{\mathbb{F}} = \mathbb{F} \vee \mathbb{H}^{k+1} \vee \cdots \vee \mathbb{H}^n$ be the filtration generated by the reference filtration $\mathbb{F}$ and the observations of defaults of secondary firms. Then:

- The default times $\tau_1, \ldots, \tau_k$ of primary firms are conditionally independent with respect to $\mathbb{F}$.
- The default times $\tau_1, \ldots, \tau_k$ of primary firms are no longer conditionally independent when we replace the filtration $\mathbb{F}$ by $\hat{\mathbb{F}}$.
- In general, the default intensity of a primary firm with respect to the filtration $\hat{\mathbb{F}}$ differs from the intensity λ^i with respect to $\mathbb{F}$.

We conclude that defaults of primary firms are also 'dependent' of defaults of secondary firms.

Case of Two Firms

To illustrate the present model, we now consider only two firms, A and B say, and we postulate that A is a primary firm, and B is a secondary firm. Let the constant process $\lambda_t^1 \equiv \lambda_1$ represent the $\mathbb{F}$-intensity of default for firm A, so that

$$\tau_1 = \inf\left\{ t \in \mathbb{R}_+ \: : \: \int_0^t \lambda_u^1 \, du = \lambda_1 t \geq -\ln \xi_1 \right\},$$

where ξ_1 is a random variable independent of $\mathbb{F}$, with the uniform law on $[0, 1]$. For the second firm, the 'intensity' of default is assumed to satisfy

$$\lambda_t^2 = \lambda_2 \mathbb{1}_{\{\tau_1 > t\}} + \alpha_2 \mathbb{1}_{\{\tau_1 \leq t\}}$$

for some positive constants λ_2 and α_2, and thus

$$\tau_2 = \inf\left\{ t \in \mathbb{R}_+ \: : \: \int_0^t \lambda_u^2 \, du \geq -\ln \xi_2 \right\}$$

where ξ_2 is a random variable with the uniform law, independent of $\mathbb{F}$, and such that ξ_1 and ξ_2 are mutually independent. Then the following properties hold:

- λ^1 is the intensity of τ_1 with respect to $\mathbb{F}$,
- λ^2 is the intensity of τ_2 with respect to $\mathbb{F} \vee \mathbb{H}^1$,
- λ^1 is not the intensity of τ_1 with respect to $\mathbb{F} \vee \mathbb{H}^2$.

Bond Valuation

The following result was established in Jarrow and Yu (2001), who assumed the fractional recovery of Treasury value scheme with the fixed recovery rates δ_1 and δ_2. Let $\lambda = \lambda_1 + \lambda_2$. For $\lambda \neq \alpha_2$, we denote

$$c_{\lambda_1, \lambda_2, \alpha_2}(u) = \frac{1}{\lambda - \alpha_2}\left(\lambda_1 e^{-\alpha_2 u} + (\lambda_2 - \alpha_2)e^{-\lambda u} \right).$$

For $\lambda = \alpha_2$, we set

$$c_{\lambda_1, \lambda_2, \alpha_2}(u) = \left(1 + \lambda_1 u \right)e^{-\lambda u}.$$

Proposition 4.5. *For the bond issued by the primary firm we have*

$$D_1(t, T) = B(t, T)\left(\delta_1 + (1 - \delta_1)e^{-\lambda_1(T - t)} \mathbb{1}_{\{\tau_1 > t\}} \right).$$

The value of a zero-coupon bond issued by the secondary firm equals, on the set $\{\tau_1 > t\}$, that is, prior to default of the primary firm

$$D_2(t, T) = B(t, T)\left(\delta_2 + (1 - \delta_2)c_{\lambda_1, \lambda_2, \alpha_2}(T - t) \mathbb{1}_{\{\tau_2 > t\}} \right).$$

On the set $\{\tau_1 \leq t\}$, that is, after default of the primary firm, it equals

$$D_2(t, T) = B(t, T)\left(\delta_2 + (1 - \delta_2)e^{-\alpha_2(T - t)} \mathbb{1}_{\{\tau_2 > t\}} \right).$$

Special Case: Zero Recovery

Assume that $\lambda_1 + \lambda_2 - \alpha_2 \neq 0$ and the bond is subject to the zero recovery scheme. For the sake of brevity, we set $r = 0$ so that $B(t,T) = 1$ for $t \leq T$. Under the present assumptions:

$$D_2(t,T) = \mathbb{Q}^* \{ \tau_2 > T \mid \mathcal{H}_t^1 \vee \mathcal{H}_t^2 \}$$

and the general formula yields

$$D_2(t,T) = \mathbb{1}_{\{\tau_2 > t\}} \frac{\mathbb{Q}^* \{ \tau_2 > T \mid \mathcal{H}_t^1 \}}{\mathbb{Q}^* \{ \tau_2 > t \mid \mathcal{H}_t^1 \}} .$$

If we set $\Lambda_t^2 = \int_0^t \lambda_u^2 \, du$ then

$$D_2(t,T) = \mathbb{1}_{\{\tau_2 > t\}} \, \mathbb{E}_{\mathbb{Q}^*} (e^{\Lambda_t^2 - \Lambda_T^2} \mid \mathcal{H}_t^1).$$

Finally, we have the following explicit result.

Corollary 4.1. *If $\delta_2 = 0$ then $D_2(t,T) = 0$ on $\{\tau_2 \leq t\}$. On the set $\{\tau_2 > t\}$ we have*

$$D_2(t,T) = \mathbb{1}_{\{\tau_1 \leq t\}} \, e^{-\alpha_2(T-t)}$$
$$+ \mathbb{1}_{\{\tau_1 > t\}} \frac{1}{\lambda - \alpha_2} \Big(\lambda_1 e^{-\alpha_2(T-t)} + (\lambda_2 - \alpha_2) e^{-\lambda(T-t)} \Big).$$

4.5 Extension of the Jarrow and Yu Model

We shall now argue that the assumption that some firms are primary while other firms are secondary is not relevant. For simplicity of presentation, we assume that:

- We have $n = 2$, that is, we consider two firms only.
- The interest rate r is zero, so that $B(t,T) = 1$ for every $t \leq T$.
- The reference filtration $\mathbb{F}$ is trivial.
- Corporate bonds are subject to the zero recovery scheme.

Since the situation is symmetric, it suffices to analyze a bond issued by the first firm. By definition, the price of this bond equals

$$D_1(t,T) = \mathbb{Q}^* \{ \tau_1 > T \mid \mathcal{H}_t^1 \vee \mathcal{H}_t^2 \}.$$

For the sake of comparison, we shall also evaluate the following values, which are based on partial observations,

$$\tilde{D}_1(t,T) = \mathbb{Q}^* \{ \tau_1 > T \mid \mathcal{H}_t^2 \}$$

and

$$\hat{D}_1(t,T) = \mathbb{Q}^* \{ \tau_1 > T \mid \mathcal{H}_t^1 \}.$$

Kusuoka's Construction

We follow here Kusuoka (1999). Under the original probability measure $\mathbb{Q}$ the random times τ_i, $i = 1, 2$ are assumed to be mutually independent random variables with exponential laws with parameters λ_1 and λ_2, respectively.

Girsanov's theorem. For a fixed $T > 0$, we define a probability measure $\mathbb{Q}^*$ equivalent to $\mathbb{Q}$ on $(\Omega, \mathcal{G})$ by setting

$$\frac{d\mathbb{Q}^*}{d\mathbb{Q}} = \eta_T, \quad \mathbb{Q}\text{-a.s.}$$

where the Radon-Nikodým density process η_t, $t \in [0, T]$, satisfies

$$\eta_t = 1 + \sum_{i=1}^{2} \int_{]0,t]} \eta_{u-} \kappa_u^i \, dM_u^i$$

where in turn

$$M_t^i = H_t^i - \int_0^{t \wedge \tau_i} \lambda_i \, du$$

Here $H_t^i = \mathbb{1}_{\{\tau_i \leq t\}}$ and the processes κ^1 and κ^2 are given by

$$\kappa_t^1 = \mathbb{1}_{\{\tau_2 < t\}} \left(\frac{\alpha_1}{\lambda_1} - 1 \right), \quad \kappa_t^2 = \mathbb{1}_{\{\tau_1 < t\}} \left(\frac{\alpha_2}{\lambda_2} - 1 \right).$$

It can be checked that the martingale intensities of τ_1 and τ_2 under $\mathbb{Q}^*$ are

$$\lambda_t^1 = \lambda_1 \mathbb{1}_{\{\tau_2 > t\}} + \alpha_1 \mathbb{1}_{\{\tau_2 \leq t\}},$$
$$\lambda_t^2 = \lambda_2 \mathbb{1}_{\{\tau_1 > t\}} + \alpha_2 \mathbb{1}_{\{\tau_1 \leq t\}}.$$

Main features. We focus on τ_1 and we denote $\Lambda_t^1 = \int_0^t \lambda_u^1 \, du$. Let us make a few observations. First, the process λ^1 is $\mathbb{H}^2$-predictable, and the process

$$M_t^1 = H_t^1 - \int_0^{t \wedge \tau_1} \lambda_u^1 \, du = H_t^1 - \Lambda_{t \wedge \tau_1}^1$$

is a $\mathbb{G}$-martingale under $\mathbb{Q}^*$. Next, the process λ^1 is not the intensity of the default time τ_1 with respect to $\mathbb{H}^2$ under $\mathbb{Q}^*$. Indeed, in general, we have

$$\mathbb{Q}^*\{\tau_1 > s \,|\, \mathcal{H}_t^1 \vee \mathcal{H}_t^2\} \neq \mathbb{1}_{\{\tau_1 > t\}} \, \mathbb{E}_{\mathbb{Q}^*} \left(e^{\Lambda_t^1 - \Lambda_s^1} \,|\, \mathcal{H}_t^2 \right).$$

Finally, the process λ^1 represents the intensity of the default time τ_1 with respect to $\mathbb{H}^2$ under a probability measure $\mathbb{Q}^1$ equivalent to $\mathbb{Q}$, where

$$\frac{d\mathbb{Q}^1}{d\mathbb{Q}} = \tilde{\eta}_T, \quad \mathbb{Q}\text{-a.s.}$$

and the Radon-Nikodým density process $\tilde{\eta}_t$, $t \in [0, T]$, satisfies

$$\tilde{\eta}_t = 1 + \int_{]0,t]} \tilde{\eta}_{u-} \kappa_u^2 \, dM_u^2.$$

For $s > t$ we have

$$\mathbb{Q}^1\{\tau_1 > s \,|\, \mathcal{H}_t^1 \vee \mathcal{H}_t^2\} = \mathbb{1}_{\{\tau_1 > t\}} \, \mathbb{E}_{\mathbb{Q}^1}\left(e^{\Lambda_t^1 - \Lambda_s^1} \,|\, \mathcal{F}_t\right)$$

but also

$$\mathbb{Q}^*\{\tau_1 > s \,|\, \mathcal{H}_t^1 \vee \mathcal{H}_t^2\} = \mathbb{Q}^1\{\tau_1 > s \,|\, \mathcal{H}_t^1 \vee \mathcal{H}_t^2\}.$$

Interpretation of Intensities

Recall that the processes λ_1 and λ_2 have jumps if $\alpha_i \neq \lambda_i$. The following result shows that the intensities λ^1 and λ^2 are 'local intensities' of default with respect to the information available at time t. It shows also that the model can in fact be reformulated as a two-dimensional Markov chain (see Lando (1998b)).

Proposition 4.6. *For $i = 1, 2$ and every $t \in \mathbb{R}_+$ we have*

$$\lambda_i = \lim_{h \downarrow 0} h^{-1} \mathbb{Q}^* \{t < \tau_i \leq t + h \,|\, \tau_1 > t, \, \tau_2 > t\}. \tag{38}$$

Moreover:

$$\alpha_1 = \lim_{h \downarrow 0} h^{-1} \mathbb{Q}^* \{t < \tau_1 \leq t + h \,|\, \tau_1 > t, \, \tau_2 \leq t\}.$$

and

$$\alpha_2 = \lim_{h \downarrow 0} h^{-1} \mathbb{Q}^* \{t < \tau_2 \leq t + h \,|\, \tau_2 > t, \, \tau_1 \leq t\}.$$

Bond Valuation

Proposition 4.7. *The price $D_1(t, T)$ on $\{\tau_1 > t\}$ equals*

$$D_1(t, T) = \mathbb{1}_{\{\tau_2 \leq t\}} \, e^{-\alpha_1(T-t)}$$
$$+ \mathbb{1}_{\{\tau_2 > t\}} \frac{1}{\lambda - \alpha_1} \left(\lambda_2 e^{-\alpha_1(T-t)} + (\lambda_1 - \alpha_1) e^{-\lambda(T-t)}\right).$$

Furthermore

$$\tilde{D}_1(t, T) = \mathbb{1}_{\{\tau_2 \leq t\}} \frac{(\lambda - \alpha_2)\lambda_2 e^{-\alpha_1(T-\tau_2)}}{\lambda_1 \alpha_2 e^{(\lambda-\alpha_2)\tau_2} + \lambda(\lambda_2 - \alpha_2)}$$
$$+ \mathbb{1}_{\{\tau_2 > t\}} \frac{\lambda - \alpha_2}{\lambda - \alpha_1} \frac{(\lambda_1 - \alpha_1) e^{-\lambda(T-t)} + \lambda_2 e^{-\alpha_1(T-t)}}{\lambda_1 e^{-(\lambda-\alpha_2)t} + \lambda_2 - \alpha_2}$$

and

$$\hat{D}_1(t, T) = \mathbb{1}_{\{\tau_1 > t\}} \frac{\lambda_2 e^{-\alpha_1 T} + (\lambda_1 - \alpha_1) e^{-\lambda T}}{\lambda_2 e^{-\alpha_1 t} + (\lambda_1 - \alpha_1) e^{-\lambda t}}.$$

Observe that:

- The formula for $D_1(t,T)$ coincides with the Jarrow and Yu formula for the bond issued by a secondary firm.
- The processes $D_1(t,T)$ and $\hat{D}_1(t,T)$ represent ex-dividend values of the bond, and thus they vanish after default time τ_1.
- The latter remark does not apply to the process $\tilde{D}_1(t,T)$.

4.6 Dependent Intensities of Credit Migrations

We present here a contribution to the dynamic theory of dependence between credit events. Specifically, we discuss here an approach towards modeling of dependent credit migrations based on the theory of continuous-time conditional Markov chains.[6] The goal is to extend the previous analysis to the case of multiple credit ratings. Assume that the current financial standing of the ith firm is reflected through the credit ranking process C^i with values in a finite set of credit grades $\mathcal{K}_i = \{1,\ldots,k_i\}$. For simplicity, we assume that the reference filtration $\mathbb{F}$ is trivial, and we consider the case of two firms.

Let $\mathbb{F}^i = \mathbb{F}^{C^i}$, $i = 1,2$, denote the filtration generated by C^i and let $\mathbb{G} = \mathbb{F}^1 \vee \mathbb{F}^2$. We examine the two following Markovian properties under the martingale measure $\mathbb{Q}^*$. The Markov property of $C = (C^1, C^2)$:

$$\mathbb{Q}^*\{C_s^1 = k, C_s^2 = l \,|\, \mathcal{G}_t\} = \mathbb{Q}^*\{C_s^1 = k, C_s^2 = l \,|\, C_t^1, C_t^2\}.$$

The $\mathbb{F}^j$-conditional Markov property of C^i for $i \neq j$:

$$\mathbb{Q}^*\{C_s^1 = k \,|\, \mathcal{G}_t) = \mathbb{Q}^*\{C_s^1 = k \,|\, \sigma(C_t^1) \vee \mathcal{F}_t^2\},$$

$$\mathbb{Q}^*\{C_s^2 = l \,|\, \mathcal{G}_t\} = \mathbb{Q}^*\{C_s^2 = l \,|\, \sigma(C_t^2) \vee \mathcal{F}_t^1\}.$$

Extension of Kusuoka's Construction

Assume that $k_1 = k_2 = 3$ (three rating grades). We consider the two independent Markov chains C^i, $i = 1,2$ defined on $(\Omega, \mathcal{G}, \mathbb{Q})$ and taking values in $\mathcal{K} = \{1,2,3\}$ with generators:

$$\Lambda^i = \begin{pmatrix} -\lambda_{12}^i - \lambda_{13}^i & \lambda_{12}^i & \lambda_{13}^i \\ \lambda_{21}^i & -\lambda_{21}^i - \lambda_{23}^i & \lambda_{23}^i \\ 0 & 0 & 0 \end{pmatrix}$$

The state $k = 3$ is the only absorbing state for each chain. We assume that $(C_0^1, C_0^2) = (1,1)$. In addition, we are given the following matrices:

---

[6] We refer to Bielecki and Rutkowski (2002) for information regarding conditional Markov chains.

$$\Lambda^{i|l} = \begin{pmatrix} -\lambda_{12}^{i|l} - \lambda_{13}^{i|l} & \lambda_{12}^{i|l} & \lambda_{13}^{i|l} \\ \lambda_{21}^{i|l} & -\lambda_{21}^{i|l} - \lambda_{23}^{i|l} & \lambda_{23}^{i|l} \\ 0 & 0 & 0 \end{pmatrix}$$

for $i = 1, 2$ and $l = 2, 3$. It should be observed that formally $\Lambda^i = \Lambda^{i|1}$ for $i = 1, 2$. In general, the intensities λ_{km}^i and $\lambda_{km}^{i|l}$ may follow $\mathbb{F}$-predictable stochastic processes.

Auxiliary Processes and Associated Martingales

We define a probability measure $\mathbb{Q}^*$ equivalent to $\mathbb{Q}$. To this end, we introduce auxiliary processes κ_{km}^i, by setting

$$\kappa_{km}^i(t) = \sum_{l=2}^{3} \mathbf{H}_{kl}^j(t-) \left(\frac{\lambda_{km}^{i|l}}{\lambda_{km}^i} - 1 \right)$$

for $i = 1, 2$, $j \neq i$, $k = 1, 2$, $m = 1, 2, 3$, $k \neq m$, where for $j = 1, 2$ and $k = 1, 2, 3$,

$$\mathbf{H}_{kl}^j(t) = H_l^j(t) H_k^j(t)$$

with $H_k^j(t) = \mathbb{1}_{\{C_t^j = k\}}$. We also define, for $i = 1, 2$ and $k \neq m$, the transition counting process

$$H_{km}^i(t) = \sum_{0 < u \leq t} H_k^i(u-) H_m^i(u).$$

For $i = 1, 2$ and $k \neq m$, the process M_{km}^i given by the expression

$$M_{km}^i(t) = H_{km}^i(t) - \int_0^t \lambda_{km}^i H_k^i(u) \, du$$

is known to follow an $\mathbb{F}^i$-martingale under $\mathbb{Q}$, and thus also a $\mathbb{G}$-martingale under $\mathbb{Q}$ where $\mathbb{G} = \mathbb{F}^1 \vee \mathbb{F}^2$.

Equivalent Probability Measure

We define a strictly positive martingale under $\mathbb{Q}$:

$$\eta_t = 1 + \sum_{i=1}^{2} \int_{]0,t]} \sum_{k=1}^{2} \sum_{m=1, m \neq k}^{3} \eta_{u-} \kappa_{km}^i(u) \, dM_{km}^i(u).$$

The process η plays the role of the Radon-Nikodým density process. For any fixed, but otherwise arbitrary, date T we define the probability measure $\mathbb{Q}^*$ equivalent to $\mathbb{Q}$ by setting:

$$\frac{d\mathbb{Q}^*}{d\mathbb{Q}} = \eta_T, \quad \mathbb{Q}\text{-a.s.}$$

The following result describes the properties of migration processes C^1 and C^2 under $\mathbb{Q}^*$. Recall that under the present convention: $\lambda_{km}^i = \lambda_{km}^{i|1}$.

Proposition 4.8. *For each* $i \neq j$ *the migration process* C^i *follows an* $\mathbb{F}^j$-*conditional Markov chain under* $\mathbb{Q}^*$. *For any* $k \neq m$ *the* $\mathbb{F}^j$-*conditional transition intensity of* C^i *under* $\mathbb{Q}^*$ *equals:*

$$\lambda_{km}^{*i}(t) = (1 + \kappa_{km}^i(t))\lambda_{km}^i = \lambda_{km}^i H_1^j(t) + \sum_{l=2}^{3} H_l^j(t)\lambda_{km}^{i|l}.$$

Conditional Markov Property

The $\mathbb{F}^j$-conditional Markov property of C^i under the equivalent probability measure $\mathbb{Q}^*$ established in Proposition 4.8 is a consequence of:

- The fact that the Radon-Nikodým density process η depends only on $C = (C^1, C^2)$.
- The fact that the migration process C has the Markov property under the original probability $\mathbb{Q}$.

Let us summarize the properties of our model under $\mathbb{Q}^*$. First, for $i = 1$, and $j \neq i$, the process $\lambda_{km}^{*i}(t)$ is the corresponding $\mathbb{F}^j$-martingale intensity. In other words, the processes M_{km}^{*i} defined as

$$M_{km}^{*i}(t) = H_{km}^i(t) - \int_0^t \lambda_{km}^{*i}(u)H_k^i(u)\,du$$

for $k \neq m$ are $\mathbb{G}$-martingales under $\mathbb{Q}^*$. Second, as we shall see soon, the intensities λ_{km}^{*i} have the natural interpretation as the 'local intensities' of credit migrations (in the special case of a trivial reference filtration C is a Markov chain under $\mathbb{Q}^*$).

Interpretation of Intensities

Let us explain the intuitive meaning of intensity parameters. For original intensities we have

$$\lambda_{kk'}^1 = \lim_{h \downarrow 0} h^{-1}\, \mathbb{Q}\{C_{t+h}^1 = k' \,|\, C_t^1 = k,\, C_t^2 = 1\},$$

but also for $l = 2, 3$

$$\lambda_{kk'}^1 = \lim_{h \downarrow 0} h^{-1}\, \mathbb{Q}^*\{C_{t+h}^1 = k' \,|\, C_t^1 = k,\, C_t^2 = l\},$$

The modified intensities satisfy, for $l = 2, 3$,

$$\lambda_{kk'}^{1;l} = \lim_{h \downarrow 0} h^{-1}\, \mathbb{Q}^*\{C_{t+h}^1 = k' \,|\, C_t^1 = k,\, C_t^2 = l\}.$$

Let us recall that model's inputs are: the original generators Λ^1, Λ^2, and the modified matrices:

$$\Lambda^{i;l} \begin{pmatrix} -\lambda_{12}^{i;l} - \lambda_{13}^{i;l} & \lambda_{12}^{i;l} & \lambda_{13}^{i;l} \\ \lambda_{21}^{i;l} & -\lambda_{21}^{i;l} - \lambda_{23}^{i;l} & \lambda_{23}^{i;l} \\ 0 & 0 & 0 \end{pmatrix}$$

for $i = 1, 2$ and $l = 2, 3$.

First-to-Change Swap

Let $C = (C^1, \ldots, C^n)$. We assume that the payoff occurs at the first change of the credit rating of the firm 1 or 2. The payoff is digital, specifically, if we set $\tau = \tau_1 \wedge \tau_2$ then the payoff at time τ equals

$$Z_\tau = K_1 \mathbb{1}_{\{\tau = \tau_1 \leq T\}} + K_2 \mathbb{1}_{\{\tau = \tau_2 \leq T\}}.$$

for some constant K_1, K_2. Let us summarize the basic steps of the valuation procedure:

- Introduce an auxiliary probability measure $\mathbb{Q}^{1,2}$ equivalent to $\mathbb{Q}^*$.
- Verify that any martingale under $\mathbb{Q}^{1,2}$ with respect to $\mathbb{G}^{1,2} = \mathbb{F} \vee \mathbb{H}^3 \vee \cdots \vee \mathbb{H}^n$ is also a martingale under $\mathbb{Q}^{1,2}$ with respect to $\mathbb{G} = \mathbb{F} \vee \mathbb{H}^1 \vee \cdots \vee \mathbb{H}^n$.
- Use the standard formula to find the $\mathbb{G}^{1,2}$-conditional laws of τ_1 and τ_2 under $\mathbb{Q}^*$, through conditional expectations with respect to $\mathbb{Q}^{1,2}$.
- Use the fact that τ_1 and τ_2 are $\mathbb{G}^{1,2}$-conditionally independent under $\mathbb{Q}^*$ in order to value the swap.

We argue that in some cases a high-dimensional (unconditional) expectation can be efficiently evaluated as a low-dimensional conditional expectation under an equivalent probability measure.

4.7 Dynamics of Dependent Credit Ratings

Let us denote by $C_t = (C_t^1, \ldots, C_t^n)$ the vector of credit ratings at time t of all relevant obligors (credit names). Some authors focus directly on specification of dynamics of the process C. Note that the assumption that the state space for C is finite is not always imposed.

Continuous Time Setup

Indirect approach. Structural/factor models (KMV, CreditMetrics, etc.) are based on the assumption that $C_t = \Psi(\xi_t)$, where ξ is a (multivariate) factor process (representing, for instance, the values of the firms). Dynamics of ξ are typically modeled as an Ito process. Note that ξ involves both idiosyncratic risks and systemic risks.

Direct approach. Models proposed by Hull and White (2001), Douady and Jeanblanc (2002) and Albanese et al. (2003) postulate that the credit ratings process is a multi-dimensional diffusion, specifically,

$$dC_t = \mu_t \, dt + \Sigma_t \, dW_t.$$

The dependence between rating migrations is introduced here through the judicious choice of the diffusion matrix Σ_t.

Discrete Time Setup

Discrete-time Markov models of credit migrations were studied by Kijima et al. (2002) and Bielecki (2002). Credit ratings are modeled as

$$C_{t+1}^n = \Theta(C_t^n, Z_{t+1}^n, B_{t+1}^n, Y_{t+1}) =$$

$$\begin{cases} \theta(C_t^n + Z_{t+1}^n + B_{t+1}^n Y_{t+1}), & \text{if} C_t^n \leq K - 1, \\ K, & \text{if} C_t^n = K, \end{cases} \tag{39}$$

where Z_t^n and B_t^n represent idiosyncratic risks, and Y_t represents systemic risks, and where $\theta(k)$ is a cut-off function.

Main practical issues arising in the context of a model's implementation are: the estimation and calibration of the model, the structure of the pricing measure, the effect of change of measures on the dependence structure. As soon as the model is estimated and calibrated, it can be easily used for risk management purposes, as well as for pricing purposes (via Monte Carlo simulation).

4.8 Defaultable Term Structure

It this section, we shall summarize the model of defaultable term structure of interest rates developed by Bielecki and Rutkowski (2000) and Schönbucher (2000a), and further generalized by Eberlein and Özkan (2003). Essentially, the model extends the Heath-Jarrow-Morton (HJM) model of term structure of default-free rates to the case of defaultable bonds. Although we do not consider here dependence between term structures of several corporate bonds, the approach presented here lends itself for such dependence analysis (see Section 13.2.7 in Bielecki and Rutkowski (2002)).

Standing Assumptions

Standard intensity-based models (as, for instance, in Jarrow and Turnbull (1995) or Jarrow et al. (1997)) rely on the following assumptions:
- Existence of the martingale measure $\mathbb{Q}^*$ is postulated.
- Relationship between the statistical probability $\mathbb{P}$ and the risk-neutral probability $\mathbb{Q}^*$ is derived via calibration.
- Credit migrations process is modeled as a Markov chain.
- Market and credit risks are separated (independent).

The HJM-type model of defaultable term structure with multiple ratings was proposed independently by Bielecki and Rutkowski (2000) and Schönbucher (2000a). The main features of this approach are:
- The model formulates sufficient consistency conditions that tie together credit spreads and recovery rates in order to construct a risk-neutral probability $\mathbb{Q}^*$ and the corresponding risk-neutral intensities of credit events.

- Statistical probability $\mathbb{P}$ and the risk-neutral probability $\mathbb{Q}^*$ are connected via the market price of interest rate risk and the market price of credit risk.
- Market and credit risks are combined in a flexible way.

Term Structure of Credit Spreads

Suppose that we are given a filtered probability space $(\Omega, \mathbb{F}, \mathbb{P})$ endowed with a d-dimensional standard Brownian motion W. We assume that the reference filtration satisfies $\mathbb{F} = \mathbb{F}^W$. For any fixed maturity $0 < T \leq T^*$, the price of a zero-coupon Treasury bond equals

$$B(t,T) = \exp\left(-\int_t^T f(t,u)\,du\right),$$

where the default-free instantaneous forward rate $f(t,T)$ process is subject to the standard (HJM) assumption.

(HJM) Dynamics of the instantaneous forward rate $f(t,T)$ are given by the expression

$$f(t,T) = f(0,T) + \int_0^t \alpha(u,T)\,du + \int_0^t \sigma(u,T)\,dW_u$$

for some function $f(0,\cdot) : [0,T^*] \to \mathbb{R}$, and some $\mathbb{F}$-adapted processes $\alpha : A \times \Omega \to \mathbb{R}$, $\sigma : A \times \Omega \to \mathbb{R}^d$, where $A = \{(u,t)\,|\,0 \leq u \leq t \leq T^*\}$.

Credit Classes

Suppose there are $K \geq 2$ credit rating classes, where the Kth class corresponds to the default-free bond. Essentially, credit rating classes are distinguished by the yields on the corresponding bonds. In other words, for any fixed maturity $0 < T \leq T^*$, the *defaultable instantaneous forward rate* $g_i(t,T)$ corresponds to the rating class $i = 1, \ldots, K - 1$. We assume that:

(HJM^i) Dynamics of the instantaneous defaultable forward rates $g_i(t,T)$ are given by

$$g_i(t,T) = g_i(0,T) + \int_0^t \alpha_i(u,T)\,du + \int_0^t \sigma_i(u,T)\,dW_u$$

for some deterministic functions $g_i(0,\cdot) : [0,T^*] \to \mathbb{R}$, and some $\mathbb{F}$-adapted processes $\alpha_i : A \times \Omega \to \mathbb{R}$, $\sigma_i : A \times \Omega \to \mathbb{R}^d$.

Credit Spreads

It is natural (although not necessary for further developments) to assume that

$$g_{K-1}(t,T) > g_{K-2}(t,T) > \cdots > g_1(t,T) > f(t,T)$$

for every $t \leq T$.

Definition 4.3. For every $i = 1, 2, \ldots, K - 1$, the ith *forward credit spread* equals $s_i(\cdot, T) = g_i(\cdot, T) - f(\cdot, T)$.

Martingale Measure $\mathbb{P}^*$

It is known from the HJM theory that the following condition (M) is sufficient to exclude arbitrage across default-free bonds for all maturities $T \leq T^*$ and the savings account.

Condition (M) There exists an $\mathbb{F}$-adapted $\mathbb{R}^d$-valued process β such that

$$\mathbb{E}_{\mathbb{P}}\left\{ \exp\left(\int_0^{T^*} \beta_u \, dW_u - \frac{1}{2} \int_0^{T^*} |\beta_u|^2 \, du \right) \right\} = 1$$

and, for any maturity $T \leq T^*$, we have

$$\alpha^*(t,T) = \tfrac{1}{2}|\sigma^*(t,T)|^2 - \sigma^*(t,T)\beta_t$$

where

$$\alpha^*(t,T) = \int_t^T \alpha(t,u) \, du$$

$$\sigma^*(t,T) = \int_t^T \sigma(t,u) \, du.$$

Let β be some process satisfying Condition (M). Then the probability measure $\mathbb{P}^*$, given by the formula

$$\frac{d\mathbb{P}^*}{d\mathbb{P}} = \exp\left(\int_0^{T^*} \beta_u \, dW_u - \frac{1}{2} \int_0^{T^*} |\beta_u|^2 \, du \right), \quad \mathbb{P}\text{-a.s.},$$

is a *martingale measure* for the default-free term structure. We will see that for any T the price $B(t,T)$ is a martingale under the measure $\mathbb{P}^*$, when discounted with the savings account B_t.

Zero-Coupon Bonds

The price of the T-maturity default-free zero-coupon bond is given by the equality

$$B(t,T) = \exp\left(-\int_t^T f(t,u) \, du \right).$$

Formally, such Treasury bond corresponds to credit class K. Similarly, the 'conditional value' of T-maturity defaultable zero-coupon bond belonging at time t to the credit class $i = 1, 2, \ldots, K-1$, equals

$$D_i(t,T) = \exp\left(-\int_t^T g_i(t,u) \, du \right).$$

We consider discounted price processes

$$Z(t,T) = B_t^{-1} B(t,T), \quad Z_i(t,T) = B_t^{-1} D_i(t,T),$$

where B is the savings account

$$B_t = \exp\left(\int_0^t f(u,u)\,du\right).$$

Let us define a Brownian motion W^* under $\mathbb{P}^*$ by setting

$$W_t^* = W_t - \int_0^t \beta_u\,du, \quad \forall t \in [0, T^*].$$

Conditional Dynamics of the Bond Price

Lemma 4.2. *Under the martingale measure $\mathbb{P}^*$, for any fixed $T \leq T^*$, the discounted price processes $Z(t,T)$ and $Z_i(t,T)$ satisfy*

$$dZ(t,T) = Z(t,T)b(t,T)\,dW_t^*,$$

where $b(t,T) = -\sigma^*(t,T)$, and

$$dZ_i(t,T) = Z_i(t,T)\big(\lambda_i(t)\,dt + b_i(t,T)\,dW_t^*\big)$$

where

$$\lambda_i(t) = a_i(t,T) - f(t,t) + b_i(t,T)\beta_t$$

and

$$a_i(t,T) = g_i(t,t) - \alpha_i^*(t,T) + \tfrac{1}{2}|\sigma_i^*(t,T)|^2$$
$$b_i(t,T) = -\sigma_i^*(t,T).$$

Observe that usually the process $Z_i(t,T)$ is not a martingale under the martingale measure $\mathbb{P}^*$. This feature is related to the fact that it does not represent the (discounted) price of a tradeable security.

Credit Migration Process

Recall that we assumed that the set of rating classes is $\mathcal{K} = \{1,\ldots,K\}$, where the class K corresponds to default. The *migration process C* is constructed in Bielecki and Rutkowski (2000a) as a (nonhomogeneous) conditionally Markov process on $\mathcal{K}$, with the state K as the unique *absorbing* state for this process. The process C is constructed on some enlarged probability space $(\Omega^*, \mathbb{G}, \mathbb{Q}^*)$, where the probability measure $\mathbb{Q}^*$ is the extended martingale measure. The reference filtration $\mathbb{F}$ is contained in the extended filtration $\mathbb{G}$. For simplicity of presentation, we summarize the results for the case $K = 3$.

Given some non-negative and $\mathbb{F}$-adapted processes $\lambda_{1,2}(t)$, $\lambda_{1,3}(t)$, $\lambda_{2,1}(t)$ and $\lambda_{2,3}(t)$, a migration process C is constructed as a conditional Markov process with the conditional intensity matrix (infinitesimal generator)

$$\Lambda(t) = \begin{pmatrix} \lambda_{1,1}(t) & \lambda_{1,2}(t) & \lambda_{1,3}(t) \\ \lambda_{2,1}(t) & \lambda_{2,2}(t) & \lambda_{2,3}(t) \\ 0 & 0 & 0 \end{pmatrix}$$

where $\lambda_{i,i}(t) = -\sum_{j\neq i}\lambda_{i,j}(t)$ for $i = 1, 2$.

The conditional Markov property (with respect to the reference filtration $\mathbb{F}$) means that if we denote by $\mathcal{F}_t^C$ the σ-field generated by C up to time t then for arbitrary $s \geq t$ and $i, j \in \mathcal{K}$ we have

$$\mathbb{Q}^*\{C_{t+s} = i \mid \mathcal{F}_t \vee \mathcal{F}_t^C\} = \mathbb{Q}^*\{C_{t+s} = i \mid \mathcal{F}_t \vee \{C_t = j\}\}.$$

The formula above provides the risk-neutral conditional probability that the defaultable bond is in class i at time $t + s$, given that it was in the credit class C_t at time t. For any date t, we denote by $\hat{C}_t$ the previous bond's rating; we shall need this notation later.

Finally, the default time τ is introduced by setting

$$\tau = \inf\{t \in \mathbb{R}_+ : C_t = 3\}.$$

Let $H_i(t) = \mathbb{1}_{\{C_t=i\}}$ for $i = 1, 2$, and let $H_{i,j}(t)$ represent the number of transitions from i to j by C over the time interval $(0, t]$. It can be shown that the process

$$M_{i,j}(t) = H_{i,j}(t) - \int_0^t \lambda_{i,j}(s)H_i(s)\,ds, \quad \forall t \in [0, T],$$

for $i = 1, 2$ and $j \neq i$, is a martingale on the enlarged probability space $(\Omega^*, \mathbb{G}, \mathbb{Q}^*)$. Let us emphasize that due to the judicious construction of the migration process C, appropriate version of the hypotheses (H.1)-(H.3) remain valid here.

Defaultable Term Structure

We maintain the simplified framework with $K = 3$. We assume the fractional recovery of Treasury value scheme. To be more specific, to each credit rating $i = 1, \ldots, K - 1$, we associate the recovery rate $\delta_i \in [0, 1)$, where δ_i is the fraction of par paid at bond's maturity, if a bond belonging to the ith class defaults prior to its maturity. Thus, the cash flow at maturity is

$$X = \mathbb{1}_{\{\tau > T\}} + \delta_{\hat{C}_\tau}\mathbb{1}_{\{\tau \leq T\}}.$$

In order to provide the model with arbitrage free properties, Bielecki and Rutkowski (2000) postulate that the risk-neutral intensities of credit migrations $\lambda_{1,2}(t)$, $\lambda_{1,3}(t)$, $\lambda_{2,1}(t)$ and $\lambda_{2,3}(t)$ are specified by the *no-arbitrage condition* (also termed the *consistency condition*):

$$\lambda_{1,2}(t)\big(Z_2(t,T)\!-\!Z_1(t,T)\big) + \lambda_{1,3}(t)\big(\delta_1 Z(t,T) - Z_1(t,T)\big)$$
$$+\lambda_1(t)Z_1(t,T) = 0,$$
$$\lambda_{2,1}(t)\big(Z_1(t,T)\!-\!Z_2(t,T)\big) + \lambda_{2,3}(t)\big(\delta_2 Z(t,T) - Z_2(t,T)\big)$$
$$+\lambda_2(t)Z_2(t,T) = 0.$$

Martingale Dynamics of a Defaultable Bond

First, we introduce the process $\hat{Z}(t,T)$ as a solution to the following SDE

$$
\begin{aligned}
d\hat{Z}(t,T) = {} & \big(Z_2(t,T) - Z_1(t,T)\big)\, dM_{1,2}(t) \\
& + \big(Z_1(t,T) - Z_2(t,T)\big)\, dM_{2,1}(t) \\
& + \big(\delta_1 Z(t,T) - Z_1(t,T)\big)\, dM_{1,3}(t) \\
& + \big(\delta_2 Z(t,T) - Z_2(t,T)\big)\, dM_{2,3}(t) \\
& + H_1(t) Z_1(t,T) b_1(t,T)\, dW_t^* \\
& + H_2(t) Z_2(t,T) b_2(t,T)\, dW_t^* \\
& + \big(\delta_1 H_{1,3}(t) + \delta_2 H_{2,3}(t)\big) Z(t,T) b(t,T)\, dW_t^*,
\end{aligned}
$$

with the initial condition $\hat{Z}(0,T) = H_1(0) Z_1(0,T) + H_2(0) Z_2(0,T)$.

It appears that the process $\hat{Z}(t,T)$ follows a martingale on $(\Omega^*, \mathbb{G}, \mathbb{Q}^*)$, so that it is justified to refer to $\mathbb{Q}^*$ as the *extended martingale measure*. The proof of the next result employs the no-arbitrage condition.

Lemma 4.3. *For any maturity $T \leq T^*$ and for every $t \in [0,T]$ we have*

$$
\hat{Z}(t,T) = \mathbb{1}_{\{C_t \neq 3\}} Z_{C_t}(t,T) + \mathbb{1}_{\{C_t = 3\}} \delta_{\hat{C}_t} Z(t,T).
$$

Next, we define the price process of a T-maturity defaultable zero-coupon bond by setting

$$
D_C(t,T) = B_t \hat{Z}(t,T)
$$

for any $t \in [0,T]$. In view of Lemma 4.3, we have that

$$
D_C(t,T) = \mathbb{1}_{\{C_t \neq 3\}} D_{C_t}(t,T) + \mathbb{1}_{\{C_t = 3\}} \delta_{\hat{C}_t} B(t,T).
$$

The defaultable bond price $D_C(t,T)$ satisfies the following properties:

- The process $D_C(t,T)$ is a $\mathbb{G}$-martingale under $\mathbb{Q}^*$, when discounted by the savings account.
- In contrast to the 'conditional price' $D_i(t,T)$, the process $D_C(t,T)$ admits discontinuities. Jumps are directly associated with changes in credit quality (ratings migrations).
- The process $D_C(t,T)$ represents the price of a tradeable security: the corporate zero-coupon bond of maturity T.

Risk-Neutral Representations

Recall that $\delta_i \in [0,1)$ is the recovery rate for a bond which was in the ith rating class just prior to default.

Proposition 4.9. *The price process $D_C(t,T)$ of a T-maturity defaultable zero-coupon bond equals*

$$D_C(t,T) = \mathbb{1}_{\{C_t \neq 3\}} B(t,T) \exp\left(-\int_t^T s_{C_t}(t,u)\,du\right)$$
$$+ \mathbb{1}_{\{C_t=3\}} \delta_{\hat{C}_t} B(t,T)$$

where $s_i(t,u) = g_i(t,u) - f(t,u)$ is the ith credit spread.

Proposition 4.10. *The price process* $D_C(t,T)$ *satisfies the risk-neutral valuation formula*

$$D_C(t,T) = B_t\,\mathbb{E}_{\mathbb{Q}^*}\big(\delta_{\hat{C}_T} B_T^{-1} \mathbb{1}_{\{\tau \leq T\}} + B_T^{-1} \mathbb{1}_{\{\tau > T\}} \mid \mathcal{G}_t\big).$$

It is also clear that

$$D_C(t,T) = B(t,T)\,\mathbb{E}_{\mathbb{Q}_T}\big(\delta_{\hat{C}_T} \mathbb{1}_{\{\tau \leq T\}} + \mathbb{1}_{\{\tau > T\}} \mid \mathcal{G}_t\big),$$

where $\mathbb{Q}_T$ stands for the T-forward measure associated with the extended martingale measure $\mathbb{Q}^*$.

Let us end this section by mentioning that Eberlein and Özkan (2003) have generalized the model presented above to the case of term structures driven by Lévy processes.

Premia for Interest Rate and Credit Event Risks

We shall now change, using a suitable version of Girsanov's theorem, the measure $\mathbb{Q}^*$ to the equivalent probability measure $\mathbb{Q}$. In the financial interpretation, the probability measure $\mathbb{Q}$ will play the role of the statistical probability (i.e., the real-world probability). It is thus natural to postulate that the restriction of the probability measure $\mathbb{Q}$ to the original probability space Ω necessarily coincides with the statistical probability $\mathbb{P}$ for the default-free market. From now on, we shall assume that the following condition holds.

Condition (P) We postulate that

$$\frac{d\mathbb{Q}}{d\mathbb{Q}^*} = \hat{\eta}_{T^*}, \quad \mathbb{Q}^*\text{-a.s.},$$

where the positive $\mathbb{Q}^*$-martingale $\hat{\eta}$ is given by the formula

$$d\hat{\eta}_t = -\hat{\eta}_t \beta_t\,dW_t^* + \hat{\eta}_{t-}\,dM_t, \quad \eta_0 = 1,$$

for some $\mathbb{R}^d$-valued $\mathbb{F}$-predictable process β, where the $\mathbb{Q}^*$-local martingale M equals

$$dM_t = \sum_{i \neq j} \kappa_{i,j}(t)\,dM_{i,j}(t)$$
$$= \sum_{i \neq j} \kappa_{i,j}(t)\big(dH_{i,j}(t) - \lambda_{i,j}(t)H_i(t)\,dt\big).$$

for some $\mathbb{F}$-predictable processes $\kappa_{i,j} > -1$.

Assume that for any $i \neq j$

$$\int_0^{T^*} \left(\kappa_{i,j}(t) + 1\right)\lambda_{i,j}(t)\, dt < \infty, \quad \mathbb{Q}^*\text{-a.s.}$$

In addition, we postulate that $\mathbb{E}_{\mathbb{Q}^*}(\hat{\eta}_{T^*}) = 1$, so that the probability measure $\mathbb{Q}$ is indeed well defined on $(\Omega^*, \mathcal{G}_{T^*})$. The financial interpretation of processes β and κ is similar as in Section 3.2, namely,

- The vector-valued process β corresponds to the *premium for the interest rate risk*.
- The matrix-valued process κ represents the *premium for the credit event risk*.

Statistical Default Intensities

We define processes $\lambda_{i,j}^{\mathbb{Q}}$ by setting, for $i \neq j$,

$$\lambda_{i,j}^{\mathbb{Q}}(t) = (\kappa_{i,j}(t) + 1)\lambda_{i,j}(t), \quad \lambda_{i,i}^{\mathbb{Q}}(t) = -\sum_{j \neq i} \lambda_{i,j}^{\mathbb{Q}}(t).$$

Proposition 4.11. *Under the equivalent probability $\mathbb{Q}$ given by condition (P), the process C is a conditionally Markov process. The matrix of conditional intensities of C under $\mathbb{Q}$ equals*

$$\Lambda_t^{\mathbb{Q}} = \begin{pmatrix} \lambda_{1,1}^{\mathbb{Q}}(t) & \cdots & \lambda_{1,K}^{\mathbb{Q}}(t) \\ \cdot & \cdots & \cdot \\ \lambda_{K-1,1}^{\mathbb{Q}}(t) & \cdots & \lambda_{K-1,K}^{\mathbb{Q}}(t) \\ 0 & \cdots & 0 \end{pmatrix}$$

If the market price for credit risk depends only on the current rating i (and not on the rating j after jump), so that $\kappa_{i,j} = \kappa_{i,i}$ for every $j \neq i$. Then $\Lambda_t^{\mathbb{Q}} = \Phi_t \Lambda_t$, where $\Phi_t = \text{diag}\,[\phi_i(t)]$ with $\phi_i(t) = \kappa_{i,i}(t) + 1$ is the diagonal matrix (this case was examined, e.g., by Jarrow et al. (1997)).

Defaultable Coupon Bond

Consider a defaultable coupon bond with the face value L that matures at time T and promises to pay coupons c_i at times $T_1 < \cdots < T_n < T$. The coupon payments are only made prior to default, and the recovery payment is made at maturity T, and is proportional to the bond's face value. Notice that the migration process C introduced in Section 4.8 may depend on both the maturity T and on recovery rates. Therefore, it is more appropriate to write $C_t = C_t(\delta, T)$, where $\delta = (\delta_1, \ldots, \delta_K)$. Similarly, we denote the price of a defaultable zero-coupon bond $D_{C(\delta,T)}(t, T)$, rather than $D_C(t, T)$.

A defaultable coupon bond can be treated as a portfolio consisting of:

- Defaultable coupons – that is, defaultable zero-coupon bonds with maturities $T_1, \ldots, T_n$, which are subject to zero recovery.
- Defaultable face value – that is, a T-maturity defaultable zero-coupon bond with a constant recovery rate δ.

We conclude that the arbitrage price of a defaultable coupon bond equals

$$D_c(t,T) = \sum_{i=1}^{n} c_i D_{C(0,T_i)}(t,T_i) + L D_{C(\delta,T)}(t,T),$$

where, by convention, we set $D_{C(0,T_i)}(t,T_i) = 0$ for $t > T_i$.

Examples of Credit Derivatives

Credit Default Swap

Consider first a basic credit default swap, as described, e.g., in Section 1.3.1 of Bielecki and Rutkowski (2002). In the present setup, the contingent payment is triggered by the event $\{C_t = K\}$. The contract is settled at time $\tau = \inf\{t < T : C_t = K\}$, and the payoff equals

$$Z_\tau = \left(1 - \delta_{\hat{C}_T} B(\tau,T)\right).$$

Notice the dependence of Z_τ on the initial rating C_0 through the default time τ and the recovery rate $\delta_{\hat{C}_T}$. The following two market conventions are common in practice:

- The buyer pays a lump sum at contract's inception (*default option*).
- The buyer pays annuities up to default time (*default swap*).

In the first case, the value at time 0 of a default option equals

$$S_0 = \mathbb{E}_{\mathbb{Q}^*}\left(B_\tau^{-1}\left(1 - \delta_{\hat{C}_T} B(\tau,T)\right)\mathbb{1}_{\{\tau \leq T\}}\right).$$

In the second case, the annuity κ can be found from the equation

$$S_0 = \kappa\, \mathbb{E}_{\mathbb{Q}^*}\left(\sum_{i=1}^{T} B_{t_i}^{-1} \mathbb{1}_{\{t_i < \tau\}}\right).$$

Notice that both the price S_0 and the annuity κ depend on the initial bond's rating C_0.

Total Rate of Return Swap

As a reference asset we take the coupon bond with the promised cash flows c_i at times T_i. Suppose the contract maturity is $\hat{T} \leq T$. In addition, suppose that the *reference rate payments* (the *annuity payments*) are made by the investor at fixed scheduled times $t_i \leq \hat{T}$, $i = 1, 2, \ldots, m$. The owner of a total rate of return swap is entitled not only to all coupon payments during the life of the contract, but also to the change in the value of the underlying bond. By convention, we assume that the default event occurs when $C_t(\delta, T) = K$. According to this convention, the reference rate κ to be paid by the investor satisfies

$$\mathbb{E}_{\mathbb{Q}^*}\left(\sum_{i=1}^{n} c_i B_{T_i}^{-1} \mathbb{1}_{\{T_i \leq \hat{T}\}} \right) + \mathbb{E}_{\mathbb{Q}^*}\left(B_\tau^{-1}\left(D_c(\tau, T) - D_c(0, T) \right) \right)$$

$$= \kappa \, \mathbb{E}_{\mathbb{Q}^*}\left(\sum_{i=1}^{m} B_{t_i}^{-1} \mathbb{1}_{\{C_{t_i}(\delta, T) \neq K\}} \right),$$

where $\tau = \inf\{t \geq 0 : C_t(\delta, T) = K\} \wedge \hat{T}$.

4.9 Concluding Remarks

It should be acknowledged that we have not discussed in the present text any results or techniques related to hedging of credit risk. Let us conclude, however, by listing the most important issues arising in practical and theoretical approaches to this problem, and giving some references that may be consulted by the interested reader.

Simplified approaches. In most practical implementations of credit risk models (see, for instance, Greenfield (2000)), it is common to impose at least some of the following simplifying assumptions:

- Only a pure credit risk instrument (e.g., a basic credit default swap) is considered.
- One deals with a one-sided counterparty risk with a fixed recovery rate (the same for a derivative product and for a corporate bond).
- The mark-to-market value of the contract is assumed to be non-negative to a non-defaultable counterparty (thus, for instance, defaultable loans and bonds or vulnerable options are covered, but defaultable swaps are excluded).
- Independence of market and credit risks is frequently postulated.
- Existence of a non-defaultable version of the contract and of a liquid market in corporate bonds and other related instruments of various maturities is assumed.

Theoretical results. More sophisticated mathematical techniques, which have a potential to be useful in hedging credit risk, have been developed in recent years, in particular:

- Suitable versions of a predictable representation theorem with respect to discontinuous martingales associated with the default event, or with credit migrations, were established (see, for instance, Bélanger et al. (2001) or

Blanchet-Scalliet and Jeanblanc (2003)). Unfortunately, the general formulae obtained through this technique seem to be very difficult to implement. A more straightforward approach to the replication of credit derivatives was proposed by Vaillant (2001) (see also Jeanblanc and Rutkowski (2003) in this regard).

- A utility-based approach to hedging of credit risk and valuation of credit derivatives was examined. In this approach, which is based on the idea of indifference pricing, hedging strategies are constructed as solutions to appropriate stochastic control problems (see, for instance, Collin-Dufresne and Hugonnier (2002) or Lukas (2001)).

- An alternative approach to hedging of credit risk, in the spirit of Markowitz mean-variance methodology, was recently developed. It involves, in general, constructing of hedging strategies in terms of solutions of certain backward stochastic differential equations as well as in terms of certain orthogonal projections (see Bielecki et al. (2004)).

References

1. C. Albanese, J. Campolieti, O. Chen and A. Zavidonov (2003) Credit barrier models. *Risk Magazine* 16(6).
2. M. Ammann (1999) *Pricing Derivative Credit Risk.* Springer-Verlag, Berlin Heidelberg New York.
3. A. Arvanitis and J. Gregory (2001) *Credit: The Complete Guide to Pricing, Hedging and Risk Management.* Risk Books, London.
4. S. Babbs and T.R. Bielecki (2003) A note on short spreads. Working paper.
5. G. Bakshi, D. Madan and F. Zhang (2001) Understanding the role of recovery in default risk models: Empirical comparisons and implied recovery rates. Working paper.
6. A. Bélanger, S.E. Shreve and D. Wong (2001) A general framework for pricing credit risk. Forthcoming in *Mathematical Finance.*
7. T.R. Bielecki (2002) A multivariate Markov model for simulating dependent migrations. Working paper.
8. T.R. Bielecki, M. Jeanblanc and M. Rutkowski (2004a) Pricing and hedging of credit risk: Replication and mean-variance approaches. Forthcoming in *Proceedings of the AMS-IMS-SIAM Summer Conference on Mathematics of Finance. Snowbird, UT, June 22-26, 2003.*
9. T.R. Bielecki, M. Jeanblanc and M. Rutkowski (2004b) On Hedging of Credit Risk and Credit Derivatives. Working paper.
10. T.R. Bielecki and M. Rutkowski (2000) Multiple ratings model of defaultable term structure. *Mathematical Finance* 10 125–139.
11. T.R. Bielecki and M. Rutkowski (2001) Credit risk modelling: Intensity based approach. In: *Handbook in Mathematical Finance: Option Pricing, Interest Rates and Risk Management,* eds. E. Jouini, J. Cvitanić, M. Musiela, Cambridge University Press.
12. T.R. Bielecki and M. Rutkowski (2002) *Credit Risk: Modelling, Valuation and Hedging.* Springer-Verlag, Berlin.

13. T.R. Bielecki and M. Rutkowski (2003) Dependent defaults and credit migrations. *Applicationes Mathematicae* 30, 121–145.
14. T.R. Bielecki and M. Rutkowski (2004) Defaultable term structure: Conditionally Markov approach. *IEEE Transactions on Automatic Control, Special Issue on Stochastic Control Methods in Financial Engineering*, in press.
15. F. Black and M. Scholes (1973) The pricing of options and corporate liabilities. *Journal of Political Economy* 81, 637–654.
16. F. Black and J.C. Cox (1976) Valuing corporate securities: Some effects of bond indenture provisions. *Journal of Finance* 31, 351–367.
17. C. Blanchet-Scalliet and M. Jeanblanc (2003) Hazard rate for credit risk and hedging defaultable contingent claims. Forthcoming in *Finance & Stochastics*.
18. M.J. Brennan and E.S. Schwartz (1977) Convertible bonds: Valuation and optimal strategies for call and conversion. *Journal of Finance* 32, 1699–1715.
19. M.J. Brennan and E.S. Schwartz (1980) Analyzing convertible bonds. *Journal of Financial and Quantitative Analysis* 15, 907–929.
20. D. Brigo and A. Alfonsi (2003) A two-dimensional shifted square-root diffusion model for credit derivatives: calibration, pricing and the impact of correlation. Working paper.
21. E. Briys and F. de Varenne (1997) Valuing risky fixed rate debt: An extension. *Journal of Financial and Quantitative Analysis* 32, 239–248.
22. L. Chen and D. Filipović (2003a) A simple model for credit migration and spread curves. Working paper.
23. L. Chen and D. Filipović (2003b) Pricing credit default swaps with default correlation and counterparty risk. Working paper.
24. P.O. Christensen, C.R. Flor, D. Lando and K.R. Miltersen (2002) Dynamic capital structure with callable debt and debt renegotiations. Working paper.
25. P. Collin-Dufresne, R.S. Goldstein and J.-N. Hugonnier (2003) A general formula for valuing defaultable securities. Working paper.
26. P. Collin-Dufresne and J.-N. Hugonnier (2002) On the pricing and hedging of contingent claims in the presence of extraneous risks. Working paper.
27. D. Cossin and H. Pirotte (2000) *Advanced Credit Risk Analysis*. J. Wiley, Chichester.
28. M. Crouhy, D. Galai and R. Mark (1998) Credit risk revisited. *Risk – Credit Risk Supplement,* March, 40–44.
29. M. Davis and V. Lo (2001) Infectious defaults. *Quantitative Finance* 1, 382–386.
30. C. Dellacherie (1972) *Capacités et processus stochastiques*. Springer-Verlag, Berlin Heidelberg New York.
31. R. Douady and M. Jeanblanc (2002) A rating-based model for credit derivatives. *European Investment Review* 1, 17–29.
32. G. Duffee (1999) Estimating the price of default. *Review of Financial Studies* 12, 197–226.
33. D. Duffie (1998a) Defaultable term structure models with fractional recovery of par. Working paper.
34. D. Duffie (1998b) First-to-default valuation. Working paper.
35. D. Duffie and D. Lando (2001) The term structure of credit spreads with incomplete accounting information. *Econometrica* 69, 633–664.
36. D. Duffie, M. Schroder and C. Skiadas (1996) Recursive valuation of defaultable securities and the timing of resolution of uncertainty. *Annals of Applied Probability* 6, 1075–1090.

37. D. Duffie and K. Singleton (1998) Simulating correlated defaults. Working paper.

38. D. Duffie and K. Singleton (1999) Modeling term structures of defaultable bonds. *Review of Financial Studies* 12, 687–720.

39. D. Duffie and K. Singleton (2003) *Credit Risk: Pricing, Measurement and Management*. Princeton University Press, Princeton.

40. D. Duffie and R. Stanton (1992) Pricing continuously resettled contingent claims. *Journal of Econom. Dynamics Control* 16, 561–573.

41. E. Eberlein and F. Özkan (2003) The defaultable Lévy term structure: ratings and restructuring. *Mathematical Finance* 13, 277–300.

42. R.J. Elliott (1982) *Stochastic Calculus and Applications*. Springer-Verlag, Berlin.

43. R.J. Elliott, M. Jeanblanc and M. Yor (2000) On models of default risk. *Mathematical Finance* 10, 179–195.

44. H. Geman, N. El Karoui and J.-C. Rochet (1995) Changes of numeraire, changes of probability measures and pricing of options. *Journal of Applied Probability* 32, 443–458.

45. K. Giesecke (2002) Default compensator, incomplete information, and the term structure of credit spreads. Working paper.

46. Y.M. Greenfield (2000) Hedging of the credit risk embedded in derivative transactions. PhD dissertation.

47. R. Guha (2003) Recovery of face value at default: Theory and empirical evidence. Working paper.

48. B. Hilberink and L.C.G. Rogers (2002) Optimal capital structure and endogenous default. *Finance and Stochastics* 6, 237–263.

49. J. Hull and A. White (2001) Valuing credit default swaps II: Modeling default correlations. *Journal of Derivatives* 8, 12–22.

50. F. Jamshidian (1997) LIBOR and swap market models and measures. *Finance and Stochastics* 1, 293–330.

51. F. Jamshidian (2002) Valuation of credit default swaps and swaptions. Working paper.

52. R.A. Jarrow and S.M. Turnbull (1995) Pricing derivatives on financial securities subject to credit risk. *Journal of Finance* 50, 53–85.

53. R.A. Jarrow, D. Lando and S.M. Turnbull (1997): A Markov model for the term structure of credit risk spreads. *Review of Financial Studies* 10, 481–523.

54. R.A. Jarrow, D. Lando and F. Yu (2002) Default risk and diversification: Theory and applications. Working paper.

55. R.A. Jarrow and F. Yu (2001) Counterparty risk and the pricing of defaultable securities. *Journal of Finance* 56, 1756–1799.

56. M. Jeanblanc and M. Rutkowski (2000) Modelling of default risk: An overview. In: *Mathematical Finance: Theory and Practice*, Beijing, pp. 171–269.

57. M. Jeanblanc and M. Rutkowski (2001) Default risk and hazard process. In: *Mathematical Finance – Bachelier Congress 2000*. H. Geman, D. Madan, S.R. Pliska and T. Vorst, eds., Springer-Verlag, Berlin, 2002, pp. 281–312.

58. M. Jeanblanc and M. Rutkowski (2003) Modelling and hedging of default risk. In: *Credit Derivatives: The Definitive Guide*. J. Gregory, ed., Risk Books, pp. 385–416.

59. M. Kijima and K. Komoribayashi (1998) A Markov chain model for valuing credit risk derivatives. *Journal of Derivatives* 6, Fall, 97–108.

60. M. Kijima, K. Komoribayashi and E. Suzuki (2002) A multivariate Markov model for simulating correlated defaults. Working paper.
61. M. Kijima and Y. Muromachi (2000) Credit events and the valuation of credit derivatives of basket type. *Rev. Derivatives Res.* 4, 55 – 79.
62. I.J. Kim, K. Ramaswamy and S. Sundaresan (1993) The valuation of corporate fixed income securities. Working paper.
63. S. Kusuoka (1999) A remark on default risk models. *Advances in Mathematical Economics* 1, 69–82.
64. D. Lando (1998a) On Cox processes and credit-risky securities. *Review of Derivatives Research* 2, 99-120.
65. D. Lando (1998b) On rating transition analysis and correlation. *Risk Publications.*
66. D. Lando (2000a) Some elements of rating-based credit risk modeling. In: *Advanced Fixed-Income Valuation Tools*, J. Wiley, Chichester, pp. 193–215.
67. D. Lando (2000b) On correlated defaults in a rating-based model: Common state variables versus simultaneous defaults. Working paper.
68. J.-P. Laurent and J. Gregory (2002) Basket defaults swaps, CDOs and factor copulas. Working paper.
69. H. Leland (1994) Corporate debt value, bond covenants, and optimal capital structure. *Journal of Finance* 49, 1213–1252.
70. H. Leland and K. Toft (1996) Optimal capital structure, endogenous bankruptcy, and the term structure of credit spreads, *Journal of Finance* 51, 987–1019.
71. F.A. Longstaff and E.S. Schwartz (1995) A simple approach to valuing risky fixed and floating rate debt. *Journal of Finance* 50, 789–819.
72. S. Lukas (2001) On pricing and hedging defaultable contingent claims. Thesis.
73. D. Madan and H. Unal (1998) Pricing the risk of default. *Review of Derivatives Research* 2, 121–160.
74. M. Musiela and M. Rutkowski (1997) *Martingale Methods in Financial Modelling.* Springer-Verlag, Berlin.
75. T.N. Nielsen, J. Saá-Requejo and P. Santa-Clara (1993) Default risk and interest rate risk: The term structure of default spreads. Working paper.
76. P. Protter (2003) *Stochastic Integration and Differential Equations.* 3rd edition, Springer-Verlag, Berlin.
77. D. Revuz and M. Yor (1999) *Continuous Martingales and Brownian Motion.* 3rd edition, Springer-Verlag, Berlin.
78. J. Saá-Requejo and P. Santa-Clara (1999) Bond pricing with default risk. Working paper.
79. P.J. Schönbucher (2000a) Credit risk modelling and credit derivatives. PhD dissertation.
80. P.J. Schönbucher (2000b) A Libor market model with default risk. Working paper.
81. P.J. Schönbucher (2003) *Credit Derivatives Pricing Models.* J.Wiley, Chichester.
82. P.J. Schönbucher and D. Schubert (2001) Copula-dependent default risk in intensity models. Working paper.
83. N. Vaillant (2001) A beginner's guide to credit derivatives. Working paper.

Stochastic Control with Application in Insurance

Christian Hipp

Institute for Finance, Banking and Insurance, University of Karlsruhe, Kronenstr. 34, 76133 Karlsruhe, Germany
christian.hipp@wiwi.uni-karlsruhe.de

1 Preface

In a talk given at the Royal Statistical Society of London, Karl Borch in 1967 made the following statement (see Taksar [44]):

> The theory of control processes seems to be *tailor made* for the problems which actuaries have struggled to formulate for more than a century. It may be interesting and useful to meditate a little how the theory would have developed if actuaries and engineers had realized that they were studying the same problems and joined forces over 50 years ago. A little reflection should teach us that a *highly specialized* problem may, when given the proper mathematical formulation, be identical to a series of other, seemingly unrelated problems.

It took some more time until (in 1994 and 1995) the first papers on stochastic control in insurance appeared (e.g. Martin-Löf [32], Brockett and Xia [3], or Browne [4]). Since then we can see a rapid development of this field with a series of papers written by Soren Asmussen, Michael Taksar, Bjarne Hoejgaard, Hanspeter Schmidli and others. It is the purpose of the following parts to give an introduction into this field and present a survey of recent results and their possible applications. The five parts are

1) Introduction into insurance risk
2) Possible control variables and stochastic control
3) Optimal investment for insurers
4) Optimal reinsurance and new business
5) Asymptotic behavior for value functions and strategies
6) Control problems with constraints: dividends and ruin.

Since we shall mostly consider optimization for a first insurer we shall concentrate on problems with infinite planning horizon. The main objective function will be the infinite time ruin (survival) probability, but most of the techniques presented here also work for other objective functions.

One major recent trend in risk management is replacement of (parts of) risk capital by sophisticated risk control. Here one will use control of investment into risky assets (capital market), control of reinsurance, of underwriting, of new business, and of setting premia. In this sense the mathematical field presented here is (or will be) part of asset liability management, of dynamic financial analysis, and of holistic risk management in insurance.

Stochastic control is well established in the finance world since the seminal papers of Merton ([33] and [34]). The books by Fleming and Rishel [8], by Fleming and Soner [9], and by Karatzas and Shreve [30] cover most of today's problems and methods in this field.

2 Introduction Into Insurance Risk

2.1 The Lundberg Risk Model

Here we consider the technical risk which is generated by the randomness of claim sizes and claim occurrence times. A classical model is the Lundberg model [31] for the risk process which uses a compound Poisson process for the claims:

$$R(t) = s + ct - S(t),$$
$$S(t) = X_1 + ... + X_{N(t)},$$

with a homogeneous Poisson process $N(t)$ having constant intensity λ and independent claim sizes $X_1, X_2, ...$ with distribution Q (the claim size distribution) which are independent of $N(t), t \geq 0$. The initial surplus is s, and c is the constant premium intensity. This process is generated by independent random variables $X_1, X_2, ..., W_1, W_2, ...$ with $X_i \sim Q$, $W_i \sim Exp(\lambda)$, where W_i is the inter-arrival time between claim X_{i-1} and X_i if $i \geq 2$, and W_1 is the waiting time until the first claim. Then $N(t)$ can be written as

$$N(t) = \max\{k : W_1 + ... + W_k \leq t\}.$$

Claim X_i occurs at time $T_i = W_1 + ... + W_i$, $i \geq 1$. The process $R(t)$ has independent stationary increments, in particular the process is Markov with respect to the natural filtration F_t generated by $R(t)$, in the following sense: for any set A in the sigma-field generated by $R(u), u \geq t$, the conditional probability $P\{A \mid F_t\}$ depends on $R(t)$ alone,

$$P\{A \mid F_t\} = P\{A \mid R(t)\}.$$

The Lundberg risk process is the standard model for nonlife insurance, simple enough to calculate probabilities of interest, but too simple to be realistic. It does not include interest earned on the surplus, no long tail business with claims which are settled a long time after occurrence of the claim, no time dependence or even randomness of premium income and of the size of the portfolio (which would lead to stochastic processes $c(t)$ and $\lambda(t)$, respectively). But the Lundberg model is still attractive because it separates and models the two major reasons for big losses: frequent claims and large claims. Most of the techniques developed for the Lundberg model are useful for more realistic and more general risk processes like the Sparre-Andersen model or the Markov modulated risk process.

2.2 Alternatives

Other risk models discussed in the insurance context are the Sparre-Andersen model and the Markov-modulated risk process. In both classes of models, the claims X_i stay iid independent of the claims arrival process $N(t)$, which in the Sparre-Andersen model is a renewal process

$$N(t) = \max\{k : W_1 + ... + W_k \leq t\}$$

with iid positive random variables W_i which are independent of the sequence of claim sizes X_i. A Sparre-Andersen risk model has the parameters s, c, Q, R where R is the distribution of the inter-arrival times W_i. In this model, the process $R(t)$ is no longer Markovian; to obtain a Markov process one has to enlarge the state space. If $T(t)$ is the time elapsed since the last claim, then $(R(t), T(t))$ is a Markov process.

In the Markov-modulated risk model one considers a continuous time homogeneous Markov process $M(t)$ on the state space $\{1, ..., I\}$, and with fixed intensities $0 \leq \lambda_1 < \lambda_2 < ... < \lambda_I$ one uses the process $\lambda(t) = \lambda_{M(t)}$ as stochastic intensity of an inhomogeneous Poisson process $N(t)$. Here we have the parameters $s, c, Q, \lambda_1, ..., \lambda_I, b_{ij}, i, j = 1, ..., I$, where b_{ij} are the transition intensities of the Markov process $M(t)$. Also in this model, $R(t)$ is not Markovian, while the process $(R(t), \lambda(t))$ is a Markov process. These risk models can be found, e.g., in Rolski et al. ([39], chapters 6 and 12.3).

A simple extension of the Lundberg risk process is the implementation of constant interest, in which the reserve earns interest at a constant rate r. In this process, the jumps are the same as in the Lundberg process, and between claims the process evolves with the dynamics

$$R'(t) = c + rR(t);$$

see Paulsen [36] for the (non-trivial!) computation of ruin probabilities in this model.

2.3 Ruin Probability

A classical risk measure is the infinite time ruin probability

$$\psi(s) = P\{R(t) < 0 \text{ for some } t \geq 0\}$$

which equals one as long as $c \leq \lambda E[X_1]$ (no safety loading), and in the case with safety loading

$$c > \lambda E[X_1]$$

(a condition which we tacitly assume throughout the paper) we have $R(t) \to \infty$, and the ruin probability satisfies the following first order integro-differential equation

$$0 = \lambda E[\psi(s - X) - \psi(s)] + c\psi'(s), s \geq 0 \qquad (1)$$

where $X \sim Q$ is a generic claim size (for technicalities and more details, e.g. for (non-)differentiability of $\psi(s)$, see Grandell [16], or Gerber [14], or Rolski et al. [39]). For exponential claim sizes with density

$$f(x) = \theta \exp(-\theta x), \; x > 0,$$

the ruin probability equals

$$\psi(s) = \frac{\lambda \mu}{c} \exp(-Rs), \qquad (2)$$

where $\mu = E[X_1] = 1/\theta$ is the mean claim size, and $R = (c - \lambda\mu)/(c\mu)$ is the adjustment coefficient of the problem which is the positive solution r of the Lundberg equation

$$\lambda + rc = \lambda E[\exp(rX)]. \qquad (3)$$

For the following parts it might help to recall how equation (2) can be derived from (1). Consider the survival probability $\delta(s) = 1 - \psi(s)$, for which $\delta(s) = 0$ for $s < 0$ and

$$0 = \lambda(g(s) - \delta(s)) + c\delta'(s), s \geq 0, \qquad (4)$$

where

$$g(s) = E[\delta(s - X)] = \int_0^s \delta(s - x)\theta e^{-\theta x} dx = \int_0^s \delta(x)\theta e^{-\theta(s - x)} dx.$$

It is easy to see that on the set $\{s \geq 0\}$, $g(s)$ satisfies the differential equation

$$g'(s) = \theta(\delta(s) - g(s)),$$

and hence on the set $\{s \geq 0\}$ the function $\delta(s)$ has a continuous second derivative $\delta''(s)$ for which

$$\begin{aligned}
0 &= \lambda(g'(s) - \delta'(s)) + c\delta''(s) \\
&= \lambda\theta(\delta(s) - g(s)) - \lambda\delta'(s) + c\delta''(s) \\
&= c\theta\delta'(s) - \lambda\delta'(s) + c\delta''(s).
\end{aligned}$$

This linear differential equation with constant coefficients has a general solution of the form

$$\delta(s) = C_1 + C_2 \exp(-Rs)$$

since $z = 0$ and $z = -R$ are the solutions to the characteristic equation

$$0 = (c\theta - \lambda)z + cz^2.$$

Using $\delta(s) \to 1$ for $s \to \infty$ we get $C_1 = 1$. From (4) at the point $s = 0$ we obtain $\lambda\delta(0) = c\delta'(0)$ or $\lambda(1 + C_2) = -cRC_2$ or finally

$$-C_2 = \frac{\lambda}{cR + \lambda} = \frac{\lambda\mu}{c}.$$

In the Markov modulated situation, the ruin probability $\psi(s,i)$ depends on the initial surplus s and the initial value of the process $\lambda(t) : \lambda(0) = \lambda_i$. The functions $\psi(s,i)$ satisfy the following interacting system of first order integro differential equations:

$$0 = \lambda_i E[\psi(s - X, i) - \psi(s,i)] + c\psi_s(s,i) + \sum_{j=1}^{I} b_{ij}\psi(s,j), \ s \geq 0, \ i = 1, ..., I.$$

In the Sparre-Andersen model the ruin probability $\psi(s) = \psi(s,0)$ is derived from a function $\psi(s,t)$, where s is the initial surplus and t is the current time since the last claim. If the waiting times W_i have a continuous density $f(x)$, then the function $\psi(s,t)$ satisfies the following integro-differential equation:

$$0 = \frac{f(t)}{1 - F(t)} E[\psi(s - X, t) - \psi(s,t)] + c\psi_s(s,t) + \psi_t(s,t), \ s \geq 0, \ t \geq 0.$$

In the Lundberg risk process with constant interest rate r, the ruin probability $\psi(s)$ satisfies the integro-differential equation

$$0 = \lambda E[\psi(s - X) - \psi(s)] + (c + rs)\psi'(s), \ s \geq 0.$$

These integro-differential equations are derived with the infinitesimal generators of the underlying risk processes (see below). In the remainder of this section we shall restrict ourselves to Lundberg risk processes.

2.4 Asymptotic Behavior For Ruin Probabilities

Equation (2) shows the typical behavior of ruin probabilities for small claim sizes for which the adjustment coefficient (see equation (3)) exists,

$$\psi(s) \sim C \exp(-Rs) \tag{5}$$

with $C > 0$, and where $a(s) \sim b(s)$ means $a(s)/b(s) \to 1$. This relation holds, e.g., if

$$r_0 = \sup\{r : E[\exp(rX)] < \infty\} > 0$$

and $\lim_{r \to r_0} E[\exp(rX)] = \infty$. See Rolski et al. ([39], chapter 5.4).

For large claims with heavy tailed distributions (for which $r_0 = 0$) the behavior is totally different: e.g., for Pareto claims with density $f(x) = ax^{-(a+1)}$, $x > 1$, $a > 1$, we have

$$\psi(s) \sim Cs^{-(a-1)}, \ s \to \infty, \tag{6}$$

with a positive constant C. Also this behavior is typical: For heavy tailed claim size distributions Q (more precisely: for all *subexponential* distributions, see Embrechts et al. ([7], chapter 1.3)) we have

$$\psi(s) \sim C \int_s^\infty Q(x, \infty) dx,$$

where $Q(t, \infty) = P\{X_1 > t\}$ is the tail probability of the claim size at the point $t \geq 0$.

The difference between the two cases will become apparent when one tries to increase the initial surplus s to a new surplus s_1 in order to halve the ruin probability. In the exponential claims case the new initial surplus s_1 equals $s_1 = s + \ln(2)/R$, while in the Pareto claims case

$$s_1 \sim 2^{1/(a-1)} s.$$

A complete survey on infinite time ruin probabilities for the Lundberg model can be found in Rolski et al. [39].

3 Possible Control Variables and Stochastic Control

We consider an insurance company managing the risk in a portfolio with claims modelled by a Lundberg risk process with parameters c, λ and Q. There is a collection of possible actions well suited for risk management: reinsurance, investment, volume control (via setting of premia), portfolio selection (via combination of the given risk portfolio with other risks which can be written), and the combination of all these actions. Here we deal with dynamic risk management, the actions are selected and changed at each point in time - according to the risk position of the company. We will treat actions involving one control variable only, and try to find the optimal (with respect to some given objective) dynamic strategy for the selected control variable, i.e. for each control variable we define a stochastic control problem which we try to solve.

3.1 Possible Control Variables

Investment, One Risky Asset

Here we consider a risky asset in which the insurer can invest, and a riskless asset, a bank account, which pays interest r. At each point in time t the insurer with current wealth $R(t)$ will invest an amount $A(t)$ into the risky asset, and what is left is on the bank account earning (costing) interest r if $R(t) - A(t) > 0$ (if $R(t) - A(t) < 0$). For simplicity we take the classical Samuelson model (logarithmic Brownian motion) for the dynamics of the asset prices $Z(t)$:

$$dZ(t) = aZ(t)dt + bZ(t)dW(t), \ Z(0) = z_0,$$

where $W(t)$ is a standard Wiener process. If $\theta(t) = A(t)/Z(t)$ is the number of shares held at time t, then the total position of the insurer has the following dynamics:

$$dR(t) = rR(t)dt + cdt - dS(t) + \theta(t)dZ(t) - r\theta(t)Z(t)dt, R(0) = s,$$

or

$$dR(t) = rR(t)dt + cdt - dS(t) + A(t)((a - r)dt + bdW(t)), R(0) = s.$$

To simplify the setup and the notation we shall restrict ourselves to the case $r = 0$.

We shall allow for all possible trading strategies $\theta(t)$ which - as stochastic processes - are (F_t)−predictable, where (F_t) is the filtration generated by the two processes $Z(t)$ and $S(t), t \geq 0$. So for the selection of $\theta(t)$ we may use the knowledge of all stock prices and claims before time t, but not the knowledge at time t which might be the size of a claim happening at time t. There is no budget constraint such as $\theta(t)Z(t) \leq R(t)$, one can borrow an arbitrary amount of money and invest it into the risky asset. We shall also neglect transaction costs and allow for shares of any (up to infinitesimal) size.

Investment, Two or More Risky Assets

Assume that the insurer can invest his money into d risky assets $Z_1(t), ...,Z_d(t)$, and the dynamics of these prices is given by the following system of stochastic differential equations

$$dZ_i(t) = Z_i(t)(a_i dt + \sum_{j=1}^{d} \sigma_{ij}dW_j(t), i = 1, ..., d, \tag{7}$$

where a_i and σ_{ij} are constants with nonsingular matrix

$$\Sigma = (\sigma_{ij})_{i,j=1,...,d},$$

and where $W_1(t), ..., W_d(t)$ are independent standard Wiener processes. If the interest is zero, $r = 0$, and if $A_i(t)$ is the amount invested into stock i at time t, then the total position of the insurer has the following dynamics:

$$dR(t) = cdt - dS(t) + a_i A_i(t) + A_i(t) \sum_{j=1}^{d} \sigma_{ij}dW_j(t), R(0) = s.$$

The stochastic process $Z(t) = (Z_1(t), ..., Z_d(t))$ is called *d-variate logarithmic Brownian motion*.

Proportional Reinsurance

In a proportional reinsurance contract each individual claim of size X is divided between first insurer and reinsurer according to a proportionality factor a : the insurer pays aX, the reinsurer pays $(1-a)X$. For this the insurer pays a reinsurance premium $h(a)$ to the reinsurer. We allow a continuous adjustment of the proportionality factor: $a(t)$ is (F_t)−predictable. Under the strategy $a(t)$ the risk process of the first insurer is given by

$$R(t) = s + ct - \int_0^t h(a(v))dv - \sum_{i=1}^{N(t)} a(T_i)X_i, \ t \geq 0.$$

The usual premium rule $h(a)$ is the expectation principle:

$$h(a) = a\rho\lambda E[X]$$

with $\rho > 1$. If $c \geq \rho\lambda E[X]$ and the first insurer wants to minimize his risk then he would choose $a(t) = 0$ and give all risk to the reinsurer. To exclude this uninteresting situation we shall always assume that reinsurance is expensive: $c < \rho\lambda E[X]$.

Unlimited XL Reinsurance

In excess of loss (XL) reinsurance each claim of size X is divided between the first insurer and the reinsurer according to a priority $0 \leq b \leq \infty$: the insurer pays $\min(X, b)$, and the reinsurer pays $(X - b)^+ = \max\{X - b, 0\}$. For this the insurer pays a reinsurance premium $h(b)$ to the reinsurer. We allow a continuous adjustment of the proportionality factor: $b(t)$ is (F_t)−predictable. Under the strategy $b(t)$ the risk process of the first insurer is given by

$$R(t) = s + ct - \int_0^t h(b(v))dv - \sum_{i=1}^{N(t)} \min\{b(T_i), X_i\}, \ t \geq 0.$$

One possible rule $h(b)$ is again the expectation principle:

$$h(b) = \rho\lambda E[(X - b)^+]$$

with $\rho > 1$. Also here we shall assume that reinsurance is expensive: $c < \rho\lambda E[X]$. Other premium principles would be the variance principle

$$h(b) = \lambda E[(X - b)^+] + \beta\lambda E[((X - b)^+)^2],$$

which puts more weight to the tail of the distribution of the claim size, or the standard deviation principle

$$h(b) = \lambda E[(X - b)^+] + \beta\sqrt{\lambda E[((X - b)^+)^2]}.$$

In general, expensive reinsurance is the situation in which $c < h(0)$.

XL-Reinsurance

In practical situations, XL-reinsurance contracts are limited by some constant $0 \leq L \leq \infty$, which leads to the following division of a claim of size X : the reinsurer pays $\min\{(X - b)^+, L\}$, and the first insurer pays what is left: $g(X, b, L) = \min\{X, b\} + (X - b - L)^+$. For this the insurer pays a reinsurance premium $h(b, L)$. Under a dynamic XL-reinsurance contract with strategy $(b(t), L(t))$ the insurer has the following risk process:

$$R(t) = s + ct - \int_0^t h(b(v), L(v)) dv - \sum_{i=1}^{N(t)} g(b(T_i), L(T_i)).$$

Reinsurance is expensive if $c < h(0, \infty)$. Possible premium schemes are the expectation principle

$$h(b, L) = \rho \lambda E[\min\{(X - b)^+, L\}],$$

the variance principle

$$h(b, L) = \lambda E[\min\{(X - b)^+, L\}] + \beta \lambda E[\min\{(X - b)^+, L\}^2]$$

or the standard deviation principle.

To minimize his risk, an insurer will choose $L(t) = \infty$ if he can afford it. So $L(t) < \infty$ will be a reasonable choice for him only if the tail of the claim size distribution matters for the reinsurance premium, as is the case in the variance or in the standard deviation principle.

Premium Control

An insurer can control the volume of his business by setting the premium c. The higher the premium rate c, the smaller the number of contracts in his portfolio, and this in turn will decrease the claims intensity λ. This will be modelled by a non-increasing function $\lambda(c)$: if $c(t)$ is the instantaneous premium rate charged, then $\lambda(c(t))$ will be the instantaneous intensity of the claims process. A realistic model has $\lambda(\infty) = 0$, and in order to get a non trivial risk minimization problem one has to change the framework a bit since otherwise the insurer would reduce his risk to zero by the choice of an infinite premium rate. One possibility is the introduction of cost of capital, i.e. for the initial surplus an interest rate ρ has to be paid continuously.

Control of New Business

This is a control problem for an insurer who controls the risk in one given portfolio by writing an appropriate proportion of business in a second indepen-dent portfolio. If $R(t)$ and $R_1(t)$ are the two independent insurance portfolios which are both modelled as Lundberg risk processes with parameters λ, c, Q and λ_1, c_1, Q_1, respectively, and if $b(t)$ is the proportion written at time t in portfolio $R_1(t)$, then the total position of the insurer consists of premium income up to time t equal to

$$ct + \int_0^t \lambda_1 b(u)du,$$

and the claims paid up to time t are $S(t) = X_1 + ... + X_{N(t)}$ for the first portfolio with distribution compound Poisson with parameters λt and Q, and $S_1(t)$ for the second portfolio with instantaneous claims intensity $\lambda_1 b(t)$ and claim size distribution Q_1. For practical applications one has to assume that $b(t) \geq 0$ (no short selling of insurance business) and $b(t) \leq 1$ (the maximum possible volume written is $R_1(t)$).

3.2 Stochastic Control

One of the most investigated classical problems in finance is the Merton optimal investment and consumption problem. In its simplest form, it reads as follows. An investor has initial wealth r_0, he can dynamically consume part of his wealth and invest dynamically another part of it in a risky asset with price process modelled as logarithmic Brownian motion:

$$dX(t) = X(t)(adt + bdW(t)), \ X(0) = x_0.$$

What is left is on a bank account earning interest at a constant rate r. If $A(t)$ is the amount invested and $c(t)$ the rate of consumption at time t, then the wealth $R(t)$ has the dynamics

$$dR(t) = A(t)(adt + bdW(t)) + (R(t) - A(t))rdt - c(t)dt, \ R(0) = r_0.$$

One is interested in the optimal strategy $(A(t), c(t))$ which maximizes expected accumulated utility of consumption

$$E\left[\int_0^\tau e^{-\rho t} u(c(t))dt\right]$$

where $\tau = \inf\{t : \ R(t) < 0\}$ is the ruin time for the investor, $\rho > 0$ is a subjective interest rate (appreciation rate), and $u(x) = x^\gamma$, $\gamma < 1$, is a special utility function. This problem is specified by the dynamics of the risky asset, by the two control variables $A(t)$ and $c(t)$, and by the objective function which is maximized.

Objective Functions

In order to properly define an optimization problem one needs to specify the planning horizon and the quantity which should be maximized. There is the finite horizon case where optimization is done over a finite interval $[0, T]$, and the infinite horizon case. In the finite horizon case, a general objective function can be written as the sum of two components: the running cost and the final cost. If $\sigma(t)$ is the strategy with values in an action space Σ, then the general objective function to be maximized reads

$$E\left[\int_0^{\tau \wedge T} u(R(t), \sigma(t), t)dt + U(R(T), \sigma(T))\right],$$

where τ is some stopping time (such as ruin time or first entry time into a certain region). In the infinite horizon case a general objective function consists of running costs and terminal costs

$$E\left[\int_0^\tau u(R(t), \sigma(t), t)dt + U(R(\tau), \sigma(\tau), \tau)\right],$$

where τ is an unlimited stopping time and $U(R(\tau), \sigma(\tau), \tau)$ could be bank-ruptcy cost when τ is time of ruin. For the above mentioned infinite horizon Merton problem, we have $U = 0$ and $u(r, \sigma, t) = \exp(-\rho t)(c(t))^\gamma, \gamma < 1$. If $c(t)$ is a dividend rate, then the quantity

$$E\left[\int_0^\tau \exp(-\rho t)c(t)^\gamma dt + U(R(\tau))\right]$$

could be interpreted as the value of the company if $c(t)$ is the dividend rate and $U(s)$ is the cost of default when the final capital is s.

We shall mainly be concerned with ruin (survival) probabilities, i.e. running cost is zero, and $U = 1$ if $\tau = \infty$, $U = 0$ elsewhere.

Infinitesimal Generators

Infinitesimal generators L are defined for Markov processes $R(t)$ and for (sufficiently smooth) functions $f(s)$ on the state space via

$$L_t f(s) = \lim_{h \searrow 0} \frac{1}{h} E\left[f(R(t+h)) - f(s) \mid R(t) = s\right],$$

where the function $f(s)$ is restricted to the domain D of L for which this limit exists. If the Markov process is time homogeneous, then the infinitesimal generator does not depend on t. Obviously, D is linear, and L_t is a linear operator. In the following examples, all processes are stationary. The domains of the generator will not be specified precisely, but it should be clear in each case that it contains the set of all functions $f(s)$ having bounded derivatives of all orders.

1) $R(t) = a + bt$: $Lf(s) = bf'(s)$;
2) $dR(t) = a(R(t))dt + b(R(t))dW(t)$: $Lf(s) = a(s)f'(s) + \frac{1}{2}b^2(s)f''(s)$;

3) $R(t) = s + ct - S(t)$, the Lundberg risk process: $Lf(s) = \lambda E[f(s - X) - f(s)] + cf'(s)$;

4) $R(t)$ the Lundberg risk process with constant interest r : $Lf(s) = \lambda E[f(s - X) - f(s)] + (c + rs)f'(s)$;

5) $(R(t), M(t))$ from the Markov modulated risk process:

$$Lf(s, i) = \lambda_i E[f(s - X, i) - f(s, i)] + cf_s(s, i) + \sum_{j=1}^{I} b_{ij} f(s, j);$$

6) $(R(t), T(t))$ from the Sparre-Andersen model:

$$Lf(s, t) = \frac{f(t)}{1 - F(t)} E[f(s - X, t) - f(s, t)] + cf_s(s, t) + f_t(s, t);$$

In the following we shall also need the infinitesimal generator for a controlled risk process, where the control strategy is *constant*. So, e.g., for optimal investment with a constant amount A invested into the risky asset, the total position $R(t)$ of the insurer has the dynamics

$$dR(t) = cdt - dS(t) + A(adt + bdW(t)),$$

and so the infinitesimal generator for the process $(R(t), X(t))$ (which is Markov) equals

$$Lf(s, x) = \lambda E[f(s - X, x) - f(s, x)] + cf_s(s, x) + Aaf_s(s, x) + \frac{1}{2} A^2 b^2 f_{ss}(s, x).$$

One can see that the infinitesimal generator is independent of x (for logarithmic Brownian motion, the initial value has no influence on the return of an investment), and so we use the notation

$$Lf(s) = \lambda E[f(s - X) - f(s)] + cf'(s) + Aaf'(s) + \frac{1}{2} A^2 b^2 f''(s).$$

For proportional reinsurance with constant proportion a the risk process of the insurer reads

$$R(t) = s + (c - h(a))t - a \sum_{i=1}^{N(t)} X_i,$$

and the corresponding infinitesimal generator is

$$Lf(s) = \lambda E[f(s - aX) - f(s)] + (c - h(a))f'(s).$$

For unlimited XL-reinsurance with constant priority $b \in [0, \infty]$ the generator equals

$$Lf(s) = \lambda E[f(s - X \wedge b) - f(s)] + (c - h(b))f'(s),$$

and for the general reinsurance contract $g(X, a)$ with reinsurance premium $h(a)$ and fixed decision vector a the generator is

$$Lf(s) = \lambda E[f(s - g(X, a)) - f(s)] + (c - h(a))f'(s).$$

The integro-differential equations for the ruin probability $\psi(s)$ are all of the form

$$L\psi(s) = 0, \ s \geq 0,$$

where L is the infinitesimal generator of the underlying risk process. It is by no means obvious that the function $\psi(s)$ is in the domain of L, but this problem can be dealt with using the so called *verification argument*.

Hamilton-Jacobi-Bellman Equations

The computation of the maximized objective function and - if it exists - of the corresponding optimal strategy is a non-trivial task: the space of possible strategies is too large (the set of all F_t−predictable processes) for a complete search. An indirect method will be used. The principle behind this method (for the case of finite horizon) is based on two observations: a) the optimal strategy depends only on the initial state (and the time to maturity), and b) the optimal strategy is specified by its value at the initial time point for each initial state (and each time to maturity). Since the concept is quite classical now and part of each book on stochastic control, the HJB equation will just be given without describing how it is derived heuristically from the optimization problem.

If A is a fixed action from the action space, which is regarded as a constant strategy $a(t) \equiv A$, then the controlled process $R^a(t)$ should be a time homogeneous Markov process with infinitesimal generator L^A. The HJB equation for an optimization problem considered in this survey, i.e.with value function $V(s)$ of the form

$$V(s) = \max_{a(.)} E[U(R^a(\tau), a(\tau), \tau)]$$

where $\tau = \inf\{t : R^a(t) < 0\}$ is the ruin time of the controlled process (other stopping times are possible, too) equals

$$\max_A L^A V(s) = 0, \ s \geq 0. \tag{8}$$

The maximizer $A = A(s)$ in this problem defines the optimal strategy: if the controlled process is in state s, then the optimal action is $A(s)$.

Consider, as a first example without optimization, a Wiener process with positive drift a and diffusion constant $b \neq 0$,

$$dR(t) = a dt + b dW(t), t \geq 0, \ R(0) = s.$$

We want to determine the ruin probability

$$\psi(s) = P\{R(t) < 0 \text{ for some } t \mid R(0) = s\}, \ s > 0.$$

The process $R(t)$ has the infinitesimal generator

$$Lf(s) = af'(s) + \frac{1}{2}b^2 f''(s). \tag{9}$$

For a short time interval from 0 to dt, there might be ruin in $[0, dt]$ - which happens with probability $o(dt)$ - or ruin occurs after dt with probability $\psi(R(dt))$. Integrating over all possible values for $R(dt)$, we obtain

$$\psi(s) = E[\psi(R(dt))] + o(dt).$$

Assuming that the function $\psi(s)$ is in the domain D of L, we get the differential equation (which corresponds to our integro-differential equations for the ruin probabilities in models with jumps)

$$0 = a\psi'(s) + \frac{1}{2}b^2\psi''(s), \; s > 0.$$

The general solution to this linear differential equation with constant coefficients reads

$$\psi(s) = C_1 + C_2 \exp(-2as/b^2).$$

For $s \to \infty$ we should have $\psi(s) \to 0$, so $C_1 = 0$. At $s = 0$ we are ruined immediately because of the fluctuation of the Wiener process, so $C_2 = 1$. As a conclusion, $\psi(s) = \exp(-2a/b^2 \, s)$. Of course this is not a rigorous proof for the ruin formula since $\psi(s) \in D$ was just assumed. For this we will use the verification argument below.

Consider next the optimal investment problem of Browne [4]. For a given investment strategy $A(t)$ ($A(t)$ is the amount invested at time t) the risk process of an investor is given by

$$dR^A(t) = \alpha dt + \beta dV(t) + A(t)(a dt + b dW(t)),$$

where $V(t), W(t)$ are two independent standard Wiener processes. The first part might model the return in an insurance portfolio modelled by a Brownian motion with drift, and the second the investment return in a risky asset with price process modelled by logarithmic Brownian motion. The problem is to find the optimal investment strategy $A(t)$ which maximizes survival probability

$$\delta(s) = P\{R^A(t) \ge 0 \text{ for all } t \mid R^A(0) = s\}.$$

According to (8) the HJB-equation for this problem reads

$$0 = \max_A\{\alpha V'(s) + \frac{1}{2}\beta^2 V''(s) + AaV'(s) + \frac{1}{2}A^2 b^2 V''(s)\}, \; s > 0.$$

A maximizing A exists only if $V''(s) \le 0$, and in this case it is given by

$$A = A(s) = -\frac{a}{b^2}\frac{V'(s)}{V''(s)}.$$

Plugging in we obtain

$$0 = \alpha V'(s) + \frac{1}{2}\beta^2 V''(s) + \frac{1}{2}\frac{a^2}{b^2}\frac{V'(s)^2}{V''(s)}, \ s > 0.$$

After dividing by $V'(s)$ and computing the negative solution $-k$ of the equation

$$0 = \alpha + \frac{1}{2}\beta^2 \frac{1}{z} + \frac{1}{2}\frac{a^2}{b^2} z$$

we obtain a solution of the form

$$A(s) = \frac{a}{b^2}k, \ V(s) = 1 - \exp(-ks), \ s > 0.$$

This tells us that a constant amount ak/b^2 is optimal, and the resulting ruin probability $\exp(-ks)$ is smaller - as it should be - than the ruin probability without investment $\exp(-2\alpha/\beta^2 \ s)$ for $s > 0$. Notice that the optimal strategy is not *buy and hold* but an anticyclic strategy: if prices go up then shares are sold, and if prices go down then shares are bought. Again, the above computations do not yet solve our maximization problem since we do not know wether $V(s)$ is the unique solution of the HJB equation and wether it is the maximal possible survival probability $\delta(s)$. For this we use the verification argument below.

Verification Argument

The verification argument closes the gap between a solution of an integro-differential equation or a HJB equation and the given problem of computing ruin probabilities or maximizing an objective function. For ease of exposition we reconsider the ruin probability $\psi(s)$ for the Brownian motion with positive drift. We had found a solution $V(s) = \exp(-2a/b^2 \ s)$ of the equation (9). Let τ be the ruin time of the process $R(t) = at + bW(t)$, and define the process $Y(t) = V(R(t \wedge \tau))$. From equation (9) one can read that $Y(t)$ is a martingale for which

$$E[Y(t)] = Y(0) = V(s).$$

For $t \to \infty$ we have $Y(t) \to 1$ on the set $\{\tau < \infty\}$, and $Y(t) \to 0$ on the set $\{\tau = \infty\}$. From bounded convergence we obtain that

$$V(s) = \lim_{t\to\infty} E[Y(t)] = P\{\tau < \infty\} = \psi(s),$$

which proves that $V(s)$ is indeed the ruin probability for initial surplus s.

As a second example we consider the optimization problem of Browne [4]. We have seen that the HJB (8) has a smooth bounded solution $V(s)$ with the property that for all possible actions A we have

$$L^A V(s) \geq 0, \ s > 0. \tag{10}$$

Let $A^*(t) = ak/b^2$ be the strategy constructed with the optimizer $A(s)$ of the HJB equation (the constant amount invested), and $A(t)$ any arbitrary admissible strategy. Let $R^*(t)$ and $R(t)$ be the corresponding risk processes and τ^* and τ the corresponding ruin times, and define the processes $Y^*(t) = V(R^*(t \wedge \tau^*))$ and $Y(t) = V(R(t \wedge \tau))$. From the HJB equation we see that $Y^*(t)$ is a martingale, and according to (10) the process $Y(t)$ is a supermartingale, and both are starting at the value $V(s)$. So for all $t \geq 0$

$$V(s) = E[Y^*(t)] \geq E[Y(t)].$$

For the further reasoning we need boundary values which are derived from the optimization problem and which are satisfied by the solution $V(s)$. In our optimization problem we wanted to maximize the survival probability $\delta(s)$ of the (controlled) risk process. The *natural boundary conditions* for the value function $\delta(s)$ are $\delta(\infty) = 0$ (and $\delta(s) = 1$ for $s < 0$). Our solution $V(s)$ of (8) satisfies the same boundary conditions (for the second condition, observe that $V(s)$ can be arbitrary for $s \leq 0$). For $t \to \infty$ we have

$$Y(t) \to 0 \text{ on } \{\tau < \infty\}, \; Y^*(t) \to 0 \text{ on } \{\tau^* < \infty\},$$

and, since $R^*(t)$ is a Brownian motion with positive drift, $Y^*(t) \to 1$ on $\{\tau^* = \infty\}$. So, $V(s) = \lim_{t \to \infty} E[Y^*(t)] = P\{\tau^* < \infty\}$. For the process $Y(t)$ the asymptotic behavior for $t \to \infty$ is less clear. For $\varepsilon > 0$ we therefore introduce the process $R_1(t)$ with investment strategy $A(t) + \varepsilon^2$ and initial surplus $s + \varepsilon$, the corresponding ruin time τ_1 and the process $Y_1(t) = V(R_1(t \wedge \tau_1))$. We have $R_1(t) = \varepsilon + R(t) + \varepsilon^2(at + bW(t))$ and hence $R_1(t) \to \infty$ on the set $\{\tau_1 = \tau = \infty\}$. As above, we have $V(s + \varepsilon) \geq P\{\tau_1 = \tau = \infty\}$. Furthermore,

$$\begin{aligned} P\{\tau_1 &< \infty \text{ and } \tau = \infty\} \\ &\leq P\{\varepsilon + \varepsilon^2(at + bW(t)) < 0 \text{ for some } t\} \\ &= \exp(-2a\varepsilon^2/(b^2\varepsilon^4)\, \varepsilon), \end{aligned}$$

and hence

$$\begin{aligned} P\{\tau = \infty\} &\leq P\{\tau_1 = \tau = \infty\} + \exp(-2a\varepsilon^2/(b^2\varepsilon^4)\, \varepsilon) \\ &\leq V(s + \varepsilon) + \exp(-2a\varepsilon^2/(b^2\varepsilon^4)\, \varepsilon). \end{aligned}$$

With $\varepsilon \to 0$ we obtain $P\{\tau = \infty\} \leq V(s)$. So, for arbitrary investment strategy $A(t)$ the corresponding survival probability $\delta(s)$ is bounded by $V(s)$,

$$\delta(s) \leq V(s),$$

and the maximum is attained by the strategy $A^*(t)$.

In the following optimization problems, the verification argument is similar to the one in Browne's problem, so we omit it and refer to the literature whenever it does *not* follow the same pattern (as, e.g., in the case of optimal reinsurance).

Steps for Solution

For the solution of a stochastic control problem via the HJB equation we will go through the following steps: write down the controlled risk process for a constant control A and its infinitesimal generator, and from this write down the HJB equation. Then show that this equation has a smooth solution satisfying the natural boundary conditions derived from the optimization problem. Then use the verification argument to show that the solution of the HJB equation is the value function of the optimization problem, and the maximizer in the equation determines the optimal strategy in feedback form. The most difficult problem is step two: an explicit solution to a HJB equation in the framework considered here is never possible; the best one can hope for is an existence proof which renders a good numerical method for computations.

4 Optimal Investment for Insurers

4.1 HJB and its Handy Form

Here we consider investment strategies $A(t)$ (the amount invested into the risky asset) which are predictable processes with respect to the natural filtration generated by the processes $S(t)$ (the claims) and $Z(t)$ (the stock price). This means that for the strategy we may use all information available just before time t, so $A(t)$ may not depend on the information that there is a claim at time t or on the size of that claim. The HJB equation for the problem to maximize survival probability by investment is

$$\sup_{A}\{\lambda E[V(s-X)-V(s)]+(c+aA)V'(s)+\frac{1}{2}b^2A^2V''(s)\}=0,\ s\geq 0.$$

Solving for A which is possible whenever $V''(s)<0$ we obtain

$$A=A(s)=-\frac{a}{b^2}\frac{V'(s)}{V''(s)}.$$

If $A(0)\neq 0$ then the fluctuations of the Wiener process would lead to immediate ruin, i.e. $V(0)=0$, which cannot be optimal since without investment we have $\delta(0)=1-\lambda E[X]/c>0$ if $\lambda E[X]<c$. Hence, $A(0)=0$ or $V''(0)=-\infty$. If we plug in the optimal $A(s)$ we obtain the integro-differential equation

$$\lambda E[V(s-X)-V(s)]+cV'(s)=\frac{1}{2}\frac{a^2}{b^2}\frac{V'(s)^2}{V''(s)},\ s\geq 0. \qquad (11)$$

The natural boundary conditions are $V(s)=0$ for $s<0$, $V(\infty)=1$, and $V''(0)=-\infty$. This equation is of second order with a singularity at 0 ($V''(0)=-\infty$); it is of little use even for numerical solutions since for the integral term $g(s)=E[V(s-X)]$ the values of $V(u)$ are needed for $0\leq u\leq s$, and for this the singularity at zero is disturbing. We will replace the equation by a system of interacting integro-differential equations which lead to a stable numerical algorithm, to an elementary proof for the existence of a solution, and to an almost explicit solution for the case of exponential claim sizes. To simplify the notation we shall first divide both sides of the equation by a^2/b^2 and denote the new claims intensity and the new premium intensity again by λ and c, respectively. This leads to an equation with $a = b = 1$. Next we introduce the function $U(s) = A(s)^2$ and rewrite (11) as

$$\lambda(g(s) - V(s)) + cV'(s) = -\frac{1}{2}V'(s)\sqrt{U(s)}, \ s \geq 0, \tag{12}$$

where $\sqrt{U(s)}$ denotes always the positive root of $U(s)$. Assuming that

$$X \text{ has a continuous density } h(x)$$

we see that the function $g(s)$ has a continuous derivative for $s \geq 0$, and so we can differentiate once more and obtain

$$\lambda(g'(s) - V'(s)) + cV''(s) = -\frac{1}{2}V''(s)A(s) - \frac{1}{2}V'(s)A'(s), \ s \geq 0.$$

Using $V''(s)A(s) = -V'(s)$ and multiplying both sides of the equation by $A(s)$ we arrive at

$$\sqrt{U(x)}\left((\lambda + \frac{1}{2})V'(x) - \lambda g'(x)\right) + cV'(x) = \frac{1}{4}U'(x)V'(x), \ s \geq 0. \tag{13}$$

The corresponding boundary conditions are $V(s) = 0$ for $s < 0$, $U(0) = 0$, and $V(\infty) = 1$. The two interaction differential equations (12) and (13) are equivalent to equation (11) in the sense that (11) has a smooth concave solution $V(s)$ satisfying the natural boundary conditions iff the system (12) and (13) has a solution $(V(s), U(s))$ with $V(s)$ concave, $U(s) = (V'(s)/V''(s))^2$, satisfying the natural boundary conditions. The system of differential equations can be used for numerical computation, and the following resulting algorithm is stable. Start with $U_0(s) = 0$ and compute the function $V_0(s)$ from (12) and the natural boundary conditions (which yields the survival probability without investment). With $(V_0(s), U_0(s))$ as starting points, define the sequence of functions $(V_n(s), U_n(s))$ recursively by $g_n(s) = E[V_n(s - X)]$,

$$\frac{\lambda(V_{n+1}(s) - g_n(s))}{c + \frac{1}{2}\sqrt{U_n(s)}} = V'_{n+1}(s), \ s \geq 0, V_{n+1}(\infty) = 1, \tag{14}$$

and

$$\left[\lambda + \frac{1}{2} - \lambda\frac{g'_n(s)}{V'_n(s)}\right]\sqrt{U_{n+1}(s)} + c = \frac{1}{4}U'_{n+1}(s), \ s \geq 0, U_{n+1}(0) = 0.$$

The condition $V_{n+1}(\infty) = 1$ can be satisfies by homogeneity of the system: if $h(s)$ is a solution to (14), then $\alpha h(s)$ is a solution, too. Hence starting with $h(0) = 1$ and norming we obtain with $V_{n+1}(s) = h(s)/h(\infty)$ a solution satisfying $V_{n+1}(\infty) = 1$. The sequence of functions $(V_n(s), U_n(s))$ converges, and the limit is a solution of the system (12) and (13) satisfying the natural boundary conditions.

4.2 Existence of a Solution

There are two papers with an existence proof for the equation (11), one based on more classical methods as in [48] (in [18]), the other with a monotonicity proof (in [19]). For the proof in [18] one assumes a locally bounded density $h(x)$ of the claim size distribution, for the proof in [19] one needs a continuous density $h(x)$. The monotonicity proof does not only work for the optimal investment problem in the Lundberg model but also for the multivariate setup needed for Markov modulated risk processes.

The monotonicity proof works as follows: first one solves the problem for a fixed given function $g(s)$ which is increasing, bounded and continuously differentiable. The corresponding equation is the HJB equation for the following optimization problem: for a given utility function $g(s)$ maximize the expected accumulated discounted wealth

$$E\left[\int_0^\tau \exp(-\lambda t)g(R(t))dt\right]$$

by the choice of an optimal investment strategy $A(t)$. One can show that the HJB equation of this problem has a smooth solution $V_g(s)$, the maximizer of the HJB equation defines the optimal investment strategy, and $V_g(s)$ is the value function of the problem. The existence proof is based on a monotonicity argument using an iteration scheme similar to the numerical algorithm above, it makes use of differential inequalities studied in [47].

Second, a monotone sequence of functions is defined starting with $V_0(s)$ the survival probability without investment, and solving recursively for the value function $V_{n+1}(s)$ of the above optimization problem with utility function $g_n(s)$. The functions $V_n(s)$ are the value function of the optimization problem of optimal investment up to the n-th claim. If $V_{n+1}(s) \geq V_n(s)$ then the same is true for the corresponding functions g, and vice versa. So we can show that we have a monotone sequence $V_n(s)$ which converges, and the limit turns out to be a solution of the original optimization problem.

The *verification argument* in the optimal investment problem follows exactly the pattern described in section 3.2.

4.3 Exponential Claim Sizes

For exponential claim sizes with density $h(x) = \theta \exp(-\theta x)$, $x > 0$, the system of interacting differential equations separates, and one obtains a differential equation for $A(s)$ alone. This phenomenon is present not only for exponential distributions but for arbitrary phasetype distributions with a density satisfying a higher order linear differential equation with constant coefficients. These distributions have the nice property that the non-local operator $g(s) = E[V(s - X)]$ can be replaced by a local operator involving derivatives of $g(s)$ and $V(s)$.

To simplify the notation we assume that X has mean 1. If the density of X is $e^{-x}, x > 0$, then the function $g(s)$ satisfies the differential equation

$$g'(s) = V(s) - g(s), \ s > 0.$$

From (12) we can see that $g'(s)$ can be represented with the factor $V'(s)$,

$$g'(s) = V(s) - g(s) = \frac{1}{\lambda} \left(c + \frac{1}{2}\sqrt{U(s)} \right) V'(s), \ s \geq 0,$$

and so we obtain a differential equation involving $U(s)$ alone:

$$\sqrt{U(x)} \left\{ \lambda + \frac{1}{2} - c - \frac{1}{2}\sqrt{U(x)} \right\} + c = \frac{1}{4}U'(x), \ s \geq 0. \tag{15}$$

This equation is closely related to the Lundberg equation below, with which one can obtain the adjustment coefficient for the exponential bound for the ruin probability of the controlled risk process:

$$\lambda + rc + \frac{1}{2} = \lambda E[\exp(rX)].$$

This equation in the exponential case reads

$$\lambda + rc + \frac{1}{2} = \frac{\lambda}{1 - r}$$

or

$$cr^2 + r \left(\lambda + \frac{1}{2} - c \right) - \frac{1}{2} = 0.$$

The equation has two solutions, $R > 0$ and $-\gamma < 0$. Since the coefficients in this equation and those in (15) coincide, we can write (15) as

$$A'(s) = 2cA(s) \left(\frac{1}{A(s)} - R \right) \left(\frac{1}{A(s)} + \gamma \right), \ s > 0,$$

and this differential equation has a solution with $A(0) = 0$ satisfying the following transcendental equation

$$\left(\frac{1}{A(s)} - R \right)^{\gamma} \left(\frac{1}{A(s)} + \gamma \right)^{R} = \exp(-(R + \gamma)s). \tag{16}$$

A function $u(s)$ with $u(0) = 1$ satisfying $-u(s)/u'(s) = A(s)$ is

$$u(s) = \frac{\exp(-Rs)}{(1 + \gamma A(s))^R}.$$

The function $u(s)$ is related to $V'(s)$ and $V(s)$ via

$$V'(s) = \frac{\lambda}{c} V(0) u(s),$$

and

$$1 - V(s) = \frac{\lambda \int_s^\infty u(y) dy}{c + \lambda \int_0^\infty u(y) dy}. \tag{17}$$

For an explicit expression of the constant $\int_0^\infty u(y) dy$ see [22].

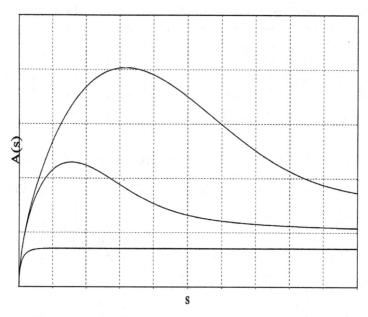

Fig. 1. Optimal Investment Strategies for Erlang Claims

In Figure 1 the optimizer $A(s)$ is shown for claim size distributions which are exponential with mean 1, Exp(1) (lowest curve), Erlang(2) (convolution of two Exp(1), middle curve) and Erlang(3) (convolution of three Exp(1), top curve).

4.4 Two or More Risky Assets

If there are two or more risky assets in which investment is possible, then the optimal investment problem is an optimal portfolio problem combined with the problem of asset allocation (i.e. in equity on the one hand, and in the money market on the other hand). If the price process of the risky assets is modelled by the $d-$variate logarithmic Brownian motion (see (7), then these two problems can be separated:

The HJB of our control problem reads

$$0 = \sup_A \left(\lambda E[V(s - X) - V(s)] + (c + a^T A)V'(s) + A^T \Sigma \Sigma^T A V''(s)\right), \quad (18)$$

where the maximum is taken over all d-vectors $A = (A_1, ..., A_d)$, and $a = (a_1, ..., a_d)$. The maximizer is

$$A = -\frac{V'(s)}{V''(s)}(\Sigma \Sigma^T)^{-1}a$$

which is a scalar multiple of a vector which does not depend on s, only the scalar is state dependent. The techniques developed in the univariate case yield that the maximizer $A(s)$ defines the optimal investment strategy in feedback form, which is investment into a fixed (state independent) portfolio with portfolio weights given by $(\Sigma \Sigma^T)^{-1}a$, only the total amount invested into risky assets varies with the wealth s of the insurer. Hence the methods in the univariate case, in particular the handy HJB, can be used also here for the computation of the value function and the optimal strategy.

5 Optimal Reinsurance and Optimal New Business

For *optimal reinsurance* we introduce a general setup as in Vogt (2003). Let A be the space of possible actions, and for $a \in A$ let $g(X, a)$ be the part of the claim X paid by the insurer, and let $X - g(X, a)$ be paid by the reinsurer. We shall assume that $0 \leq g(x, a) \leq x$ and that $x \rightarrow g(x, a)$ is non-decreasing for all $a \in A$. Furthermore, A should be a compact topological space for which $a \rightarrow g(x, a)$ is continuous for all x. Finally, we assume that there is $a_0 \in A$ for which $g(x, a_0) = x$ for all x. The reinsurance premium for active a equals $h(a)$, a continuous function on A. Then the risk process of the insurer under a dynamic strategy $a(t)$ evolves as

$$R(t) = s + ct - \int_0^t h(a(u))du - \sum_{i=1}^{N(t)} g(X_i, a(T_i)), \quad t \geq 0.$$

The problem is to find the optimal admissible dynamic strategy $a(t), t \geq 0$, for which the survival probability of the insurer is maximal. Admissible strategies are all predictable processes $a(t)$ (with respect to the filtration generated by $S(t)$). The HJB equation for this problem is

$$0 = \sup_{a \in A} \{\lambda E[V(s - g(X, a)) - V(s)] + (c - h(a))V'(s)\},$$

and the natural boundary conditions are $V(s) = 0$ for $s < 0$ and $V(\infty) = 1$. There is no example for which this equation has an explicit solution, not even the maximizer $a(s)$ can be computed in closed form. The program in this section is to prove existence of a smooth solution for this equation, and to derive a numerical algorithm for its computation. In some cases, general properties of the optimal solution can be given.

The general approach starts with the observation that $V' > 0$ and hence the sup is attained at some a for which $c - h(a) > 0$. Solving for $V'(s)$ we obtain

$$V'(s) = \inf_{a \in A_0} \left\{ \frac{\lambda E[V(s) - V(s - g(X, a))]}{c - h(a)} \right\}, \tag{19}$$

where $A_0 = \{a \in A : h(a) < c\}$. This equation can be solved via the iteration

$$V'_{n+1}(s) = \inf_{a \in A_0} \left\{ \frac{\lambda E[V_n(s) - V_n(s - g(X, a))]}{c - h(a)} \right\}, s \geq 0,$$

starting with $V_0(s)$ the survival probability without reinsurance, i.e. with contract specification a_0. The sequence of functions $V_n(s)$ is non-decreasing, so it converges to a limiting function $V(s)$ which turns out to be a solution of equation (19). So the existence of a solution can be derived in a most general framework. For numerical computation, the equation (19) yields a stable algorithm. The boundary condition can again be achieved using homogeneity of the equation.

There are several possible pricing rules $h(a)$ for the reinsurance contract. One possible rule is the expectation principle

$$h(a) = \lambda \rho E[X - g(X, a)]$$

with $\rho > 1$. Reinsurance is expensive under this rule if $\lambda \rho E[X] > c$. Another possible rule would be the variance principle

$$h(a) = \lambda E[X - g(X, a)] + \rho \lambda E[(X - g(X, a))^2],$$

which is expensive if $\lambda E[X] + \rho \lambda E[X^2] > c$. We shall, however, not allow that the pricing rule for the reinsurer can change with time, which is a bit unrealistic, it neglects liquidity cycles.

For *optimal new business* we obtain a simpler HJB equation for which the optimal strategy is bang-bang (since the HJB equation is linear in the control variable): The HJB equation reads

$$0 = \sup_{0 \leq b \leq 1} \{\lambda E[V(s - X) - V(s)] + cV'(s) + b\lambda_1 E[V(s - Y) - V(s)] + bc_1 V'(s)\}$$

which has an optimizer $b = b(s)$ which is zero if $\lambda_1 E[V(s - Y) - V(s)] + c_1 V'(s) \leq 0$ and is equal to one if $\lambda_1 E[V(s - Y) - V(s)] + c_1 V'(s) > 0$. This yields the following formula for the derivative $V'(s)$:

$$V'(s) = \min\{ \frac{\lambda E[V(s) - V(s - X)]}{c}, \tag{20}$$

$$\frac{\lambda E[V(s) - V(s - X)] + \lambda_1 E[V(s) - V(s - X)]}{c + c_1}.$$

This equation can be solved via the iteration

$$V'_{n+1}(s) = \min\{\frac{\lambda E[V_n(s) - V_n(s - X)]}{c},$$

$$\frac{\lambda E[V_n(s) - V_n(s - X)] + \lambda_1 E[V_n(s) - V_n(s - Y)]}{c + c_1}\}$$

which yields a non decreasing sequence of functions $V_n(s)$ converging to a function $V(s)$ which is a solution to (20). For numerical computation, formula (20) produces a stable algorithm.

5.1 Optimal Proportional Reinsurance

Here, $A = [0, 1]$ and $g(x, a) = ax$. We consider the reinsurance premium rule $h(a) = \lambda\rho(1 - a)E[X]$ with $\lambda\rho E[X] > c$ The HJB equation in this case reads

$$V'(s) = \inf_{a \in [0,1]} \frac{\lambda E[V(s) - V(s - aX)]}{(c - \lambda\rho(1 - a)E[X])}, \quad s \geq 0. \tag{21}$$

Not even in the exponential claim size case we can compute the value function or the optimizer in explicit form. But for exponential [or phasetype] distributions we can replace the non local operator $g(s, a) = E[V(s - aX)]$ by a local one: for claims with density $h(x) = \exp(-x)$, $x > 0$, we have

$$g_s(s, a) = \frac{1}{a}(V(s) - g(s, a)), \quad s > 0.$$

Notice that the minimum in (21) is over $a \in (a_1, 1]$ with $a_1 > 0$ from $\lambda\rho(1 - a_1)E[X] = c$.This implies that for the optimizer we have $a(s) \geq a_1$. Hence in the controlled risk process $R^*(t)$ under the optimizer strategy we have claims bounded from below by some positive constant. Then one can easily show that $R^*(t) \to \infty$ on the set of no ruin $\{\tau^* = \infty\}$. This lower bound can not be assumed for arbitrary strategies $a(t)$, and hence for the verification argument one needs a different reasoning. In the optimal investment problem we had replaced the controlled process $R(t)$ by a second one with slightly changed control and initial surplus. Here one can replace the controlled process $R(t)$ by one in which the premium rate c of the first insurer is slightly changed, and then use continuity of the solution of (21) with respect to c. For details see Schmidli [41] and Vogt [45].

In Figure 2 we show the optimizer functions $a(s)$ for the case of exponential claims with different means. Surprisingly, we see different types of functions: no reinsurance for all s ($a(s) \equiv 1$), continuous non-increasing, and finally with jump and then increasing. In each case we see convergence of $a(s)$ which is investigated in more detail in section 6.4.

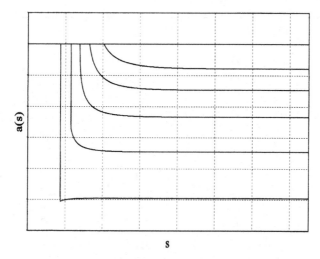

Fig. 2. Optimal Proportional Re Strategies

5.2 Optimal Unlimited XL Reinsurance

Here, $A = [0, \infty]$ and $g(x, a) = x \wedge a$. We first consider $h(a) = \lambda \rho E[(X - a)^+]$, the expectation principle. The HJB equation reads

$$V'(s) = \inf_{0 \leq a \leq \infty} \frac{\lambda E[V(s) - V(s - X \wedge a)]}{c - h(a)}. \qquad (22)$$

The function to be minimized is discontinuous:

$$g(s, a) = E[V(s - X \wedge a)]$$

has a jump at $a = s$.

In Figure 3 we show the functions $g(s, a)$ for exponential claim size. We see that the maximizer is $a = \infty$ whenever the maximum is not in $[0, s]$ (since for $a > s$ we have $E[V(s - X \wedge a)] = E[V(s - X)]$ and since $h(a)$ is non-increasing). So the maximum will be at $a = \infty$, or at $a = s$, or in the interval (a_1, s), where a_1 is derived from $h(a_1) = c$.

For exponential claims the numerical computation of the value function $V(s)$ can be simplified: for $h(x) = \theta e^{-x}$, $x > 0$, we have

$$
\begin{aligned}
g(s, a) &= \int_0^a V(s - x)h(x)dx + V(s - a)e^{-\theta a} \\
&= \int_0^s V(x)h(s - x)dx - e^{-\theta a} \int_0^{s-a} (x)h(s - a - x)dx + V(s - a)e^{-\theta a} \\
&= g(s) - e^{-\theta a}g(s - a) + V(s - a)e^{-\theta a},
\end{aligned}
$$

where $g(s) = E[V(s - X)]$. In Figure 4 we show the maximizer $a(s)$ for exponential claim sizes; for small s we have $a(s) = \infty$, then $a(s) = s$ which

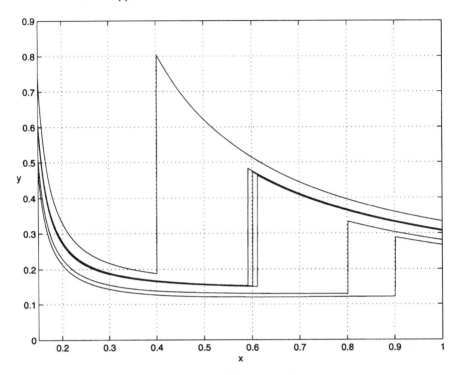

Fig. 3. Integral Term with Jumps

means that under this reinsurance specification the next claim will not cause ruin, and then $a(s) < s$ with converging $a(s)$.

For the verification argument - which again uses variation of the premium intensity - and for other technical details see [24] as well as [45].

5.3 Optimal XL Reinsurance

For XL-reinsurance with limit we have a control problem with two decision variables: $A = [0, \infty] \times (0, \infty]$, where $(b, L) \in A$ denotes the vector with priority b and limit L. No reinsurance will be identified by the pair $(0, \infty)$. For this case little more can be said about the value function and the optimal reinsurance strategy besides existence and verification statements, and asymptotic properties in the small claims case. This is due to computational intractability of the bivariate decision variable (which is not as bad as a bivariate state variable, but nevertheless).

One might conjecture that the insurer would always choose $L = \infty$ since his survival will be caused by extremely large events, and so he will reinsure the unlimited tail of the claim size distribution, as long as the reinsurance premium does not give much weight to the tail. This would be the case, e.g., for Pareto claims and for the expectation principle. However, Vogt [45] showed

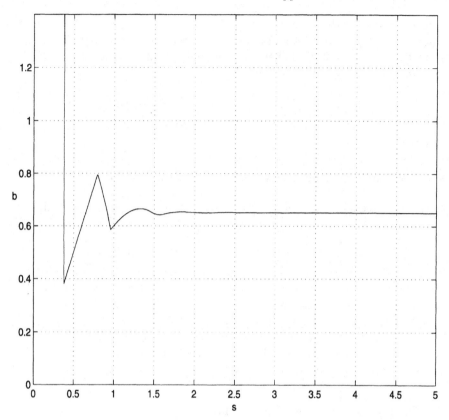

Fig. 4. Optimal XL-Re Strategy

that $L = \infty$ is never optimal under the expectation principle and for large claim sizes with density $f(x)$, and with hazard rate $r(x) = f(x)/(1 - F(x))$ for which $r(x) \to 0$ when $x \to \infty$ (which is true for Pareto claims). And for exponential claims and the expectation principle, $L = \infty$ is always optimal. This is against intuition, though true.

5.4 Optimal New Business

Here we have $A = [0, 1]$, and the optimizer $a(s)$ satisfies $a(s) = 1$ for $s \leq s_0$, and $a(s) = 0$ for $s > s_0$, where s_0 is the intersection point of the two curves from the HJB equation

$$s \to \frac{\lambda E[V(s) - V(s - X)]}{c}$$

and

$$s \to \frac{\lambda E[V(s) - V(s - X)] + \lambda_1 E[V(s) - V(s - Y)]}{c + c_1}.$$

At the point $s = 0$ the choice $b = 1$ is optimal whenever $c/\lambda < c/\lambda_1$, i.e. even non-profitable business (without a safety loading) would be written. Of course, such strategies work only if there is a market on which (also non-profitable) insurance business can be sold (see [20]).

6 Asymptotic Behavior for Value Function and Strategies

6.1 Optimal Investment: Exponential Claims

In the case of exponential claims the following equation characterizes $A(s)$:

$$\left(\frac{1}{A(s)} - R\right)^\gamma \left(\frac{1}{A(s)} + \gamma\right)^R = \exp(-(R + \gamma)s),$$

where $A(s) \geq 0$ (see 16). This implies $A(s) \to 1/R$ for $s \to \infty$. Formula (17) and the asymptotic relation $u(s) \sim \exp(-Rs)$ together imply that the asymptotic behavior of the value function $V(s)$ is given by $1 - V(s) \sim C\exp(-Rs)$, $s \to \infty$, where C is a positive constant. The same asymptotics can be obtained by a constant strategy: $A(s) \equiv 1/R$ yields a survival probability $V_1(s)$ satisfying $1 - V_1(s) \sim C_1 \exp(-Rs), s \to \infty$.

6.2 Optimal Investment: Small Claims

The ruin probability in Lundberg's risk model is exponentially bounded in the case of small claims,

$$\psi^0(s) \leq \exp(-R_0 s), s \geq 0,$$

where R_0 - the adjustment coefficient - is the solution of Lundberg's equation (3). For this we need a safety loading and finite exponential moments $E[\exp(rX)], r > 0$, of the claims. Furthermore, this exponential bound is sharp in the sense that

$$\psi^0(s) \sim C \exp(-R_0 s), s \to \infty$$

(see (5)). With investment the same is true in the case of *optimal* investment in which the adjustment coefficient $R > R_0$ is defined via the modified Lundberg equation

$$\lambda + cr + \frac{a^2}{2b^2} = \lambda E[\exp(rX)].$$

For the inequality see [11], and for the asymptotics [17] and [22]. Furthermore, the maximizer $A(s)$ satisfies $A(s) \to 1/R$, and the constant strategy based on $A(s) = 1/R$ has a ruin probability of the same exponential order $\exp(-Rs)$.

The asymptotics is derived in [22] under the assumption that $E[\exp(rX)] < \infty$ for some $r > R$, that X has a continuous density, and the condition

$$\sup_y E[\exp(R(X - y)) \mid X > y] < \infty.$$

The asymptotic behavior of the value function implies that optimal investment improves the situation because of $R_0 < R$, but investment of a constant amount does the same. This is disappointing: our optimal strategy is - at least for large values of s - not much better than a very simple strategy. However, the case of small claims is not the one important for applications.

Notice that investment of a constant proportion into the risky asset cannot be recommended in the small claims case. In Norberg and Kalashnikov [29] and in Frolova et al. [10] it is shown that the ruin probability in this case behaves as Cs^{-a} for some $C > 0, a > 0$, which is as large as the ruin probability without investment with Pareto claims (see (6)). So, investment can improve the situation of the insurer, but investment of a constant proportion will not, it will lead to a much more risky position. See also Paulsen and Gjessing [38].

6.3 Optimal Investment: Large Claims

For large claims (such as the family of subexponential distributions) a general theory on the asymptotics of the value function in the optimal investment problem is not available. One case has recently been considered and solved by Karamata's theory of regularly varying functions. It is the case in which the claim size distribution has a tail which is regularly varying with exponent $\rho < -1$, i.e.

$$P\{X > t\} = t^\rho L(t), \ t \to \infty$$

with a slowly varying function $L(t)$. In this case, the asymptotics of the ruin probability without investment is given by

$$\psi^0(s) \sim C_0 s^{\rho+1} L_0(s), \ s \to \infty,$$

with a different slowly varying function $L_0(s)$. Gaier and Grandits [12] could prove that with optimal investment, the ruin probability $\psi(s) = 1 - V(s)$ satisfies

$$\psi(s) \sim C_1 s^\rho L_1(s), \ s \to \infty,$$

and the optimizer $A(s)$ is asymptotically linear,

$$A(s) \sim \frac{a}{b^2(1 - \rho)} s, \ s \to \infty.$$

With the strategy derived from the linear function $A(s) = (a/(b^2(1 - \rho)))s$ one can obtain almost the same asymptotic behavior of the corresponding ruin probability $\psi^l(s)$:

$$\psi^l(s) \sim C_2 s^\rho L_2(s), \ s \to \infty.$$

So also in this large claims case, insurance improves the situation of the insurer: $s^{\rho+1}$ is replaced by s^ρ, and this can be achieved - at least to the same order, as one can show using the methods in [12] - by the investment of a specific constant proportion of current wealth.

6.4 Optimal Reinsurance

For the general reinsurance problem there is an asymptotic behavior of the value function and the optimal strategy in the small claims case: one can show that for arrangements $g(x, a)$, $a \in A$, satisfying the assumptions

- A is a compact topological space,
- $x \to g(x, a)$ is continuous for all $x \in \mathbb{R}$,
- $x \to g(x, a)$ is monotone for all $a \in A$,
- $|g(x, a) - g(y, a)| \le |x - y|$ for all $a \in A$,
- the adjustment coefficient $R = \sup_{a \in A} R(a)$ exists, where $R(a)$ is the positive solution to the equation

$$\lambda + (c - h(a))r = \lambda E[\exp(rg(X, a))],$$

and for claim size distributions satisfying

- X has a continuous density $h(x)$,
- $E[\exp(rX)] < \infty$ for some $r > R$,
- $\sup\{E[\exp(R(X - y)) \mid X > y] < \infty$

the value function of the problem satisfies the asymptotic relation

$$1 - V(s) \sim C \exp(-Rs),$$

and for the optimal strategy we have

$$a(s) \to a^*, \ t \to \infty,$$

where a^* is the action for which $R(a)$ attains its maximum (see Vogt [45]). Also for the survival probability $V_1(s)$ of the controlled process with constant strategy $a(t) \equiv a^*$ we have $1 - V(s) \sim C_1 \exp(-Rs)$.

The computation of the adjustment coefficient $R = \sup\{R(a) : a \in A\}$ can be simplified: R is the positive parameter for which the function $V(s) = 1 - e^{-Rs}$ satisfies the HJB equation, i.e.

$$0 = \sup_{a \in A}\{\lambda E[1 - \exp(Rg(X, a))] + R(c - h(a))\},$$

and a^* is the maximizer of this equation. To see this let R^* be the solution of the above equation which we assume to be attained at some $a^* \in A$. Observe first that for arbitrary $a \in A$ we have

$$\lambda E[\exp(R^* g(X, a)) - 1] \ge R^*(c - h(a)).$$

The adjustment coefficients $R(a)$ are defined via

$$\lambda E[\exp(R(a)g(X, a)) - 1] = R(a)(c - h(a)),$$

and we have

$$\lambda E[\exp(rg(X, a)) - 1] \geq r(c - h(a))$$

iff $r \geq R(a)$. Applying this for $r = R^*$ we obtain $R^* \geq R(a)$ for all $a \in A$. Since $R^* = R(a^*) \leq R$ we obtain the assertion $R = R^*$.

For further aspects of maximizing the adjustment coefficient see [46].

For large claims the situation is less transparent. For unlimited XL-reinsurance the controlled process has an adjustment coefficient $R = \sup_{a \in A} R(a)$, and with this quantity the Lundberg inequality $1 - V(s) \leq \exp(-Rs)$ holds. Numerical experiments support the conjecture that also in this case the strategies converge to a^* for which $R = R(a^*)$ (see Vogt [45]). For (limited) XL-reinsurance an adjustment coefficient does not exist, but numerical experiments still seem to indicate that the strategies converge (see [45]). In this area, there are many interesting open questions which are also relevant for practical applications.

7 A Control Problem with Constraint: Dividends and Ruin

Since the problem of stochastic control with constraints is still open for infinite horizon cases and continuous time, we consider discrete time in this chapter. The stochastic model presented is a toy model, but the method of solution seems to be transferable to more complicated models, also to continuous time models.

7.1 A Simple Insurance Model with Dividend Payments

Let $X_1, X_2, \ldots$ be the total sum of claims per period modelled by iid nonnegative integer valued random variables, let c be the total premium per period which is a positive integer, and s the initial surplus which is a non negative integer. The reserve $R(t)$ of the company without dividend payment evolves in discrete time as $R(0) = s$ and

$$R(t + 1) = R(t) + c - X_{t+1}, t \geq 0.$$

We assume that $P\{X_t > c\} > 0$ and that $c > E[X_t]$. Recall that the infinite time ruin probability

$$\psi^0(s) = P\{R(t) < 0 \text{ for some } t \geq 0\}$$

satisfies the equation

$$\psi^0(s) = E[\psi^0(s + c - X)].$$

We consider dividends $d(t)$ which are paid at the beginning of period $t + 1, t \geq 0$. If $\mathcal{F}(t)$ is the σ-field generated by $R(h), h \leq t$, then $d(t)$ is an $\mathcal{F}(t)$-measurable nonnegative random variable. As a measure of profitability we use expected accumulated discounted dividends:

$$u^d(s) = E\left[\sum_{t=0}^{\tau^d-1} v^t d(t)\right],$$

where v is a discount factor. With dividend payment the reserve is $R^d(t)$ defined by $R^d(0) = s$ and

$$R^d(t+1) = R^d(t) - d(t) + c - X_{t+1}, t \geq 0. \tag{23}$$

In the upper index of summation we use τ^d as the ruin time in the risk process $R^d(t)$. The ruin probability of the reserve $R^d(t)$ is denoted by

$$\psi^d(s) = P\{\tau^d < \infty\}.$$

There is a tradeoff between stability and profitability: Minimizing ruin probability means no dividend payment, $d(t) \equiv 0$ or $u^d(s) = 0$, and the reserve process $R(t)$ goes to $+\infty$. Maximizing $u^d(s)$ leads to a dividend payment scheme for which ruin is certain,

$$\psi^d(s) = 1 \text{ for all } s \geq 0,$$

and the reserve process $R^d(t)$ remains bounded (see Bühlmann ([5], chapter 6.4) and references given there, as well as Gerber (1979)).

We consider the problem of optimal dividend payment under a ruin constraint, i.e. for $0 < \alpha \leq 1$ and initial surplus fixed we derive an optimal dividend payment scheme $d(t)$ for which

$$\psi^d(s) \leq \alpha \tag{24}$$

and for which $u^d(s)$ is maximal in the class of all dividend payment schemes satisfying the constraint (24). This is done using a modified Hamilton-Jacobi-Bellman (HJB) equation and via the construction of the process of optimal admissible ruin probabilities.

7.2 Modified HJB Equation

The HJB equation for the value function $u(s)$ of the problem to maximize profitability without a ruin constraint is

$$u(s) = \sup_{\delta}\{\delta + vE[u(s - \delta + c - X)]\}, \tag{25}$$

where the maximum is taken over all $0 \le \delta \le s$. The optimal strategy is then defined via (23) and

$$d(t) = \delta(R^d(t)),$$

where $\delta = \delta(s)$ is the maximizer in (25).The modified HJB for the value function $u(s, \alpha)$ under the constraint (24) is

$$u(s, \alpha) = \sup_{\delta, \beta}\{\delta + vE[u(s - \delta + c - X, \beta(X))]\}, \tag{26}$$

where the maximum is taken over all $0 \le \delta \le s$ and functions $\beta(x)$ satisfying $E[\beta(X)] \le \alpha$ and

$$\psi^0(s - \delta + c - x) \le \beta(x) \le 1.$$

If there is no admissible pair (δ, β), then the maximum is interpreted as zero. Under the additional assumption $vP\{X < c\}$ one can show (via a contraction argument) that equation (26) has a solution, and that the maximum is attained at certain values $\delta = \delta(s, \alpha)$ and functions $\beta(x) = \beta(s, \alpha; x)$. With these functions the process of *optimal admissible* ruin probabilities $b(t), t \ge 0$, is defined as $b(0) = \alpha$ and

$$b(t + 1) = \beta(R^d(t), b(t); X_{t+1}), t \ge 0, \tag{27}$$

and the optimal dividend payment strategy is defined through

$$d(t) = \delta(R^d(t), b(t)), t \ge 0. \tag{28}$$

It turns out that $d(t)$ is an admissible strategy satisfying the constraint (24), and that $d(t)$ maximizes profitability under this constraint. The process $b(t), t \ge 0$, is a martingale with mean α satisfying

$$b(t) \ge \psi^0(R^d(t-1) - d(t-1) + c - X_t) = \psi^0(R^d(t)), t \ge 0.$$

At time t the value $b(t)$ is the ruin constraint which is active at t. Furthermore,

$$b(t) = P\{\tau^d < \infty \mid X_1, ..., X_t\}.$$

The strategy $d(t)$ is path dependent: for $t \ge 0$ the value $d(t)$ depends on $R^d(t)$ and $b(t)$, i.e. on the current state of the reserve and the current active ruin constraint. For details see [21].

7.3 Numerical Example and Conjectures

The numerical computation of the value function and the optimal strategy is based on a recursive solution of equation (26):

$$u_{n+1}(s,\alpha) = \sup_{\delta,\beta}\{\delta + vE[u_n(s-\delta+c-X,\beta(X))]\}.$$

Again, a contraction argument implies that the sequence of functions $u_n(s,\alpha)$ converges. As an example we consider the special case of a skip free risk process: $c = 1, P\{X_1 = 0\} = 1 - P\{X_1 = 2\} = 0.7$. Using the above iteration we computed the functions $u(s,\alpha), \beta_1(s,\alpha) = \beta(s,\alpha;1)$ (the value for $\beta_2(s,\alpha) = \beta(s,\alpha;2)$ can be derived from the martingale condition), and $\delta(s,\alpha)$. For $\alpha \le \psi^0(s)$ we know that $u(s,\alpha) = 0$, and we have set $\beta(s,\alpha;1) = 1$. From the numerical results we derive the following conjectures:

1) $\delta(s,\alpha) = \sum_{i=1}^{\infty} 1_{(\alpha > \alpha(s))}, \; \alpha(0) = \dots = \alpha(4) = 1, \; \alpha(5) = 0.1055, \; \alpha(6) = 0.041, \; \alpha(7) = 0.0185, \; \alpha(8) = 0.009.$
2) $\beta_1(s,\alpha) = \psi^0(s+1) + (\alpha - \psi^0(s))(1 - \psi^0(s+1))/(1 - \psi^0(s)), \alpha \ge \psi^0(s).$
3) $u_\alpha(s,\alpha) = \infty$ at the point $\inf\{a : u(s,a) > 0\} = \psi^0(s).$

Conjecture 2. is obvious for $s = 0$: it follows from $\beta_2(0,\alpha) = 1$ and

$$p\beta_1(0,\alpha) + (1-p)\beta_2(0,\alpha) = \alpha.$$

The corresponding problem without ruin constraint has an optimal strategy which pays dividends whenever the surplus s reaches the value 5, i.e. $\delta(s,1) = 0, s = 0, \dots, 4, \; \delta(5,1) = 1$. The value function $u(s)$ can be computed using the following system of linear equations:

$$u(0) = vpu(1)$$
$$u(i) = vpu(i+1) + v(1-p)u(i-1), i = 1, \dots, 4$$
$$u(5) = 1 + u(4);$$

we obtain

$$u(0) = 6.2752, \; u(1) = 9.2335, \; u(2) = 10.8971,$$
$$u(3) = 12.0771, \; u(4) = 13.1004, \; u(5) = 14.1004.$$

Our numerical results in the constrained case are in line with these values: $\delta(s,\alpha) = 0$ for $s = 0, \dots, 4$ and all α (see assertion (h) in the above Lemma, and $u(s,1) = u(s)$.

Figures 5 and 6 show the value function $u(s,\alpha)$ and the function $\beta(s,\alpha)$ for $s = 0, \dots, 10$, computed with a step size of $\Delta = 1/2000$ (i.e. $u(s,\alpha), \beta(s,\alpha)$ are approximated at the points $k\Delta, k = 0, \dots, 2000$), and the range of s is restricted to $s \le 20$. On the x-axis α runs from 0 to 1. For each value of s a separate curve is shown.

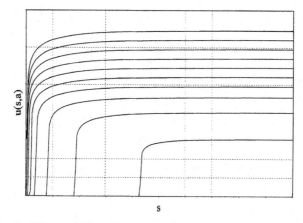

Fig. 5. Value Function for Optimal Dividends

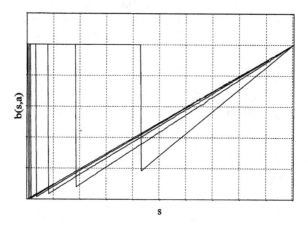

Fig. 6. Running Constraints

7.4 Earlier and Further Work

Earlier approaches to optimal dividend payment without constraints or with different constraints can be found in Bühlmann's book [5], in Gerber ([14] and [15]), and in Paulsen [36]. The Lagrange multiplier method used in Altman [1] does not seem to work in the infinite horizon situation considered above. Hipp and Schmidli [23] compute optimal dividend strategies satisfying (24) of the form

$$d(t) = \begin{cases} 0 \text{ if } R^d(t) \leq c(s, \alpha) \\ M \text{ if } R^d(t) > c(s, \alpha) \end{cases} \tag{29}$$

for compound risk processes $R(t)$ in continuous time, and for exponentially distributed claim sizes. These strategies are optimal only in the class of strategies having form (29). They show that universal optimal strategies satisfying

(24) can be derived from a modified HJB, adjusted to the Lundberg model:

$$0 = \min[\sup_{\beta}\{\lambda E[u(s + c - X, \beta(X)) - u(s, \alpha)] - \rho u(s, \alpha) \qquad (30)$$

$$+cu_s(s, \alpha) - \lambda(E[\beta(X)] - \alpha)u_\alpha(s, \alpha)\}, 1 - u_s(s, \alpha)].$$

Again, the maximum is taken over all functions $\beta(x)$ satisfying the following constraint:

$$1 \geq \beta(x) \geq \psi^0(s - x).$$

Notice that here we do not have the restriction $E[\beta(X)] = \alpha$.

8 Conclusions

The techniques and methods of risk management described in this survey paper are presented from a theoretical viewpoint (existence of optimal strategies, their properties, and the construction of numerical algorithms). Of course, the implementation of these methods in real life needs a considerable amount of additional work, such as

- parameter estimation,
- sensitivity investigations and
- real markets modelling.

All these points are related, but the main and most difficult implementation problem would be the last one. To be more specific, we consider two aspects:

- transaction costs for investment, and
- illiquid reinsurance markets.

For the first problem, one might look for strategies which are suboptimal and have low transaction costs. The second problem is based on the fact that reinsurance contracts are never adjusted in continuous time, and when they are adjusted, the reinsurance premium is fixed via a new bargaining which takes the actual market situation into account. To come closer to optimal reinsurance one would look for suboptimal strategies which could be implemented into a multi-year reinsurance contract. Stochastic control for reinsurance would help to find such suboptimal strategies (e.g. in unlimited XL-reinsurance, with three levels "no reinsurance", "priority equals surplus", and "constant (asymptotic) priority", which are active automatically when the surplus is low, intermediate, or large, respectively). Today's insurance companies are not yet using these ideas of sophisticated risk management; times of increasing costs of capital will, however, generate a demand for the methods presented in this paper.

References

1. Altman, E. (1999) Constrained Markov decision processes. Chapman&Hall, New York.
2. Asmussen, S. and M. Taksar (1997) Controlled diffusion models for optimal dividend payout. Insurance: Mathematics and Economics 20, 1-15.
3. Brockett, P., and Xia, X. (1995) Operations research in insurance: a review. Trans. Act. Soc.XLVII, 7-80.
4. Browne, S. (1995) Optimal investment policies for a firm with a random risk process: exponential utility and minimizing the probability of ruin. Math. Operations Res. 20, 937-958.
5. Bühlmann, H. (1996) Mathematical Methods of Risk Theory. Springer, Berlin.
6. Dufresne, F. and Gerber, H.U. (1991). Risk theory for the compound Poisson process that is perturbed by diffusion. Insurance: Math. Econom. 10, 51–59.
7. Embrechts, P, Klüppelberg, C., and Mikosch, T. (2001) Modelling extremal eventsfor insurance and finance. 3rd ed. Springer, Berlin.
8. Fleming, W.H. and Rishel, R.W. (1975) Deterministic and stochastic optimal control. Springer, New York.
9. Fleming, W.H. and Soner, M. (1993) Controlled Markov processes and viscosity solutions. Springer, New York.
10. Frolova, A., Kabanov, Y., Pergamenshchikov, S. (2002) In the insurance business risky investments are dangerous. Finance and Stochastics, 6, 227-235.
11. Gaier, J., Grandits, P. and Schachermeyer, W. (2003). Asymptotic ruin probabilities and optimal investment. To appear in: Annals of Applied Probability
12. Gaier, J. and Grandits, P. (2001) Ruin probabilities in the presence of regularly varying tails and optimal investment. Preprint, University of Vienna.
13. Gerber, H.U. (1969). Entscheidungskriterien für den zusammengesetzten Poisson-Prozess. Schweiz. Verein. Versicherungsmath. Mitt. 69, 185–228.
14. Gerber, H. U. (1979) An Introduction to Mathematical Risk Theory. S.S. Huebner Foundation Monographs, University of Pennsilvania.
15. Gerber, H. U. (1981) On the probability of ruin in the presence of a linear dividend barrier. Scand. Act. J. 2, 105-115.
16. Grandell, J. (1991) Aspects of Risk Theory. Springer, New York.
17. Grandits, P. (2003) An analogue of the Cramer-Lundberg approximation in the optimal investment case. Preprint, Technical University Vienna.
18. Hipp, C. and Plum, M. (2000). Optimal investment for insurers. Insurance: Math. Econom. 27, 215–228.
19. Hipp, C. and Plum, M. (2003) Optimal investment for investors with state dependent income, and for insurers. Finance and Stochastics 7, 299-321.
20. Hipp, C., and Taksar, M. (2000) Stochastic Control for Optimal New Business. Insurance: Mathematics and Economics 26, 185-192
21. Hipp, C. (2003) Optimal dividend payment under a ruin constraint: discrete time and state space. To appear in: Blätter der DGVFM.
22. Hipp, C. and Schmidli, H. (2003) Asymptotics of the ruin probability for the controlled risk process: the small claims case. To appear in: Scandinavian Actuarial J.
23. Hipp, C. and Schmidli, H. (2003) Dividend optimization under a ruin constraint: the Lundberg model. In preparation.
24. Hipp, C. and Vogt, M. (2003) Optimal dynamic XL reinsurance. To appear in ASTIN Bulletin.

25. Hoejgaard, B. and M. Taksar (1998) Optimal proportional reinsurance policies for diffusion models. Scand. Actuarial J. 166-180.
26. Højgaard, B. and Taksar, M. (1998). Optimal proportional reinsurance policies for diffusion models with transaction costs. Insurance Math. Econom. 22, 41–51.
27. Hoejgaard, B. and M. Taksar (1998) Controlling risk exposure and dividents payout schemes: Insurance company example. Mathematical Finance, to appear.
28. Højgaard, B. (2000). Optimal dynamic premium control in non-life insurance. Maximizing dividend pay-outs. Preprint, University of Aalborg.
29. Kalashnikov, V. and Norberg, R. (2002) Power tailed ruin probabilities in the presence of small claims and risky investments. Stoch. Proc. Appl. 98.
30. Karatzas, I., Shreve, S. (1997) Methods of mathematical finance. Springer, Heidelberg.
31. Lundberg, P. (1903) Approximerad framställing av sannolikhetsfunktionen. Aterförsäkring av kollektivrisker. Akad. Afhandling. Almqvist & Wiksell, Upsalla.
32. Martin-Löf, A. (1994) Lectures on the use of control theory in insurance. Scand. Actuarial J. 1-25.
33. Merton, R.C. (1969) Lifetime portfolio selection under uncertainity: The continuos-time case. Review of Economics and Statistics, 51, 247-257.
34. Merton, R.C. (1971) Optimum consumption and portfolio rules in a continous-time model. Journal of Economic Theory, 3, 373-413.
35. Øksendal, B. (1998) Stochastic Differential Equations, an Introduction with Applications. 5th ed., Springer, Berlin.
36. Paulsen, J. (1998) Ruin theory with compounding assets - a survey. Insurance: Mathematics and Economics, 22, 3-16.
37. Paulsen, J. (2003) Optimal dividend payouts for diffusions with solvency constraints. Preprint, University of Bergen.
38. Paulsen, J. and H.K. Gjessing (1997) Ruin theory with stochastic return on investments. Advances in Applied Probability, vol.29, 965-985.
39. Rolski, T. Schmidli, H. Schmidt, V. Teugels, J. (1998) Stochastic Processes for Insurance and Finance. Wiley Series in Probability and Statistics
40. Schmidli, H. (1995). Cramér-Lundberg approximations for ruin probabilities of risk processes perturbed by diffusion. Insurance: Math. Econom. 16, 135–149.
41. Schmidli, H. (2001). Optimal proportional reinsurance policies in a dynamic setting. Scand. Actuarial J., 55–68.
42. Schmidli, H. (2002). On minimising the ruin probability by investment and reinsurance. Ann. Appl. Probab., to appear.
43. Schmidli, H. (2002). Asymptotics of ruin probabilities for risk processes under optimal reinsurance policies. Preprint, Laboratory of Actuarial Mathematics, University of Copenhagen.
44. Taksar, M. (2000) Optimal Risk/Dividend Distribution Control Models: Applications to Insurance. Mathematical Methods of Operations Research, 1, 1-42.
45. Vogt, M (2003) Optimale dynamische Rückversicherung - ein kontrolltheoretischer Ansatz. Thesis, University of Karlsruhe.
46. Waters, H. R. (1983) Some Mathematical Aspects of Reinsurance. Insurance, Mathematics and Economics 2, 17-26.
47. Walter, W. (1970) Differential and Integral Inequalities. Springer, New York.
48. Walter, W. (1998) Ordinary Differential Equations. Readings in Mathematics, Springer, New York.

Nonlinear Expectations, Nonlinear Evaluations and Risk Measures

Shige Peng[*]

Institute of Mathematics, Shandong University
250100, Jinan, China
peng@sdu.edu.cn

1 Introduction

1.1 Searching the Mechanism of Evaluations of Risky Assets

We are interested in the following problem: let $(X_t)_{0 \leq t \leq T}$ be an $\mathbf{R}^d$–valued process, Y a random value depending on the trajectory of X. Assume that, at each fixed time $t \leq T$, the information available to an agent (an individual, a firm, or even a market) is the trajectory of X before t. Thus at time T, the random value $Y(\omega)$ will become known to this agent. The question is: how this agent evaluates Y at the time t? If this Y is traded in a financial market, it is called a derivative, i.e. a contract whose outcome depends on the evolution of the underlying process X. The output of this evaluation can be the maximum value the agent can accept to buy it or the minimum value to sell it. It may depend on his economic situation, his knowledge on the history of X, his risk aversion and utility function. In many situation this individual evaluation may be very different from the actual market price.

Examples of derivatives are futures and option contracts based on the underlying asset X, such as a commodity, a stock index, the interest rate, an exchange rate; or an individual stock; or a mortgage backed security. Here the term derivative is in general sense, i.e., it may be a positive or a negative number.

[*] The author would like to acknowledge the partial support from the Natural Science Foundation of China, grant No. 10131040. He would like to give his special thanks to the organizers as well as the audience of CIME–EMS school, in the beautiful town of Bressanone, for their warm hospitality and enthusiasm. This memorable Italian trip and lecture experience could never been realized without the persistence and the efforts of the organizers to overcome the author's 'Shengen–Italy–visa–paradox'. He would like to thank Li Juan as well as Xu Mingyu for their careful examinations and suggestions to the manuscript.

The well–known Black & Scholes option pricing theory (1973) has made the most significant contribution, over the last 30 years, in modeling the evaluation of derivatives in financial markets.

One of the important limitations of Black–Scholes-Merton approach is that it is heavily based on the assumption that the statistic behavior of the stochastic process X is exogenously specified. The fact that the Black–Scholes pricing of Y is independent of the preference of the involved individuals is also frequently argued. On the other hand, in the situation where Y is not traded, the main arguments of BS model, i.e. the replication of a claim in an arbitrage–free market, are no longer viable, and the evaluation of Y is often preference–dependent.

In this lecture the evaluation of Y will be treated under a new viewpoint. We will introduce an evaluation operator $\mathcal{E}_{t,T}[Y]$ to define the value of Y evaluated by the agent at time t. This operator $\mathcal{E}_{t,T}[\cdot]$ assigns an $(X_s)_{0 \le s \le T^-}$ dependent random variable Y to an $(X_s)_{0 \le s \le t}$–dependent $\mathcal{E}_{t,T}[Y]$. Although this value $\mathcal{E}_{t,T}[Y]$ is very complicated and is different from one agent to anther, we can still find some axiomatic assumptions to describe the mathematical properties of this operator. The evaluation of Y is treated as a filtration consistent nonlinear expectation or, more general, a filtration consistent nonlinear evaluation. We will prove that this expectation or evaluation is completely determined by a simple function g.

1.2 Axiomatic Assumptions for Evaluations of Derivatives

General Situations: $\mathcal{F}_t^X$–Consistent Nonlinear Evaluations

Let us give a more specific formulation to the above evaluation problem. Let $X = (X_t)_{t \ge 0}$ be a d–dimensional process, it may be the prices of stocks in a financial market, the rates of exchanges, the rates of local and global inflations etc. We assume that at each time $t \ge 0$, the information of an agent (a firm, a group of people, a financial market) is the history of X during the time interval $[0, t]$. Namely, his actual filtration is

$$\mathcal{F}_t^X = \sigma\{X_s; s \le t\}.$$

We denote the set of all real valued $\mathcal{F}_t^X$–measurable random variables by $m\mathcal{F}_t^X$. Under this notation an X–underlying derivative Y, with maturity $t \in [0, \infty)$, is an $\mathcal{F}_t^X$–measurable random variable, i.e., $Y \in m\mathcal{F}_t^X$. We will find the law of evaluation of Y at each time $s \in [0, t]$. We denote this evaluated value by $\mathcal{E}_{s,t}[Y]$. It is reasonable to assume that $\mathcal{E}_{s,t}[Y]$ is $\mathcal{F}_s^X$–measurable. We thus have the following system of evaluator: for each $Y \in m\mathcal{F}_t^X$

$$\mathcal{E}_{s,t}[Y] : m\mathcal{F}_t^X \longrightarrow m\mathcal{F}_s^X.$$

In particular

$$\mathcal{E}_{0,t}[Y] : m\mathcal{F}_t^X \longrightarrow \mathbf{R}.$$

We will make the following **Axiomatic Assumptions** for $(\mathcal{E}_{s,t}[\cdot])_{0\leq s\leq t<\infty}$:

(A1) Monotonicity: $\mathcal{E}_{s,t}[Y] \geq \mathcal{E}_{s,t}[Y']$, if $Y \geq Y'$.
(A2) $\mathcal{E}_{s,s}[Y] = Y$, if $Y \in m\mathcal{F}_s^X$, particularly $\mathcal{E}_{0,0}[c] = c$.
(A3) Time consistency: $\mathcal{E}_{s,t}[\mathcal{E}_{t,T}[Y]] = \mathcal{E}_{s,T}[Y]$, if $s \leq t \leq T$, $Y \in m\mathcal{F}_T^X$.
(A4) "Zero–one law": for each $s \leq t$, $\mathcal{E}_{s,t}[1_A Y] = 1_A \mathcal{E}_{s,t}[Y]$, $\forall A \in \mathcal{F}_s^X$.

Remark 1.1. Conditions (A1) and (A2) are obvious. Condition (A3) means that at the time $t \leq T$, $\mathcal{E}_{t,T}[Y]$ can be also treated as a derivative with the maturity t. At the time $s \leq t$, the price $\mathcal{E}_{s,t}[\mathcal{E}_{t,T}[Y]]$ of this derivative is the same as the price of the original derivative Y with maturity T, i.e., $\mathcal{E}_{s,T}[Y]$.

Remark 1.2. The meaning of condition (A4) is: at time s, the agent knows whether $X_{.\wedge s}$ is in A. If it is in A, then the value $\mathcal{E}_{s,t}[1_A Y]$ is the same as $\mathcal{E}_{s,t}[Y]$ $1_A Y = Y$. Otherwise $1_A Y$ is zero thus it costs nothing. A more generalization of (A4) is
(A4') For each $s \leq t$,

$$1_A \mathcal{E}_{s,t}[1_A Y] = 1_A \mathcal{E}_{s,t}[Y], \quad \forall A \in \mathcal{F}_s^X.$$

In this lecture we will not study this case (see Peng 2003 [Peng2003b]).

$\mathcal{F}_t^X$–Consistent Nonlinear Expectations

In many situations we assume furthermore, instead of **(A2)** , that

(A2') For each $0 \leq s \leq t$, $Y \in m\mathcal{F}_s^X$, $\mathcal{E}_{s,t}[Y] = Y$.

Remark 1.3. The meaning of condition (A2') is: the market has a zero–interesting rate, i.e., $r_t \equiv 0$. We observe that in many cases, even when $r_t \neq 0$, we can still define the following discounted evaluation

$$\mathcal{E}_{t,T}^r[Y] := \mathcal{E}_{t,T}[Y \exp(-\int_t^T r_s ds)].$$

This $\mathcal{E}_{t,T}^r[\cdot]$ satisfies (A2').

Let us fix a sufficiently large $T < \infty$ and consider $\mathcal{E}_{s,t}[Y]$ for $0 \leq s \leq t \leq T$ and $Y \in m\mathcal{F}_t^X$. By (A2')

$$\mathcal{E}_{s,t}[Y] = \mathcal{E}_{s,t}[\mathcal{E}_{t,T}[Y]] = \mathcal{E}_{s,T}[Y].$$

We then only need to treat $\mathcal{E}[Y|\mathcal{F}_s^X] := \mathcal{E}_{s,T}[Y]$:

$$\mathcal{E}[Y|\mathcal{F}_s^X] : m\mathcal{F}_T^X \to m\mathcal{F}_s^X,$$
$$\mathcal{E}[Y] = \mathcal{E}[Y|\mathcal{F}_0^X] : m\mathcal{F}_T^X \to \mathbf{R}.$$

By the Axiomatic assumptions, we have, for each $Y, Z \in m\mathcal{F}_T^X$ and $t \leq T$,

(A1) **Monotonicity:** $\mathcal{E}[Y|\mathcal{F}_t^X] \geq \mathcal{E}[Z|\mathcal{F}_t^X]$, if $Y \geq Z$;
(A2') **Constant–preserving:** $\mathcal{E}[Y|\mathcal{F}_t^X] = Y$, if $Y \in m\mathcal{F}_t^X$;
(A3) **Time consistency:** $\mathcal{E}[\mathcal{E}[Y|\mathcal{F}_t^X]|\mathcal{F}_s^X] = \mathcal{E}[Y|\mathcal{F}_s^X]$, if $s \leq t \leq T$;
(A4) **"Zero–one law":** $\mathcal{E}[1_A Y|\mathcal{F}_t^X] = 1_A\mathcal{E}[Y|\mathcal{F}_t^X]$, $\forall A \in \mathcal{F}_t^X$.

In particular, the functional $\mathcal{E}[\cdot]$ is a nonlinear expectation, i.e., it satisfies

(a1) Monotonicity: $\mathcal{E}[Y] \geq \mathcal{E}[Z]$, if $Y \geq Z$;
(a2) Constant–preserving: $\mathcal{E}[c] = c$.

From (A3) and (A4) we have, each $0 \leq T < \infty$ and $Y \in m\mathcal{F}_T^X$,

$$\mathcal{E}[1_A\mathcal{E}[Y|\mathcal{F}_t^X]] = \mathcal{E}[1_A Y], \ \forall A \in \mathcal{F}_t^X. \tag{1}$$

We recall that this is just the classical definition of the conditional expectation given $\mathcal{F}_t^X$. In the next section we will prove that in nonlinear situations we can also derive all the Axiomatic assumptions (A1), (A2'), (A3) and (A4) by this definition (1) provided $\mathcal{E}$ is strictly monotone. In this case we call $\mathcal{E}[\cdot]$ an $\mathcal{F}_t^X$–**consistent nonlinear expectation**.

Remark 1.4. From the above reasoning it is clear that the Axiomatic assumptions (A1)–(A4) are also applied in many other situations to measuring a risky value Y in a dynamical situation. In fact, an advantage is that they are also workable in the situation where the risky value Y is not exchanged in markets. For example, a result of a decision is in general not exchangeable. For example, it is applicable to an individual or a group's evaluation of a derivative Y. In some situation an agent can not have all information $\mathcal{F}_t^X$, but this formulation can be also applied to the situation of partially observation, i.e., with a smaller filtration $\mathcal{G}_t \subset \mathcal{F}_t$, $t \geq 0$.

Remark 1.5. It is clear that for the formulation of an $\mathcal{F}_t^X$–consistent evaluation it is not needed to introduce an a priori probability space. But in this lecture we will be within the framework of Brownian Motion filtration $(\mathcal{F}_t)_{t\geq 0}$. For more general situation, see [Peng2002].

1.3 Organization of the Lecture

In the next section, we will give the formulations of filtration consistent evaluations and expectations under the filtration $\mathcal{F}_t$ generated by a Brownian Motion. Then in Section 3, we present BSDE theory and introduce a large sort of filtration consistent nonlinear evaluations and expectations, i.e., g–evaluations and g–expectations. This g–evaluation is entirely determined by a simple real function g. We also present a nonlinear decomposition theorem of Doob–Meyer's type, for the related g–supermartingale. This result plays a central role in Section 4, in which we will prove that the notion of g–expectations is large enough to represent all "regular" $\mathcal{F}_t$–consistent nonlinear expectations. This result permit us to find the simple mechanism, i.e., the function g, of the above apparently very abstract evaluations. We also provide a simple method to test and then find the function g. In Section 5, we present some basic method to solve numerically BSDE such as g–expectations and g–evaluations.

The nonlinear martingale theorem in self–content in this lecture, including the related upcrossing inequalities.

2 Brownian Filtration Consistent Evaluations and Expectations

2.1 Main Notations and Definitions

In this lecture, we will study the above evaluation problem within the following standard framework. Let $(\Omega, \mathcal{F}, P)$ be a probability space equipped with a filtration $(\mathcal{F}_t)_{t\geq 0}$, $(B_t)_{t\geq 0}$ be a standard d-dimensional Brownian Motion defined on this space. We assume that $(\mathcal{F}_t)$ is the natural filtration of B:

$$\mathcal{F}_t = \sigma\{\sigma\{B_s; 0 \leq s \leq t\} \cup \mathcal{N}\}, \quad \mathcal{F}_\infty^0 := \bigcup_{t>0} \mathcal{F}_t.$$

where $\mathcal{N}$ is the collection of P–null sets in Ω. A vector valued stochastic process $X_t = X(\omega, t)$, $t \geq 0$, is said to be $\mathcal{F}_t$-adapted (or more specifically $(\mathcal{F}_t)_{0\leq t<\infty}$ -adapted), if for each $t \in [0, \infty)$, $(X_t(\cdot))$ is an $\mathcal{F}_t$–measurable random variable. $\mathcal{F}_t$ represents our information before time t. Thus the meaning that X is $\mathcal{F}_t$-adapted process is that at the current time t_0, we know all trajectories of X_t for $t \leq t_0$. All processes discussed in this lecture are assumed to be $\mathcal{F}_t$–adapted. We need the following notations. Let $p \geq 1$ and $\tau \leq T$ be a given $\mathcal{F}_t$–stopping time.

- The scalar product and norm of the Euclid space $\mathbf{R}^n$ are respectively denoted by $\langle \cdot, \cdot \rangle$ and $|\cdot|$.
- $L^p(\mathcal{F}_\tau; R^m) :=$ {the space of all real valued $\mathcal{F}_\tau$–measurable random variables such that $E[|\xi|^p] < \infty$};
- $L^p_\mathcal{F}(0, \tau; R^m) :=$ {R^m–valued and $\mathcal{F}_t$–adapted and stochastic processes such that $E \int_0^\tau |\phi_t|^p dt < \infty$};
- $D^p_\mathcal{F}(0, \tau; R^m) :=$ {all RCLL processes in $L^p_\mathcal{F}(0, \tau; R^m)$ such that $E[\sup_{0\leq t\leq \tau} |\phi_t|^p] < \infty$};
- $S^p_\mathcal{F}(0, \tau; R^m) :=$ {all continuous processes in $D^p_\mathcal{F}(0, \tau; R^m)$ };
- $\mathcal{S}_T :=$ {the collection of all $\mathcal{F}_t$–stopping times bounded by $\tau \leq T$};

- $\mathcal{S}_T^0 := \{\tau \in \mathcal{S}_T$ and $\cup_{i=1}^n \{\tau = t_i\} = \Omega$, with some deterministic $0 \le t_1 < \cdots < t_N\}$.

In the case $m = 1$, we denote them by $L^p(\mathcal{F}_\tau)$, $L^p_{\mathcal{F}}(0,\tau)$, $D^p_{\mathcal{F}}(0,\tau)$ and $S^p_{\mathcal{F}}(0,\tau)$. We observe that all elements in $D^2_{\mathcal{F}}(0,T)$ are $\mathcal{F}_t$–predictable. When $p = 2$, the above L^p are separable Hilbert spaces.

We observe the following fact: for each $\phi \in L^p_{\mathcal{F}}(0,T)$ there exists a progressively measurable process $\bar{\phi}$ which is stochastically equivalent to ϕ, i.e.,

$$P(\omega : \phi_t(\omega) = \bar{\phi}_t(\omega)) = 1, \forall t \in [0,T].$$

In this lecture, we will not distinguish the two processes.

We now give a rigorous definition of $\mathcal{F}_t$–consistent evaluations and expectations:

Definition 2.1. *The system of operators*

$$\mathcal{E}_{s,t}[\cdot] : L^2(\mathcal{F}_t) \to L^2(\mathcal{F}_s), \ 0 \le s \le t \le T \tag{2}$$

is called an $\mathcal{F}_t$–consistent nonlinear evaluation defined on $L^2(\mathcal{F}_T)$ if for each $0 \le s \le t < T$ and for each Y and $Y' \in L^2(\mathcal{F}_t)$, we have

(A1) Monotonicity: $\mathcal{E}_{t,T}[Y] \ge \mathcal{E}_{t,T}[Y']$, a.s., if $Y \ge Y'$, a.s.;
(A2) $\mathcal{E}_{t,t}[Y] = Y$, if $Y \in L^2(\mathcal{F}_t)$, a.s., particularly $\mathcal{E}_{0,0}[c] = c$;
(A3) Time consistency: $\mathcal{E}_{r,s}[\mathcal{E}_{s,t}[Y]] = \mathcal{E}_{r,t}[Y]$, a.s., if $r \le s \le t \le T$;
(A4) "Zero–one law": for each $s \le t$, $\mathcal{E}_{s,t}[1_A Y] = 1_A \mathcal{E}_{s,t}[Y]$, a.s., $\forall A \in \mathcal{F}_s$.

Remark 2.1. By (A4) it is easy to check that $\mathcal{E}_{s,t}[0] = 0$, a.s.. A condition weaker than (A4) is

(A4') For each $s \le t$, $1_A \mathcal{E}_{s,t}[1_A Y] = 1_A \mathcal{E}_{s,t}[Y]$, a.s., $\forall A \in \mathcal{F}_s$.

As we discussed in the introduction, if (A2) is strengthen to

(A2') $\mathcal{E}_{s,t}[Y] = Y$, a.s., $\forall Y \in L^2(\mathcal{F}_s)$

then we have

Proposition 2.1. *We assume (A1), (A2'), (A3) and (A4). Then, with the definition*

$$\mathcal{E}[Y|\mathcal{F}_t] := \mathcal{E}_{t,T}[Y], \ a.s., \ Y \in L^2(\mathcal{F}_T) \tag{3}$$

We have

(A1) Monotonicity: $\mathcal{E}[Y|\mathcal{F}_t] \ge \mathcal{E}[Z|\mathcal{F}_t]$, a.s., if $Y \ge Z$, a.s.;
(A2') Constant–preserving: $\mathcal{E}[Y|\mathcal{F}_t] = Y$, a.s., if $Y \in L^2(\mathcal{F}_t)$;
(A3) Time consistency: $\mathcal{E}[\mathcal{E}[Y|\mathcal{F}_t]|\mathcal{F}_s] = \mathcal{E}[Y|\mathcal{F}_s]$, a.s., if $s \le t \le T$;
(A4) "Zero–one law": for each t, $\mathcal{E}[1_A Y|\mathcal{F}_t] = 1_A \mathcal{E}[Y|\mathcal{F}_t]$, a.s., $\forall A \in \mathcal{F}_t$.

Definition 2.2. *The system of operators*

$$\mathcal{E}[\cdot|\mathcal{F}_t] : L^2(\mathcal{F}_T) \to L^2(\mathcal{F}_t), \ 0 \le t < T \qquad (4)$$

satisfying the above axiomatic assumptions (A1), (A2'), (A3) and (A4) is called an $\mathcal{F}_t$-consistent nonlinear expectation (or simply $\mathcal{F}$-expectation) defined on $L^2(\mathcal{F}_T)$.

2.2 $\mathcal{F}_t$-Consistent Nonlinear Expectations

The above $\mathcal{F}_t$-consistent nonlinear expectations can be also introduced in a classical way, beginning from the notion of nonlinear expectations:

Definition 2.3. *A nonlinear expectation defined on $L^2(\mathcal{F}_T)$ is a functional:*

$$\mathcal{E}[\cdot] : L^2(\mathcal{F}_T) \longmapsto \mathbf{R}$$

satisfying the following properties: (a1) Strict monotonicity:

$$\text{if} \ \ Y_1 \ge Y_2 \ \ a.s., \ \ \text{then} \ \ \mathcal{E}[Y_1] \ge \mathcal{E}[Y_2];$$
$$\text{if} \ \ Y_1 \ge Y_2 \ \ a.s., \ \ \mathcal{E}[Y_1] = \mathcal{E}[Y_2] \ \ \Longleftrightarrow \ \ Y_1 = Y_2 \ \ a.s.$$

(a2) preserving of constants:

$$\mathcal{E}[c] = c, \quad \text{for each constant } c.$$

Lemma 2.1. *Let $t \le T$ and $\eta_1, \eta_2 \in L^2(\mathcal{F}_t)$. If for each $A \in \mathcal{F}_t$,*

$$\mathcal{E}[\eta_1 1_A] = \mathcal{E}[\eta_2 1_A],$$

then we have

$$\eta_2 = \eta_1, \quad a.s. \qquad (5)$$

Proof. We choose $A = \{\eta_1 \ge \eta_2\} \in \mathcal{F}_t$. Since $(\eta_1 - \eta_2)1_A \ge 0$ and $\mathcal{E}[\eta_1 1_A] = \mathcal{E}[\eta_2 1_A]$, it follows that $\eta_1 1_A = \eta_2 1_A$ a.s.. Thus $\eta_2 \ge \eta_1$ a.s. With the same argument we can prove that $\eta_1 \ge \eta_2$ a.s. It follows that (5) holds. The proof is complete. □

Definition 2.4. *A nonlinear expectation is called an $\mathcal{F}$-expectation defined on $L^2(\mathcal{F}_T)$ if for each $Y \in L^2(\mathcal{F}_T)$ and $t \in [0, T]$, there exists a random variable $\eta \in L^2(\mathcal{F}_t)$, such that*

$$\mathcal{E}[Y 1_A] = \mathcal{E}[\eta 1_A], \quad \forall A \in \mathcal{F}_t. \qquad (6)$$

From Lemma 2.1, such an η is uniquely defined. We also denote it by $\eta = \mathcal{E}[Y|\mathcal{F}_t]$. $\mathcal{E}[Y|\mathcal{F}_t]$ is called the conditional $\mathcal{F}$-expectation of Y under $\mathcal{F}_t$. It is characterized by

$$\mathcal{E}[Y 1_A] = \mathcal{E}[\mathcal{E}[Y|\mathcal{F}_t]1_A], \quad \forall A \in \mathcal{F}_t. \qquad (7)$$

We will see that, in fact this definition of $\mathcal{E}[\cdot|\mathcal{F}_t]$ coincides with Definition 2.2. Indeed, we have the following lemmas. The first one checks (A3) and (A2'):

Lemma 2.2. *We have, for each $0 \leq s \leq t \leq T$ and $Y \in \mathcal{F}_T$,*

$$\mathcal{E}[\mathcal{E}[Y|\mathcal{F}_t]|\mathcal{F}_s] = \mathcal{E}[Y|\mathcal{F}_s] \quad a.s. . \tag{8}$$

In particular,

$$\mathcal{E}[\mathcal{E}[Y|\mathcal{F}_t]] = \mathcal{E}[Y]. \tag{9}$$

If $Y \in L^2(\mathcal{F}_t)$, we also have

$$\mathcal{E}[Y|\mathcal{F}_t] = Y, \quad a. s..$$

Proof. Since $A \in \mathcal{F}_s \subset \mathcal{F}_t$, thus

$$\begin{aligned}
\mathcal{E}[\mathcal{E}[\mathcal{E}[Y|\mathcal{F}_t]|\mathcal{F}_s]1_A] &= \mathcal{E}[\mathcal{E}[Y|\mathcal{F}_t]1_A] \\
&= \mathcal{E}[Y1_A] \\
&= \mathcal{E}[\mathcal{E}[Y|\mathcal{F}_s]1_A].
\end{aligned}$$

It follows from Lemma 2.1 that (8) holds. (9) follows easily from the fact that $\mathcal{F}_0$ is the trivial σ-algebra (since $B_0 = 0$). Finally, if $Y \in L^2(\mathcal{F}_t)$, then the only $\eta \in L^2(\mathcal{F}_t)$ satisfying (6) is Y itself. Thus (A2') holds.   □

The second lemma checks (A4):

Lemma 2.3. *We have*

$$\mathcal{E}[Y1_A|\mathcal{F}_t] = \mathcal{E}[Y|\mathcal{F}_t]1_A, \quad \forall A \in \mathcal{F}_t, \quad a.s. . \tag{10}$$

Proof. For each $B \in \mathcal{F}_t$, we have

$$\begin{aligned}
\mathcal{E}[\mathcal{E}[Y1_A|\mathcal{F}_t]1_B] &= \mathcal{E}[Y1_A1_B] \\
&= \mathcal{E}[\mathcal{E}[Y|\mathcal{F}_t]1_{A\cap B}] \\
&= \mathcal{E}[[\mathcal{E}[Y|\mathcal{F}_t]1_A]1_B].
\end{aligned}$$

Thus (10) holds.   □

$\mathcal{E}[\cdot|\mathcal{F}_t]$ has also the monotonicity property:

Lemma 2.4. *For any $X, Y \in L^2(\mathcal{F}_T)$, if $X \leq Y$ a.s., then we have for each $t \in [0, T]$,*

$$\mathcal{E}[X|\mathcal{F}_t] \leq \mathcal{E}[Y|\mathcal{F}_t] \quad a.s.$$

In this case, if for some $t \in [0, T)$, one has $\mathcal{E}[X|\mathcal{F}_t] = \mathcal{E}[Y|\mathcal{F}_t]$, a.s., then $X = Y$, a.s..

Proof. Define $X_t = \mathcal{E}[X|\mathcal{F}_t]$ and $Y_t = \mathcal{E}[Y|\mathcal{F}_t]$, and let $A \in \mathcal{F}_t$. Because of the monotonicity of $\mathcal{E}$, we have

$$\mathcal{E}[X_t1_A] = \mathcal{E}[X1_A] \leq \mathcal{E}[Y1_A] = \mathcal{E}[Y_t1_A].$$

Now, take $A = \{X_t > Y_t\}$. If $P(A) > 0$, the strict monotonicity of $\mathcal{E}$ implies that

$$\mathcal{E}[X_t 1_A] > \mathcal{E}[Y_t 1_A].$$

Comparing the two above inequalities, we conclude that $P(A) = 0$.

Now if for some $t \in [0, T)$, one has $\mathcal{E}[X|\mathcal{F}_t] = \mathcal{E}[Y|\mathcal{F}_t]$, then $\mathcal{E}[X] = \mathcal{E}[Y]$. It follows from the strict monotonicity of $\mathcal{E}[\cdot]$ that $X = Y$, a.s.. $\qquad \square$

We then can conclude

Proposition 2.2. *Let $\mathcal{E}[\cdot]$ be defined in Definition 2.3. If for each $0 \le t \le T < \infty$ and $Y \in L^2(\mathcal{F}_T)$, there exists a $\mathcal{E}[Y|\mathcal{F}_t] \in L^2(\Omega, \mathcal{F}_t, P)$ satisfying relation (7), then $(\mathcal{E}[Y|\mathcal{F}_t])_{0 \le t < T}$ is an $\mathcal{F}_t$-consistent nonlinear expectation defined on $L^2(\mathcal{F}_T)$.*

Proof. We have already (A1), (A3) and (A4). (A2') can be checked by a similar argument. $\qquad \square$

Lemma 2.5. *For any Y, $Y' \in L^2(\mathcal{F}_T)$ and for each $t \in [0, T]$ and $A \in \mathcal{F}_t$ we have*

$$\mathcal{E}[Y1_A + Y'1_{A^C}|\mathcal{F}_t] = \mathcal{E}[Y|\mathcal{F}_t]1_A + \mathcal{E}[Y'|\mathcal{F}_t]1_{A^C} \qquad (11)$$

Proof. We have

$$\begin{aligned}
\mathcal{E}[Y1_A + Y'1_{A^C}|\mathcal{F}_t] &= \mathcal{E}[Y1_A + Y'1_{A^C}|\mathcal{F}_t]1_A + \mathcal{E}[Y1_A + Y'1_{A^C}|\mathcal{F}_t]1_{A^C} \\
&= \mathcal{E}[(Y1_A + Y'1_{A^C})1_A|\mathcal{F}_t] + \mathcal{E}[(Y1_A + Y'1_{A^C})1_{A^C}|\mathcal{F}_t] \\
&= \mathcal{E}[Y1_A|\mathcal{F}_t] + \mathcal{E}[Y'1_{A^C}|\mathcal{F}_t] \\
&= \mathcal{E}[Y|\mathcal{F}_t]1_A + \mathcal{E}[Y'|\mathcal{F}_t]1_{A^C}.
\end{aligned}$$

$\square$

Remark 2.2. (11) is equivalent to (A4'): $1_A \mathcal{E}[Y1_A|\mathcal{F}_t] = 1_A \mathcal{E}[Y|\mathcal{F}_t]$.

2.3 $\mathcal{F}_t$-Consistent Nonlinear Evaluations

Just as in Subsection 2.2, we can also introduce $\mathcal{F}_t$-consistent nonlinear evaluations in the following way:

Definition 2.5. *An evaluation is a family of nonlinear functionals parameterized by $t \in [0, T]$*

$$\mathcal{E}_{0,t}[\cdot] : L^2(\mathcal{F}_t) \longmapsto \mathbf{R}$$

which satisfies the following strict monotonicity properties: for each $t \ge 0$ and $Y_1, Y_2 \in L^2(\mathcal{F}_t)$, we have

$$\text{if} \quad Y_1 \ge Y_2 \quad \text{a.s., then} \quad \mathcal{E}_{0,t}[Y_1] \ge \mathcal{E}_{0,t}[Y_2];$$

$$\text{if} \quad Y_1 \ge Y_2 \quad \text{a.s., then} \quad \mathcal{E}_{0,t}[Y_1] = \mathcal{E}_{0,t}[Y_2] \quad \text{iff} \quad Y_1 = Y_2 \quad \text{a.s..}$$

Lemma 2.6. *For each $t \leq T$ and $\eta_1, \eta_2 \in L^2(\mathcal{F}_t)$. If*

$$\mathcal{E}_{0,t}[\eta_1 1_A] \leq \mathcal{E}_{0,t}[\eta_2 1_A], \quad \forall A \in \mathcal{F}_t,$$

then

$$\eta_1 \leq \eta_2, \quad a.s.$$

If

$$\mathcal{E}_{0,t}[\eta_1 1_A] = \mathcal{E}_{0,t}[\eta_2 1_A], \quad \forall A \in \mathcal{F}_t,$$

then

$$\eta_2 = \eta_1, \quad a.s. \tag{12}$$

Proof. To prove the first assertion, we set $A = \{\eta_1 \geq \eta_2\} \in \mathcal{F}_t$. Since $(\eta_1 - \eta_2)1_A \geq 0$, thus the monotonicity yields $\mathcal{E}_{0,t}[\eta_1 1_A] \geq \mathcal{E}_{0,t}[\eta_2 1_A]$. With $\mathcal{E}_{0,t}[\eta_1 1_A] \leq \mathcal{E}_{0,t}[\eta_2 1_A]$, it then follows from the strict monotonicity that $\eta_1 1_A = \eta_2 1_A$ a.s.. i.e., $\eta_1 \leq \eta_2$ a.s. The second assertion is a simple consequence of the first one. $\qquad\square$

We can also define $\mathcal{F}-$ evaluation operators

Definition 2.6. *A nonlinear evaluation $(\mathcal{E}_{0,t}[\cdot])_{t\in[0,T]}$ defined on $L^2(\mathcal{F}_T)$ is called $\mathcal{F}$-evaluation if for each $0 \leq s \leq t \leq T$ and $Y \in L^2(\mathcal{F}_t)$ there exists a random variable $\eta \in L^2(\mathcal{F}_s)$, such that*

$$\mathcal{E}_{0,t}[Y 1_A] = \mathcal{E}_{0,s}[\eta 1_A], \quad \forall A \in \mathcal{F}_s.$$

From Lemma 2.6, such η is uniquely defined. We denote it by $\eta = \mathcal{E}_{s,t}[Y]$. $\mathcal{E}_{s,t}[\cdot]$ satisfies

$$\mathcal{E}_{0,t}[Y 1_A] = \mathcal{E}_{0,s}[\mathcal{E}_{s,t}[Y] 1_A], \quad \forall A \in \mathcal{F}_s. \tag{13}$$

We can prove that $(\mathcal{E}_{s,t}[\cdot])_{0 \leq s \leq t \leq T}$ is the $\mathcal{F}_t-$consistent nonlinear evaluation defined on $L^2(\mathcal{F}_T)$. We first check (A3):

Lemma 2.7. *For each $0 \leq r \leq s \leq t \leq T$ and $Y \in L^2(\mathcal{F}_t)$, we have*

$$\mathcal{E}_{r,s}[\mathcal{E}_{s,t}[Y]] = \mathcal{E}_{r,t}[Y] \quad a.s. \tag{14}$$

In particular,

$$\mathcal{E}_{0,s}[\mathcal{E}_{s,t}[Y]] = \mathcal{E}_{0,t}[Y] \quad a.s.. \tag{15}$$

Proof. Since $A \in \mathcal{F}_r \subset \mathcal{F}_s$. Thus by (13),

$$\begin{aligned} \mathcal{E}_{0,r}[\mathcal{E}_{r,s}[\mathcal{E}_{s,t}[Y]] 1_A] &= \mathcal{E}_{0,s}[\mathcal{E}_{s,t}[Y] 1_A] \\ &= \mathcal{E}_{0,t}[Y 1_A] \\ &= \mathcal{E}_{0,r}[\mathcal{E}_{r,t}[Y] 1_A]. \end{aligned}$$

It follows from Lemma 2.6 that (14) holds.

Let $r = 0$, (15) follows then easily from the fact that $\mathcal{F}_0$ is the trivial σ-algebra (since $B_0 = 0$). $\qquad\square$

We then check (A4):

Lemma 2.8. *For each $0 \leq s \leq t$, $Y \in L^2(\mathcal{F}_t)$ and $A \in \mathcal{F}_s$, we have*

$$\mathcal{E}_{s,t}[Y 1_A] = \mathcal{E}_{s,t}[Y] 1_A, \quad a.s.. \tag{16}$$

Proof. For each $0 \leq s \leq t$ and $B \in \mathcal{F}_s$, we have, by (13),

$$\begin{aligned}
\mathcal{E}_{0,s}[\mathcal{E}_{s,t}[Y 1_A] 1_B] &= \mathcal{E}_{0,t}[Y 1_A 1_B] \\
&= \mathcal{E}_{0,s}[\mathcal{E}_{s,t}[Y] 1_{A \cap B}] \\
&= \mathcal{E}_{0,s}[[\mathcal{E}_{s,t}[Y] 1_A] 1_B].
\end{aligned}$$

It follows from Lemma 2.6 that (16) holds. $\square$

We also have (A2):

Lemma 2.9. *For each $0 \leq t < T$, and $\eta \in L^2(\mathcal{F}_t)$, we have*

$$\mathcal{E}_{t,t}[\eta] = \eta, \quad a.s..$$

Proof. By (13) we have

$$\mathcal{E}_{0,t}[\mathcal{E}_{t,t}[\eta] 1_A] = \mathcal{E}_{0,t}[\eta 1_A], \quad \forall A \in L^2(\mathcal{F}_t).$$

$\square$

We then can conclude

Proposition 2.3. *Let $(\mathcal{E}_{0,t}[\cdot])_{t \in [0,T]}$ be a nonlinear evaluation defined on $L^2(\mathcal{F}_T)$. If for each $0 \leq s \leq t \leq T$ and $Y \in L^2(\mathcal{F}_t)$, there exists an $\mathcal{E}_{s,t}[Y] \in L^2(\mathcal{F}_s)$ satisfying relation (13), then $(\mathcal{E}_{s,t}[Y])_{0 \leq s \leq t < T}$ satisfies the Axiomatic assumptions (A1)–(A4) listed in Definition 2.1.*

Proof. The above three lemmas have proved (A2)–(A4). (A1) is a direct consequence of the first assertion of Lemma 2.6. $\square$

Moreover, we can prove the following strict monotonicity $\mathcal{E}_{s,t}[\cdot]$.

Lemma 2.10. *For each $0 \leq s \leq t \leq T$ and $X, Y \in L^2(\mathcal{F}_t)$ such that $X \leq Y$ a.s., if $\mathcal{E}_{s,t}[X] = \mathcal{E}_{s,t}[Y]$, a.s, then $X = Y$ a.s..*

Proof. Since $\mathcal{E}_{s,t}[X] = \mathcal{E}_{s,t}[Y]$, thus

$$\begin{aligned}
\mathcal{E}_{0,t}[X] &= \mathcal{E}_{0,s}[\mathcal{E}_{s,t}[X]] = \mathcal{E}_{0,s}[\mathcal{E}_{s,t}[Y]] \\
&= \mathcal{E}_{0,t}[Y].
\end{aligned}$$

It follows from the strict monotonicity of $\mathcal{E}[\cdot]$ that $X = Y$, a.s.. $\square$

We also have the following properties

Lemma 2.11. *For each* $0 \leq s \leq t \leq T$, X, $Y \in L^2(\mathcal{F}_t)$ *and* $A \in \mathcal{F}_s$, *we have*

$$\mathcal{E}_{s,t}[X1_A + Y1_{A^c}] = \mathcal{E}_{s,t}[X]1_A + \mathcal{E}_{s,t}[Y]1_{A^c}.$$

Proof. According to Lemma 2.8,

$$\begin{aligned}
\mathcal{E}_{s,t}[X1_A + Y1_{A^c}] &= \mathcal{E}_{s,t}[X1_A + Y1_{A^c}]1_A + \mathcal{E}_{s,t}[X1_A + Y1_{A^c}]1_{A^c} \\
&= \mathcal{E}_{s,t}[(X1_A + Y1_{A^c})1_A] + \mathcal{E}_{s,t}[(X1_A + Y1_{A^c})1_{A^c}] \\
&= \mathcal{E}_{s,t}[X1_A] + \mathcal{E}_{s,t}[Y1_{A^c}] \\
&= \mathcal{E}_{s,t}[X]1_A + \mathcal{E}_{s,t}[Y]1_{A^c}.
\end{aligned}$$

$\square$

3 Backward Stochastic Differential Equations: g–Evaluations and g–Expectations

3.1 BSDE: Existence, Uniqueness and Basic Estimates

BSDE Theory plays a central role in this lecture. A lot of $\mathcal{F}_t$–consistent non-linear expectations and evaluations are derived by BSDEs. We first consider the following form of BSDE:

$$Y_t = \xi + \int_t^T g(s, Y_s, Z_s)ds - \int_t^T Z_s dB_s. \tag{17}$$

The setting of our problem is somewhat unusual: to find a pair of $\mathcal{F}_t$-adapted processes $(Y, Z) \in L^2_{\mathcal{F}}(0, T; \mathbf{R}^m \times \mathbf{R}^{m \times d})$ satisfying BSDE (17).

Remark 3.1. The solution Y is an ordinary Itô's process:

$$Y_t = Y_0 - \int_0^t g(s, Y_s, Z_s)ds + \int_0^t Z_s dB_s.$$

To prove the existence and uniqueness of BSDE (17) we first consider a very simple case: g is a real valued process that is independent of the variable (y, z). We have

Lemma 3.1. *For a fixed* $\xi \in L^2(\mathcal{F}_T)$ *and* $g_0(\cdot)$ *satisfying*

$$\mathbf{E}\left(\int_0^T |g_0(t)|dt\right)^2 < \infty,$$

there exists a unique pair of processes $(y_\cdot, z_\cdot) \in L^2_{\mathcal{F}}(0, T; \mathbf{R}^{1+d})$, *satisfies the following BSDE*

$$y_t = \xi + \int_t^T g_0(s)ds - \int_t^T z_s dB_s. \tag{18}$$

If $g_0(\cdot) \in L^2_{\mathcal{F}}(0,T)$, then $(y_\cdot, z_\cdot) \in S^2_{\mathcal{F}}(0,T) \times L^2_{\mathcal{F}}(0,T;\mathbf{R}^d)$. We have the following basic estimate:

$$|y_t|^2 + \mathbf{E}^{\mathcal{F}_t} \int_t^T [\frac{\beta}{2}|y_s|^2 + |z_s|^2] e^{\beta(s-t)} ds \tag{19}$$

$$\leq \mathbf{E}^{\mathcal{F}_t} |\xi|^2 e^{\beta(T-t)} + \frac{2}{\beta} \mathbf{E}^{\mathcal{F}_t} \int_t^T |g_0(s)|^2 e^{\beta(s-t)} ds$$

In particular

$$|y_0|^2 + \mathbf{E} \int_0^T [\frac{\beta}{2}|y_s|^2 + |z_s|^2] e^{\beta s} ds \tag{20}$$

$$\leq \mathbf{E}|\xi|^2 e^{\beta T} + \frac{2}{\beta} \mathbf{E} \int_0^T |g_0(s)|^2 e^{\beta s} ds,$$

where β is an arbitrary constant. We also have

$$\mathbf{E}[\sup_{0 \leq t \leq T} |y_t|^2] \leq c\, \mathbf{E}[|\xi|^2 + \int_0^T |g_0(s)|^2 ds], \tag{21}$$

where the constant c depends only on T.

Proof. We define

$$M_t = \mathbf{E}^{\mathcal{F}_t}[\xi + \int_0^T g_0(s)ds].$$

M is a square integrable $(\mathcal{F}_t)$-martingale. By representation theorem of Brownian martingale (see Lemma 7.1), there exists a unique adapted process $(z_t) \in L^2_{\mathcal{F}}(0,T;\mathbf{R}^d)$ such that

$$M_t = M_0 + \int_0^t z_s dB_s.$$

Thus

$$M_t = M_T - \int_t^T z_s dB_s.$$

We denote

$$y_t = M_t - \int_0^t g_0(s)ds = M_T - \int_0^t g_0(s)ds - \int_t^T z_s dB_s.$$

Since $M_T = \xi + \int_0^T g_0(s)ds$, we have immediately (18).

The uniqueness is a simple consequence of the estimate (20). We only need to prove these two estimates. To prove (19), we first consider the case where ξ and $g_0(\cdot)$ are both bounded. Since

$$y_t = \mathbf{E}^{\mathcal{F}_t}\left[\xi + \int_t^T g_0(s)ds\right]$$

thus the process y is also bounded. We then apply Itô's formula to $|y_s|^2 e^{\beta s}$ for $s \in [t, T]$:

$$
|y_t|^2 e^{\beta t} + \int_t^T [\beta |y_s|^2 + |z_s|^2] e^{\beta s} ds
$$

$$
= |\xi|^2 e^{\beta T} + \int_t^T 2y_s g_0(s) e^{\beta s} ds - \int_t^T e^{\beta s} 2y_s z_s dB_s.
$$

We take conditional expectation under $\mathcal{F}_t$ on both sides of the above relation:

$$
|y_t|^2 e^{\beta t} + \mathbf{E}^{\mathcal{F}_t} \int_t^T [\beta |y_s|^2 + |z_s|^2] e^{\beta s} ds
$$

$$
= \mathbf{E}^{\mathcal{F}_t} |\xi|^2 e^{\beta T} + \mathbf{E}^{\mathcal{F}_t} \int_t^T 2y_s g_0(s) e^{\beta s} ds.
$$

Thus

$$
|y_t|^2 + \mathbf{E}^{\mathcal{F}_t} \int_t^T [\beta |y_s|^2 + |z_s|^2] e^{\beta (s-t)} ds
$$

$$
= \mathbf{E}^{\mathcal{F}_t} |\xi|^2 e^{\beta (T-t)} + \mathbf{E}^{\mathcal{F}_t} \int_t^T 2y_s g_0(s) e^{\beta (s-t)} ds
$$

$$
\leq \mathbf{E}^{\mathcal{F}_t} |\xi|^2 e^{\beta (T-t)} + \mathbf{E}^{\mathcal{F}_t} \int_t^T [\frac{\beta}{2} |y_s|^2 + \frac{2}{\beta} |g_0(s)|^2] e^{\beta (s-t)} ds.
$$

From this it follows (19) and (20).

We now consider the case where ξ and $g_0(\cdot)$ are possibly unbounded. We set

$$
\xi^n := (\xi \wedge n) \vee (-n), \qquad g_0^n(s) := (g_0(s) \wedge n) \vee (-n)
$$

and

$$
y_t^n := \xi^n + \int_t^T g_0^n(s) ds - \int_t^T z_s^n dB_s. \tag{22}
$$

We observe that, for each positive integers n and k, ξ^n, ξ^k, g_0^n as well as g_0^k are all bounded. We thus have

$$
|y_t^n|^2 e^{\beta t} + \mathbf{E}^{\mathcal{F}_t} \int_t^T [\beta |y_s^n|^2 + |z_s^n|^2] e^{\beta s} ds \tag{23}
$$

$$
= \mathbf{E}^{\mathcal{F}_t} |\xi^n|^2 e^{\beta T} + \mathbf{E}^{\mathcal{F}_t} \int_t^T 2y_s^n g_0^n(s) e^{\beta s} ds
$$

and

$$
|y_t^n|^2 + \mathbf{E}^{\mathcal{F}_t} \int_t^T [\frac{\beta}{2} |y_s^n|^2 + |z_s^n|^2] e^{\beta (s-t)} ds \tag{24}
$$

$$
\leq \mathbf{E}^{\mathcal{F}_t} |\xi^n|^2 e^{\beta (T-t)} + \frac{2}{\beta} \mathbf{E}^{\mathcal{F}_t} \int_t^T |g_0^n(s)|^2 e^{\beta (s-t)} ds
$$

as well as

$$\mathbf{E}\int_0^T [\frac{\beta}{2}|y_s^n - y_s^k|^2 + |z_s^n - z_s^k|^2]e^{\beta s}ds$$

$$\leq \mathbf{E}|\xi^n - \xi^k|^2 e^{\beta T} + \frac{2}{\beta}\mathbf{E}\int_0^T |g_0^n(s) - g_0^k(s)|^2 e^{\beta s}ds.$$

The last inequality implies that both $\{y^n\}$ and $\{z^n\}$ are Cauchy sequences in $L_{\mathcal{F}}^2(0,T)$. Thus (19) is proved by letting n tends to ∞ in (24).

It is clear that the solution y has continuous paths. (21) is a simple consequence of (20) together with B-D-G inequality applied to (18). Thus $y \in S_{\mathcal{F}}^2(0,T)$.   □

Remark 3.2. By passing to the limit in both sides of (22) as $n \to \infty$, we also have the relation

$$|y_t|^2 e^{\beta t} + \mathbf{E}^{\mathcal{F}_t}\int_t^T [\beta|y_s|^2 + |z_s|^2]e^{\beta s}ds \tag{25}$$

$$= \mathbf{E}^{\mathcal{F}_t}|\xi|^2 e^{\beta T} + \mathbf{E}^{\mathcal{F}_t}\int_t^T 2y_s g_0(s)e^{\beta s}ds$$

and, in particular,

$$|y_0|^2 + \mathbf{E}\int_0^T [\beta|y_s|^2 + |z_s|^2]e^{\beta s}ds \tag{26}$$

$$= \mathbf{E}|\xi|^2 e^{\beta T} + \mathbf{E}\int_0^T 2y_s g_0(s)e^{\beta s}ds.$$

With the above basic estimates we can consider the general case of BSDE (17). We assume that

$$g = g(\omega, t, y, z) : \Omega \times [0,T] \times \mathbf{R}^m \times \mathbf{R}^{m \times d} \to \mathbf{R}^m$$

satisfies the following conditions: for each $(y,z) \in \mathbf{R}^m \times \mathbf{R}^{m \times d}$, $g(\cdot, y, z)$ is an $\mathbf{R}^m$-valued and $\mathcal{F}_t$-adapted process satisfying the Lipschitz condition in (y,z), i.e., for each $y, y' \in \mathbf{R}^m$ and $z, z' \in \mathbf{R}^{m \times d}$

$$|g(t,y,z) - g(t,y',z')| \leq C(|y - y'| + |z - z'|). \tag{27}$$

We also assume

$$g(\cdot, 0, 0) \in L_{\mathcal{F}}^2(0,T). \tag{28}$$

The following is the basic result of BSDE: the existence and uniqueness theorem.

Theorem 3.1. *Assume that g satisfies (28) and (27). Then for any given terminal condition $\xi \in L^2(\mathcal{F}_T; \mathbf{R}^m)$, the BSDE (17) has a unique solution, i.e., there exists a unique pair of $\mathcal{F}_t$-adapted processes $(Y,Z) \in S_{\mathcal{F}}^2(0,T;\mathbf{R}^m) \times L_{\mathcal{F}}^2(0,T;\mathbf{R}^{m \times d})$ satisfying (17).*

Proof. In the basic estimate (19) we fix $\beta = 8(1 + C^2)$, where C is the Lipschitz constant of g in (y, z). To this β, we introduce a norm in the Hilbert space $L^2_{\mathcal{F}}(0, T; \mathbf{R}^n)$:

$$\|v(\cdot)\|_\beta \equiv \{\mathbf{E} \int_0^T |v_s|^2 e^{\beta s} ds\}^{\frac{1}{2}}.$$

Clearly this is equivalent to the original norm of $L^2_{\mathcal{F}}(0, T; \mathbf{R}^n)$. But this norm is more convenient to construct a contraction mapping in order to apply the fixed point theorem. We thus set

$$Y_t = \xi + \int_t^T g(s, y_s, z_s) ds - \int_t^T Z_s dB_s$$

We define a mapping

$$I[(y., z.)] := (Y., Z.) : L^2_{\mathcal{F}}(0, T; \mathbf{R}^m \times \mathbf{R}^{m \times d}) \to L^2_{\mathcal{F}}(0, T; \mathbf{R}^m \times \mathbf{R}^{m \times d}).$$

We need to prove that I is a contraction mapping under the norm $\| \cdot \|_\beta$. For any two elements (y, z) and (y', z') in $L^2_{\mathcal{F}}(0, T; \mathbf{R}^m \times \mathbf{R}^{m \times d})$ we set

$$(Y, Z) = I[(y, z)], \qquad (Y', Z') = I[(y', z')],$$

and denote their differences by $(\hat{y}, \hat{z}) = (y - y', z - z')$, $(\hat{Y}, \hat{Z}) = (Y - Y', Z - Z')$. By the basic estimate (20) we have

$$\mathbf{E} \int_0^T (\frac{\beta}{2}|\hat{Y}_s|^2 + |\hat{Z}_s|^2) e^{\beta s} ds \le \frac{2}{\beta} \mathbf{E} \int_0^T |g(s, y_s, z_s) - g(s, y'_s, z'_s)|^2 e^{\beta s} ds.$$

Since g satisfies Lipschitz condition, we have

$$\mathbf{E} \int_0^T [\frac{\beta}{2}|\hat{Y}_s|^2 + |\hat{Z}_s|^2] e^{\beta s} ds \le \frac{4C^2}{\beta} \mathbf{E} \int_0^T [|\hat{y}_s|^2 + |\hat{z}_s|^2] e^{\beta s} ds.$$

Since $\beta = 8(1 + C^2)$, thus

$$\mathbf{E} \int_0^T [|\hat{Y}_s|^2 + |\hat{Z}_s|^2] e^{\beta s} ds \le \frac{1}{2} \mathbf{E} \int_0^T [|\hat{y}_s|^2 + |\hat{z}_s|^2] e^{\beta s} ds,$$

or

$$\|(\hat{Y}, \hat{Z})\|_\beta \le \frac{1}{\sqrt{2}} \|(\hat{y}, \hat{z})\|_\beta.$$

Thus I is a strict contraction mapping of $L^2_{\mathcal{F}}(0, T; \mathbf{R}^m \times \mathbf{R}^{m \times d})$. It follows by the fixed point theorem that BSDE (17) has a unique solution. $(Y, Z) \in L^2_{\mathcal{F}}(0, T; \mathbf{R}^m \times \mathbf{R}^{m \times d})$. It then follows from (28) and (27) that $g(\cdot, Y., Z.) \in L^2_{\mathcal{F}}(0, T)$. Thus by Lemma 3.1 $Y \in S^2_{\mathcal{F}}(0, T)$. $\square$

The basic estimates (19) and (20) can also be applied to prove the continuous dependence theorem of BSDE (17) with respect to parameters. Let (Y^1, Z^1) and (Y^2, Z^2) be respectively the solution of the following two BSDEs:

$$Y_t^1 = \xi^1 + \int_t^T [g(s, Y_s^1, Z_s^1) + \varphi^1{}_s]ds - \int_t^T Z_s^1 dB_s. \tag{29}$$

$$Y_t^2 = \xi^2 + \int_t^T [g(s, Y_s^2, Z_s^2) + \varphi^2{}_s]ds - \int_t^T Z_s^2 dB_s. \tag{30}$$

Here the terminal condition ξ^1 and ξ^2 are given elements in $L^2(\mathcal{F}_T; \mathbf{R}^m)$ and φ^1 and φ^2 are two given processes in $L^2_{\mathcal{F}}(0, T; \mathbf{R}^m)$. Let g be the same as in Theorem 3.1. Analogue to the previous method, using Itô's formula applied to $|Y_s^1 - Y_s^2|^2 e^{\beta(s-t)}$ in the interval $[t, T]$, we can obtain the following estimate.

Theorem 3.2. *The difference of the solutions of BSDE (29) and (30) satisfies*

$$|Y_t^1 - Y_t^2|^2 + \frac{1}{2}\mathbf{E}^{\mathcal{F}_t} \int_t^T [|Y_s^1 - Y_s^2|^2 + |Z_s^1 - Z_s^2|^2]e^{\beta(s-t)}ds \tag{31}$$

$$\leq \mathbf{E}^{\mathcal{F}_t}|\xi^1 - \xi^2|^2 e^{\beta(T-t)} + \mathbf{E}^{\mathcal{F}_t} \int_t^T |\varphi_s^1 - \varphi_s^2|e^{\beta(s-t)}ds,$$

where $\beta = 16(1 + C^2)$. We also have

$$\mathbf{E}[\sup_{0 \leq t \leq T} |Y_t^1 - Y_t^2|^2] \leq c\mathbf{E}[|\xi^1 - \xi^2|^2] + c\mathbf{E}\int_0^T |\varphi_s^1 - \varphi_s^2|^2 ds. \tag{32}$$

In particular, when $\varphi_s^1 \equiv 0$, (set $\xi^2 = 0$, $\varphi_s^2 = -g(s, 0, 0)$),

$$\mathbf{E}[\sup_{0 \leq t \leq T} |Y_t^1|^2] \leq c\mathbf{E}[|\xi^1|^2] + c\mathbf{E}\int_0^T |g(s, 0, 0)|^2 ds. \tag{33}$$

where the constant c depends only on the Lipschitz constant of g and T.

Proof. We apply estimate (19) to $(y_t, z_t) = (Y_t^1 - Y_t^2, Z_t^1 - Z_t^2)$:

$$|y_t|^2 + \mathbf{E}^{\mathcal{F}_t} \int_t^T [\frac{\beta}{2}|y_s|^2 + |z_s|^2]e^{\beta(s-t)}ds$$

$$\leq \mathbf{E}^{\mathcal{F}_t}|\xi|^2 e^{\beta(T-t)} + \frac{2}{\beta}\mathbf{E}^{\mathcal{F}_t} \int_t^T |\hat{g}(s)|^2 e^{\beta(s-t)}ds,$$

where $\hat{g}(s) := g(s, Y_s^1, Z_s^1) - g(s, Y_s^2, Z_s^2) + \varphi_s^1 - \varphi_s^2$. This with $|\hat{g}(s)| \leq C(|y_t| + |z_t|) + |\varphi_s^1 - \varphi_s^2|$, yields (32). This estimate with (21) yields (33). □

For a fixed $t_0 \in [0, T]$, we denote

$$\mathcal{F}_t^{t_0} = \sigma\{\sigma(B_s - B_{t_0}; t_0 \leq s \leq t) \cup \mathcal{N}\}, \quad t \in [t_0, T].$$

The following is a simple corollary of the uniqueness of BSDE (17).

Proposition 3.1. *We still assume that g satisfies Assumptions (28) and (27). If moreover, for a fixed $t_0 \in [0,T]$ and for each $(y,z) \in \mathbf{R}^m \times \mathbf{R}^{m \times d}$, the process $g(\cdot, y, z)$ is $(\mathcal{F}_t^{t_0})$-adapted on the interval $[t_0, T]$ and $\xi \in L^2(\Omega, \mathcal{F}_T^{t_0}, P; \mathbf{R}^m)$. Then the solution $(Y., Z.)$ of BSDE (17) is also $(\mathcal{F}_t^{t_0})$-adapted on $[t_0, T]$. In particular, Y_{t_0} and Z_{t_0} are deterministic.*

Proof. Let (Y', Z') be the solution of $(\mathcal{F}_t^{t_0})$-adapted solution, on the interval $[t_0, T]$ of the BSDE

$$Y_t' = \xi + \int_t^T g(s, Y_s', Z_s')ds - \int_t^T Z_s' dB_s^0,$$

where we denote $B_t^0 \equiv B_t - B_{t_0}$. Observe that $(B_t^0)_{t_0 \leq s \leq T}$ is an $(\mathcal{F}_t^{t_0})$ - Brownian Motion on $[t_0, T]$. On the other hand the same processes $(Y_t', Z_t')_{t_0 \leq t \leq T}$ is also $\mathcal{F}_t$-adapted and

$$\int_t^T Z_s' dB_s = \int_t^T Z_s' dB_s^0, \quad t \in [t_0, T].$$

Thus from the uniqueness result of Theorem 3.1, The solution (Y, Z) of BSDE (17) coincides with (Y', Z') on $[t_0, T]$. Thus (Y, Z) is $(\mathcal{F}_t^{t_0})$-adapted. $\square$

Remark 3.3. A special situation of BSDE (17) is when ξ is deterministic and $g(t, y, z)$ is a deterministic function of (t, y, z). In this case the solution of BSDE (17) is simply $(Y., Z.) \equiv (Y_0(\cdot), 0)$, where $Y_0(\cdot)$ is the solution of the following ordinary differential equation defined on $[0, T]$:

$$-\dot{Y}_0(t) = g(t, Y_0(t), 0), \quad Y_0(T) = \xi.$$

3.2 1–Dimensional BSDE

We will see that each standard 1–dimensional BSDE on $[0, T]$ induces an $\mathcal{F}_t$-consistent evaluation, called g–evaluation, where $g = g(t, y, z)$ is the generator of the corresponding BSDE which is a simple real valued function. If (and only if) g satisfies $g(t, y, 0) \equiv 0$, then the corresponding $\mathcal{F}_t$-consistent evaluation becomes an $\mathcal{F}_t$-consistent expectation. We also notice that the present state of art of mathematical finance corresponds mostly to $m = 1$. It also covers many linear or nonlinear parabolic and elliptic PDEs. In fact $m > 1$ corresponds to systems of PDEs.

The function g is defined as follows

$$g(\omega, t, y, z) : \Omega \times [0, T] \times \mathbf{R} \times \mathbf{R}^d \longmapsto \mathbf{R}.$$

We assume, for each $y, y' \in \mathbf{R}$, $z, z' \in \mathbf{R}^d$, $t \in [0, T]$, g satisfies

$$
\begin{cases}
\text{(i)} & g(\cdot, y, z) \in L^2_{\mathcal{F}}(0, T), \quad \text{for each } y \in \mathbf{R}, \ z \in \mathbf{R}^d; \\
\text{(ii)} & |g(t, y_1, z_1) - g(t, y_2, z_2)| \leq \nu|y_1 - y_2| + \mu|z_1 - z_2|; \\
& \text{and } \mathbf{one} \text{ of the following three conditions} \\
\text{(iii)} & g(\cdot, y, z)|_{y=0, z=0} \equiv 0; \\
\text{(iii')} & g(\cdot, y, 0) \equiv 0; \\
\text{(iii'')} & g \text{ is independent of } y \text{ and } g(\cdot, z)|_{z=0} \equiv 0.
\end{cases}
\tag{34}
$$

where μ, ν are given non negative constants. It is clear that (iii") $\Rightarrow$ (iii') $\Rightarrow$ (iii).

Comparison Theorem

We first present an important property: The **Comparison Theorem** of BSDE. We will present this theorem in the case where the solution Y is possibly a RCLL (right continuous with left limit) process i.e., P-almost all of its paths of $Y.(\omega)$ are right continuous with left limit. An RCLL process $(A_t(\omega))_{t \in [0,T]}$ is called an increasing process if P-almost all of its paths are non-decreasing with $A_0(\omega) = 0$.

We first consider the following problem:
to find a solution $(Y, Z) \in L^2_{\mathcal{F}}(0, T; \mathbf{R}^{1+d})$ of the following BSDE

$$
Y_t = \xi + \int_t^T g(s, Y_s, Z_s)ds + (V_T - V_t) - \int_t^T Z_s dB_s,
\tag{35}
$$

The following is a simple corollary of Theorem 3.1.

Proposition 3.2. *We assume (34)–(i), (ii). Then, for each $\xi \in L^2(\mathcal{F}_T)$ and $V \in D^2_{\mathcal{F}}(0, T)$, there exists a unique solution $(Y, Z) \in D^2_{\mathcal{F}}(0, T) \times L^2_{\mathcal{F}}(0, T; \mathbf{R}^d)$ of the BSDE (35). Moreover $Y + V \in S^2_{\mathcal{F}}(0, T)$.*

Proof. The case $V_t \equiv 0$ corresponds to Theorem 3.1. For the general situation we let $\bar{Y}_t := Y_t + V_t$. The existence and uniqueness of BSDE (35) is equivalent to the solution $(\bar{Y}, Z) \in D^2_{\mathcal{F}}(0, T) \times L^2_{\mathcal{F}}(0, T; \mathbf{R}^d)$ of the following standard BSDE :

$$
\bar{Y}_t = \xi + V_T + \int_t^T g(s, \bar{Y}_s - V_s, Z_s)ds - \int_t^T Z_s dB_s.
$$

$\square$

For a given random variable

$$
\hat{\xi} \in L^2(\mathcal{F}_T), \ \hat{V} \in D^2_{\mathcal{F}}(0, T)
\tag{36}
$$

let $(\hat{Y}, \hat{Z}) \in L^2_{\mathcal{F}}(0, T; \mathbf{R}^{1+d})$ be the solution of the following BSDE

$$
\hat{Y}_t = \hat{\xi} + \int_t^T g(s, \hat{Y}_s, \hat{Z}_s)ds + (\hat{V}_T - \hat{V}_t) - \int_t^T \hat{Z}_s dB_s.
\tag{37}
$$

It is easy to prove that the difference $(Y - \hat{Y}, Z - \hat{Z})$ satisfies exactly the same estimate (31) given in Theorem 3.2. Using B–D–G inequality, we then derive the following estimate.

Proposition 3.3. *We assume (34)–(i), (ii). Then the difference of the solutions of BDSE (35) and (37) satisfies*

$$\mathbf{E}[\sup_{0\leq t\leq T} |Y_t + V_t - (\hat{Y}_t + \hat{V}_t)|^2] + \mathbf{E}\int_0^T |Z_s - \hat{Z}_s|^2 ds \leq cE|\xi + V_T - (\hat{\xi} + \hat{V}_T)|^2.$$
(38)

where the constant c depends only on T and the Lipschitz constant of g w.r.t. (y, z).

We now present

Theorem 3.3. *(Comparison Theorem of BSDE) We make the same assumption as in Proposition 3.2. Let* (Y', Z') *be the solution of the following simple BSDE*

$$Y_t' = \xi' + \int_t^T \bar{g}_s ds + V_T' - V_t' - \int_t^T Z_s' dB_s.$$
(39)

where $(\bar{g}_t)$, $(V_t') \in L^2_{\mathcal{F}}(0,T;\mathbf{R})$ *and* $\xi' \in L^2(\mathcal{F}_T)$ *are given such that*

$$\xi \geq \xi', \quad g(Y_t', Z_t', t) \geq \bar{g}_t, \quad a.s., \quad a.e.,$$
(40)

and such that $\hat{V} = V - V'$ *is an increasing process. We then have*

$$Y_t \geq Y_t', \quad a.e., \quad a.s..$$
(41)

We also have Strict Comparison Theorem: under the above conditions

$$Y_0 = Y_0' \iff \xi = \xi', \quad g(s, Y_s', Z_s') \equiv \bar{g}_s \quad and \; V_s \equiv V_s'.$$
(42)

Sketch of the Proof. We only consider the case $d = 1$ (i.e., B is a 1-dimensional Brownian Motion) and prove the case $t = 0$. The general situation is left to the reader as an exercise. We set $\hat{g}_s = g(s, Y_s', Z_s') - \bar{g}_s$ and

$$\hat{Y} = Y - Y', \; \hat{Z} = Z - Z', \; \hat{\xi} = \xi - \xi'.$$

The pair $(\hat{Y}, \hat{Z})$ can be regarded as the solution of the following linear BSDE:

$$\begin{cases} -d\hat{Y}_s = (a_s\hat{Y}_s + b_s\hat{Z}_s + \hat{g}_s)ds + d\hat{V}_s - \hat{Z}_s dB_s, \\ \hat{Y}_T = \hat{\xi}, \end{cases}$$

where

$$a_s := \begin{cases} \frac{g(s,Y_s,Z_s)-g(s,Y_s',Z_s)}{Y_s-Y_s'}, & \text{if } Y_s \neq Y_s', \\ 0, & \text{if } Y_s = Y_s', \end{cases}$$

$$b_s := \begin{cases} \frac{g(s,Y_s',Z_s)-g(s,Y_s',Z_s')}{Z_s-Z_s'}, & \text{if } Z_s \neq Z_s', \\ 0, & \text{if } Z_s = Z_s'. \end{cases}$$

Since g satisfies Lipschitz condition, thus $|a_s| \leq C$ and $|b_s| \leq C$. We set

$$Q_t := \exp \left[\int_0^t b_s dB_s - \frac{1}{2} \int_0^t |b_s|^2 ds + \int_0^t a_s ds \right].$$

We apply Itô's formula to $Q_t \hat{Y}_t$ on the interval $[0, T]$ and then take expectation:

$$\hat{Y}_0 = \mathbf{E}[\hat{Y}_T Q_T + \int_0^T Q_t \hat{g}_t dt + \int_0^T Q_t d\hat{V}_t] \geq 0.$$

From this we have $Y_0 \geq Y_0'$. This method also applies to prove $Y_t \geq Y_t'$ when $t > 0$.

By Girsanov Theorem,

$$\mathbf{E}[\hat{Y}_T Q_T + \int_0^T Q_t \hat{g}_t dt + \int_0^T Q_t d\hat{V}_t] = 0$$

if and only the following non negative quantities are zero: $\hat{Y}_t = 0$, $\hat{g}_t \equiv 0$ and $\hat{V}_T = 0$, a.s, a.e.. Thus we have the strict comparison.   □

Remark 3.4. In many situations the Comparison Theorem is applied to compare the following type of two BSDEs:

$$Y_t^1 = \xi^1 + \int_t^T [g(s, Y_s^1, Z_s^1) + c_s^1] ds - \int_t^T Z_s^1 dB_s, \tag{43}$$

and

$$Y_t^2 = \xi^2 + \int_t^T [g(s, Y_s^2, Z_s^2) + c_s^2] ds - \int_t^T Z_s^2 dB_s, \tag{44}$$

where $c^1(\cdot), c^2(\cdot) \in L_{\mathcal{F}}^2(0, T)$. In this case if we have

$$c_s^1 \geq c_s^2, \text{ a.s., a.e., } \quad \xi^1 \geq \xi^2, \text{ a.s..}$$

Then it is easy to apply Theorem 3.3 to derive $Y_t^1 \geq Y_t^2$, a.s., a.e..

Example 3.1. We consider a special case of BSDE (43) with $g(s, 0, 0) \equiv 0$. In this case if $c_s^2 \equiv 0$ and $\xi^2 = 0$, then the unique solution of BSDE (44) is $(Y_s^2, Z_s^2) \equiv 0$. It then follows from Remark 3.4 that if ξ^1 and $c^1(\cdot)$ are both non negative, then the solution Y^1 of (43) is also non negative. In this case we have also, by strict comparison,

$$Y_0^1 = 0 \iff c_s^1 \equiv 0 \text{ and } \xi^1 = 0.$$

An interpretation in finance is: If an investor want to obtain an opportunity of non negative return, i.e., $\xi^1 \geq 0$, then he must invest at the present time some nonnegative value, i.e., $Y_0^1 \geq 0$. If $\xi \geq 0$, a.s. and $\mathbf{E}[\xi^1] > 0$, then his investment has to be positive: $Y_0^1 > 0$.

Example 3.2. We assume that $g(s,0,0) \equiv 0$ and $\xi \geq 0$ with $\mathbf{E}[\xi] > 0$. Consider the following BSDE parameterized by $\lambda \in (0,\infty)$:

$$Y_t^\lambda = \lambda\xi + \int_t^T g(s, Y_s^\lambda, Z_s^\lambda)ds - \int_t^T Z_s^\lambda dB_s.$$

We can prove that

$$\lim_{\lambda\uparrow\infty} Y_0^\lambda = +\infty.$$

In fact we compare its solution with the one of the following BSDE

$$\bar{Y}_t^\lambda = \lambda\xi + \int_t^T C(-|\bar{Y}_s^\lambda| - |\bar{Z}_s^\lambda|)ds - \int_t^T \bar{Z}_s^\lambda dB_s,$$

where $C > 0$ is the Lipschitz constant of g with respect to (y, z). By Comparison Theorem, we have

(i) $Y_0^\lambda \geq \bar{Y}_0^\lambda$, for each $\lambda > 0$;
(ii) $\bar{Y}_0^1 > 0$, when $\lambda = 1$
We also observe that for each $\lambda \geq 0$, we have $\bar{Y}_t^\lambda \equiv \lambda\bar{Y}_t^1$ and $\bar{Z}_t^\lambda \equiv \lambda\bar{Z}_t^1$. From this and (i), (ii) it follows that

$$Y_0^\lambda \geq \bar{Y}_0^\lambda = \lambda\bar{Y}_0^1 \uparrow \infty.$$

Exercise 3.1. Prove that Y_0^λ is also bounded by:

$$Y_0^\lambda \leq \lambda\hat{Y}_0,$$

where $\hat{Y}_0$ is a constant.

Backward Stochastic Monotone Semigroups and g–Evaluations

We now discuss the backward semigroup property of the solution Y of a BSDE. We introduce the following definition: Given $t \leq T$ and $Y \in L^2(\mathcal{F}_t)$. We consider the following BSDE defined on the interval $[0,t]$

$$y_s = Y + \int_s^t g(r, y_r, z_r)dr - \int_s^t z_r dB_r, \quad s \in [0,t]. \tag{45}$$

Definition 3.1. *We define, for each $0 \leq s \leq t < \infty$ and $Y \in L^2(\mathcal{F}_t)$,*

$$\mathcal{E}_{s,t}^g[Y] := y_s. \tag{46}$$

The system $\mathcal{E}_{s,t}^g[\cdot] : L^2(\mathcal{F}_t) \to L^2(\mathcal{F}_s)$, $0 \leq s \leq t \leq T$ is called g–evaluation.

Remark 3.5. s and t can be also two uniformly bounded $\mathcal{F}_t$–stopping times.

Theorem 3.4. *Let the function g satisfies (i)–(iii) of (34). Then the g–evaluation $\mathcal{E}^g_{s,t}[\cdot]$ defined in (46) satisfies the Axiomatic assumptions (A1)–(A4) listed in Definition 2.1: it is an $\mathcal{F}_t$–consistent nonlinear evaluation operator. Furthermore, we have:*
(A5) *For each $Y_1, Y_2 \in L^2(\mathcal{F}_t)$*

$$\mathcal{E}^{-g_{\mu\nu}}_{s,t}[Y_1 - Y_2] \le \mathcal{E}^g_{s,t}[Y_1] - \mathcal{E}^g_{s,t}[Y_2] \le \mathcal{E}^{g_{\mu\nu}}_{s,t}[Y_1 - Y_2]. \qquad (47)$$

In particular, If g is independent of y, i.e., (iii") satisfies, then we have

$$\mathcal{E}^{-g_\mu}_{s,t}[Y_1 - Y_2] \le \mathcal{E}^g_{s,t}[Y_1] - \mathcal{E}^g_{s,t}[Y_2] \le \mathcal{E}^{g_\mu}_{s,t}[Y_1 - Y_2]. \qquad (48)$$

Here $g_{\mu,\nu}(y,z) := \nu|y| + \mu|z|$, $g_\mu(z) := \mu|z|$,ν and μ are the Lipschitz constants of g w.r.t. y and z, respectively.

Proof. (A1) is directly from Comparison Theorem. (A2) is obvious. As for (A4), we multiply the BSDE (45) by 1_A, $A \in \mathcal{F}_s$ on the interval $[s,t]$. Since $g(r,0,0) \equiv 0$, we have, for $u \in [s,t]$,

$$y_u 1_A = Y 1_A + \int_u^t 1_A g(r, y_r, z_r) dr - \int_u^t 1_A z_r dB_r$$
$$= Y 1_A + \int_u^t g(r, 1_A y_r, 1_A z_r) dr - \int_u^t 1_A z_r dB_r.$$

This implies that $(1_A y_r, 1_A z_r)_{r \in [s,t]}$ is the solution of the same backward equation with terminal condition $Y 1_A$. Thus

$$1_A \mathcal{E}^g_{s,t}[Y] = \mathcal{E}^g_{s,t}[1_A Y].$$

Thus we have (A4). (A3) simply follows from the uniqueness of BSDE, i.e., for each $s \le u \le t$, we have

$$\mathcal{E}^g_{s,t}[Y] = \mathcal{E}^g_{s,u}[y_u] = \mathcal{E}^g_{s,u}[\mathcal{E}^g_{u,t}[Y]]. \qquad (49)$$

(A5) is the direct consequence of the following proposition.   □

Proposition 3.4. *We assume that g_1 and g_2 satisfy (i)–(ii) of assumption (34). If g_1 is dominated by g_2 in the following sense*

$$g_1(t,y,z) - g_1(t,y',z') \le g_2(t, y - y', z - z'), \quad \forall y, y' \in \mathbf{R}, \ \forall z, z' \in \mathbf{R}^d, \ (50)$$

then $\mathcal{E}^{g_1}[\cdot]$ is also dominated by $\mathcal{E}^{g_2}[\cdot]$ in the following sense: for each $t > 0$ and $Y, Y' \in L^2(\mathcal{F}_t)$, we have

$$\mathcal{E}^{g_1}_{u,t}[Y] - \mathcal{E}^{g_1}_{u,t}[Y'] \le \mathcal{E}^{g_2}_{u,t}[Y - Y']. \qquad (51)$$

If g is dominated by itself, then $\mathcal{E}_g[\cdot]$ is also dominated by itself.

Proof. We consider the following three BSDEs

$$-dy_r = g_1(r, y_r, z_r)dr - z_r dB_r, \quad y_t = Y,$$
$$-dy'_r = g_1(r, y'_r, z'_r)dr - z'_r dB_r, \quad y'_t = Y'$$

and

$$-dY_r = g_2(r, Y_r, Z_r)dr - Z_r dB_r, \quad Y_t = Y - Y'.$$

We denote $(\hat{y}_r, \hat{z}_r) = (y_r - y'_r, z_r - z'_r)$ and $\hat{g}_r = g_1(r, y_r, z_r) - g_1(r, y'_r, z'_r)$

$$-d\hat{y}_r = \hat{g}_r dr - \hat{z}_r dB_r, \quad \hat{y}_t = Y - Y'.$$

Condition (50) implies $g_2(r, \hat{y}_r, \hat{z}_r) \geq \hat{g}_r$. It follows from Comparison Theorem that

$$\hat{y}_u \leq Y_u, \quad \forall u \in [0, t], \text{ a.s.}$$

By the definition of $\mathcal{E}^g[\cdot]$ it follows that (51) holds. $\qquad\square$

Example: Black–Scholes Evaluations

Consider a financial market consisting of $d + 1$ assets: a bond and d stocks. We denote by $P_0(t)$ the price of the bond and by $P_i(t)$ the price of i-th stock at time t. We assume that $P_0(\cdot)$ is the solution of the ordinary differential equation

$$dP_0(t) = r(t)P_0(t)dt, \quad P_0(0) = 1,$$

$\{P_i(\cdot)\}_{i=1}^d$ is the solution of the following SDE

$$dP_i(t) = P_i(t)[b_i(t)dt + \sum_{j=1}^d \sigma_{ij}(t)dB_t^j],$$
$$P_i(0) = p_i, \quad i = 1, \cdots, d.$$

Here r is the interest rate of the bond; $\{b_i\}_{i=1}^d$ is the rate of the expected return, $\{\sigma_{ij}\}_{i,j=1}^d$ the volatility of the stocks. We assume that r, b, σ and σ^{-1} are all $\mathcal{F}_t$–adapted and uniformly bounded processes on $[0, \infty)$. The problem is how a market evaluates an European type of derivative $\xi \in L^2(\mathcal{F}_T)$ with maturity T? To solve this problem we consider an investor who has, at a time $t \leq T$, $n_0(t)$ bonds and $n_i(t)$ i-stocks, $i = 1, \cdots, d$, i.e., he invests $n_0(t)P_0(t)$ in bond and $\pi_i(t) = n_i(t)P_i(t)$ in the i-th stock. $\pi(t) = (\pi_1(t), \cdots, \pi_d(t))$, $0 \leq t \leq T$ is an $\mathbf{R}^d$ valued, square-integrable and adapted process. We define by $y(t)$ the investor's wealth invested in the market at time t:

$$y(t) = n_0(t)P_0(t) + \sum_{i=1}^d \pi_i(t).$$

We make the so called self–financing assumption: in the period $[0, T]$, the investor does not withdraw his money from, or put some other person's money into his account y_t. Under this condition, his wealth y evolves according to

$$dy(t) = n_0(t)dP_0(t) + \sum\nolimits_{i=1}^{d} n_i(t)dP_i(t).$$

or

$$dy(t) = [r(t)y(t) + \sum\nolimits_{i=1}^{d}(b_i(t) - r(t))\pi_i(t)]dt + \sum\nolimits_{i,j=1}^{d}\sigma_{ij}(t)\pi_i(t)dB_t^j.$$

We denote

$$g(t, y, z) = -r(t)y - \sum\nolimits_{i,j=1}^{d}(b_i(t) - r(t))\sigma_{ji}^{-1}(t)z_j.$$

Then, by the change of variable $z_j(t) = \sum_{i=1}^{d}\sigma_{ij}(t)\pi_i(t)$, the above equation becomes

$$-dy(t) = g(t, y(t), z(t))dt - z(t)dB_t.$$

We observe that the function g satisfies (i) and (ii) of (34). It follows from the existence and uniqueness theorem of BSDE (Theorem 3.1) that for each derivative $\xi \in L^2(\mathcal{F}_T)$, there exists a unique solution $(y(\cdot), z(\cdot)) \in L_{\mathcal{F}}^2(0, T; \mathbf{R}^{1+d})$ with the terminal condition $y_T = \xi$. This meaning is significant: in order to replicate the derivative ξ, the investor needs and only needs to invest $y(t)$ at the present time t and then, during the time interval $[t, T]$, to perform the strategy $\pi_i(s) = \sigma_{ij}^{-1}(s)z_j(s)$. Furthermore, by Comparison Theorem of BSDE, if he wants to replicate a ξ' which is bigger than ξ, (i.e., $\xi' \geq \xi$, a.s., $P(\xi' \geq \xi) > 0$), then he must pay more. This means that no arbitrage–free strategy exists. This $y(t)$ is called the Black–Scholes price, or Black–Scholes evaluation, of ξ at the time t. We define, as in (46) $\mathcal{E}_{t,T}^g[\xi] = y_t$. We observe that the function g satisfies (i)–(iii) of condition (34). It follows from Theorem 3.4 that $\mathcal{E}_{t,T}^g[\cdot]$ satisfies (A1)–(A4) of $\mathcal{F}_t$–consistent evaluation.

g–Expectations

In this subsection we will consider a particularly interesting situation of the above stochastic semigroups: when g satisfies $g(s, y, z)|_{z=0} \equiv 0$, i.e., it satisfy (i), (ii) and (iii') in (34). In this situation $\mathcal{E}_{s,t}^g[Y]$ satisfies (A2'):

Proposition 3.5. *For each* $0 \leq s \leq t \leq T$*, and* $Y \in L^2(\mathcal{F}_s)$*, we have*

$$\mathcal{E}_{s,t}^g[Y] = Y. \tag{52}$$

Proof. We consider the solution (y, z) of (45) with the same terminal condition Y, but defined on $[s, t]$:

$$y_u = Y + \int_u^t g(r, y_r, z_r)dr - \int_u^t z_r dB_r, \quad u \in [s, t]. \tag{53}$$

We have $y_u = \mathcal{E}_{u,t}^g[Y]$. But by Assumption (34)–(iii'), it is easy to check $(y_u, z_u) \equiv (Y, 0)$. We thus have (52). $\qquad\square$

Thus we can define $\mathcal{F}_t$–consistent nonlinear expectation $\mathcal{E}_g[Y|\mathcal{F}_t]$:

Definition 3.2. *We define*

$$\mathcal{E}_g[Y] := \mathcal{E}_{0,T}^g[Y], \ \mathcal{E}_g[Y|\mathcal{F}_t] := \mathcal{E}_{t,T}^g[Y], \ Y \in L^2(\mathcal{F}_T). \tag{54}$$

$\mathcal{E}_g[Y]$ *is called g–expectation of* Y. *In particular, if* $g = \mu|z|$ *then we denote* $\mathcal{E}_g[Y] = \mathcal{E}^\mu[Y]$.

g–expectations is nonlinear but it satisfies all other properties of a classical linear expectation.

Proposition 3.6. *We assume that g satisfies (i), (ii) and (iii') in (34). Then the g–expectation* $\mathcal{E}_g[\cdot]$ *defined in (54) is an* $\mathcal{F}_t$–*consistent nonlinear expectation defined on* $L^2(\mathcal{F}_T)$. *That is, it satisfies (A1), (A2'), (A3) and (A4) listed in Definition 2.2. Moreover,* $\mathcal{E}_g[\cdot]$ *is dominated by* $\mathcal{E}^\mu[\cdot]$ *and* $\mathcal{E}^{g_{\mu\nu}}[\cdot]$ *in the following sense:*

$$-\mathcal{E}^\mu[-Y|\mathcal{F}_t] \le \mathcal{E}_g[Y|\mathcal{F}_t] \le \mathcal{E}^\mu[Y|\mathcal{F}_t], \ \forall Y \in L^2(\mathcal{F}_T). \tag{55}$$

and

$$\mathcal{E}_{t,T}^{-g_{\mu,\nu}}[Y_1 - Y_2] \le \mathcal{E}_g[Y_1|\mathcal{F}_t] - \mathcal{E}_g[Y_2|\mathcal{F}_t] \le \mathcal{E}_{t,T}^{g_{\mu,\nu}}[Y_1 - Y_2], \tag{56}$$
$$\forall Y_1, Y_2 \in L^2(\mathcal{F}_T).$$

Proof. Since $\mathcal{E}_{s,t}^g[\cdot]$ satisfies (A1), (A2'), (A3) and (A4), by Proposition 2.1, $\mathcal{E}_g[\cdot|\mathcal{F}_t]$ defined in (54) satisfies (A1), (A2'), (A3) and (A4) of $\mathcal{F}_t$–expectations.

(56) is directly by (47). (55) is proved from the comparison theorem of BSDE since $\mathcal{E}^\mu[\cdot] = \mathcal{E}_{g_\mu}[\cdot]$, with $g_\mu(z) = \mu|z| \ge g(t, y, z)$. □

Definition 3.3. *Let* $\tau \le T$ *be a stopping time. We also define*

$$\mathcal{E}_g[Y|\mathcal{F}_\tau] = \mathcal{E}_{\tau,T}^g[Y].$$

Definition 3.4. *(g-martingales) A process* $(Y_t)_{0 \le t \le T}$ *with* $E[Y_t^2] < \infty$ *for all t is called a g-martingale (resp. g-supermartingale, g-submartingale) if, for each* $s \le t \le T$, *we have*

$$\mathcal{E}_g[Y_t|\mathcal{F}_s] = Y_s, \quad (resp. \le Y_s, \ge Y_s).$$

The importance of this special setting follows from the following economically meaningful property.

Lemma 3.2. *Let the function g satisfies (i), (ii) and (iii") of (34). Then*

$$\mathcal{E}_g[Y + \eta|\mathcal{F}_t] = \mathcal{E}_g[Y|\mathcal{F}_t] + \eta, \quad \forall \eta \in L^2(\Omega, \mathcal{F}_t, P). \tag{57}$$

Proof. Consider the BSDE

$$-dy_s = g(s, z_s)ds - z_s dB_s, \ t \le s \le T,$$
$$y_T = Y.$$

We have by the definition $\mathcal{E}_g[Y|\mathcal{F}_t] = y_t$. On the other hand, it is easy to check that $(y_s', z_s') := (y_s + \eta, z_s)$, $s \in [t, T]$ solve the above equation with the terminal condition $y_T' = Y + \eta$. It then follows that

$$\mathcal{E}_g[Y + \eta|\mathcal{F}_t] = y_t' = y_t + \eta = \mathcal{E}_g[Y|\mathcal{F}_t] + \eta.$$

$\square$

Remark 3.6. Economically, (57) means that the nonlinearity of $\mathcal{E}_g[Y + \eta]$ is only due to the risky part of $Y + \eta$.

We will always write in the sequel $\mathcal{E}^\mu[Y] := \mathcal{E}_g[Y]$ for $g = \mu|z|$ and $\mathcal{E}^{-\mu}[Y] := \mathcal{E}_g[Y]$ for $g \equiv -\mu|z|$. Note that

$$\forall c > 0, \quad \mathcal{E}^\mu[cY|\mathcal{F}_t] = c\mathcal{E}^\mu[Y|\mathcal{F}_t] \tag{58}$$

and

$$\forall c < 0, \quad \mathcal{E}^\mu[cY|\mathcal{F}_t] = -c\mathcal{E}^\mu[-Y|\mathcal{F}_t].$$

An important feature of $\mathcal{E}^\mu[\cdot]$ is

Proposition 3.7. *Let g satisfy (i), (ii) and (iii') of Assumption (34), then $\mathcal{E}_g[\cdot]$ is dominated by $\mathcal{E}^\mu[\cdot]$ in the following sense, for each $t \geq 0$,*

$$\mathcal{E}_g[Y|\mathcal{F}_t] - \mathcal{E}_g[Y'|\mathcal{F}_t] \leq \mathcal{E}_{t,T}^{g_\mu, v}[Y - Y'], \forall Y, Y' \in L^2(\mathcal{F}_T). \tag{59}$$

If g is independent of y, i.e., (iii'') satisfies, then we have

$$\mathcal{E}_g[Y|\mathcal{F}_t] - \mathcal{E}_g[Y'|\mathcal{F}_t] \leq \mathcal{E}^\mu[Y - Y'|\mathcal{F}_t], \forall Y, Y' \in L^2(\mathcal{F}_T). \tag{60}$$

In particular, $\mathcal{E}^\mu[\cdot]$ is dominated by itself:

$$\mathcal{E}^\mu[Y|\mathcal{F}_t] - \mathcal{E}^\mu[Y'|\mathcal{F}_t] \leq \mathcal{E}^\mu[Y - Y'|\mathcal{F}_t], \forall Y, Y' \in L^2(\mathcal{F}_T). \tag{61}$$

Proof. We observe that $\mathcal{E}_{t,T}^{g_\mu, 0}[Y] = \mathcal{E}^\mu[Y|\mathcal{F}_t]$. Thus (59) as well as (61) are directly derived by (A5) of Theorem 3.4. $\square$

The self–domination property (61) of $\mathcal{E}^\mu[\cdot]$ permit us to defined a norm

Definition 3.5. *We define*

$$\|Y\|_\mu := \mathcal{E}^\mu[|Y|], \ Y \in L^2(\mathcal{F}_T).$$

Proposition 3.8. $\|\cdot\|_\mu$ *forms a norm in $L^2(\mathcal{F}_T)$.*

Proof. The triangle inequality $\|Y\|_\mu + \|Z\|_\mu \leq \|Y + Z\|_\mu$ follows from (61) with $t = 0$. By (58) we also have $\|cY\|_\mu = c\|Y\|_\mu$, $c \geq 0$. $\square$

Proposition 3.9. *Under $\|\cdot\|_\mu$, $\mathcal{E}_g[\cdot|\mathcal{F}_t]$ is a contraction mapping:*

$$\|\mathcal{E}_g[Y|\mathcal{F}_t] - \mathcal{E}_g[Y'|\mathcal{F}_t]\|_\mu \leq \|Y - Y'\|_\mu.$$

Proof. It is an easy consequence of (59).   □

Proposition 3.10. *For each $\mu > 0$, and $T > 0$, there exist a constant $c_{\mu,T}$ such that*

$$\mathbf{E}[|Y|] \leq \mathcal{E}^{\mu}[|Y|] \leq c_{\mu,T}(\mathbf{E}[|Y|^2])^{1/2}. \tag{62}$$

Proof. By definition,

$$\mathcal{E}^{\mu}[|Y||\mathcal{F}_t] = |Y| + \int_t^T \mu|Z_s|ds - \int_t^T Z_s dB_s \tag{63}$$

$$= |Y| + \int_t^T b_{\mu}(s)Z_s ds - \int_t^T Z_s dB_s,$$

where $b_{\mu}(s) = \mu\frac{Z_s}{|Z_s|}1_{\{|Z_s|>0\}}$. Let Q^{μ} be the solution of SDE

$$dQ_t^{\mu} = b_{\mu}(t)Q_t^{\mu}dB_t, \quad Q_0^{\mu} = 1.$$

Using Itô's formula to $Q_t^{\mu}\mathcal{E}^{\mu}[|Y||\mathcal{F}_t]$, we have

$$\mathcal{E}^{\mu}[|Y|] = \mathcal{E}^{\mu}[|Y||\mathcal{F}_0] = E[Q_T^{\mu}|Y|] \leq \{E[(Q_T^{\mu})^2]\}^{1/2} \cdot \{E[|Y|^2]\}^{1/2}.$$

But since $|b_{\mu}| \leq \mu$, there exists a constant $c_{\mu,T}$ depending only on μ and T, such that $E[(Q_T^{\mu})^2]^{1/2} \leq c_{\mu,T}$. We thus have the second inequality of (62). The first inequality is derived by taking $t = 0$ on both sides of (63) and then taking expectation.   □

We then have

Corollary 3.1. *Let T be fixed. Then the extension $L_{\mu}(\mathcal{F}_T)$ of $L^2(\mathcal{F}_T)$ under the norm $\|\cdot\|_{\mu}$ is a Banach space.*

Lemma 3.3. *We have for all $\mu > 0$ and $Y \in L^2(\mathcal{F}_T)$,*

$$\mathbf{E}[\mathcal{E}^{\mu}[Y|\mathcal{F}_t]^2] \leq e^{\mu^2(T-t)}E[Y^2].$$

Proof. By definition,

$$\mathcal{E}^{\mu}[Y|\mathcal{F}_t] = Y + \int_t^T \mu|Z_s|ds - \int_t^T Z_s dB_s.$$

Ito's formula gives

$$\mathcal{E}^{\mu}[Y|\mathcal{F}_t]^2 = Y^2 + \int_t^T 2\mu\mathcal{E}^{\mu}[Y|\mathcal{F}_s]|Z_s|ds - 2\int_t^T \mathcal{E}^{\mu}[Y|\mathcal{F}_s]Z_s dB_s - \int_t^T Z_s^2 ds.$$

Taking expectations, we deduce that

$$\mathbf{E}[\mathcal{E}^{\mu}[Y|\mathcal{F}_t]^2] = \mathbf{E}[Y^2] + \int_t^T \mathbf{E}[2\mu\mathcal{E}^{\mu}[Y|\mathcal{F}_s]|Z_s|]ds - \int_t^T \mathbf{E}[Z_s^2]ds$$

$$\leq \mathbf{E}[Y^2] + \mu^2 \int_t^T \mathbf{E}[\mathcal{E}^{\mu}[Y|\mathcal{F}_s]^2]ds$$

(because of $2ab \leq a^2 + b^2$). The claim follows then immediately from Gronwall's inequality.   □

Upcrossing Inequality of $\mathcal{E}^g$–Supermartingales and Optional Sampling Inequality

We begin with an easy upcrossing inequality which reveals the main idea to prove such kind of inequalities in nonlinear situation.

Proposition 3.11. *Let g satisfy (i), (ii), (iii') of (34) and let (Y_t) be a g - supermartin–gale on $[0, T]$. Let $0 = t_0 < t_1 < \cdots < t_n = T$, and $a < b$ be two constants. Then there exists a constant $c > 0$ such that the number $U_a^b[Y, n]$ of upcrossings of $[a, b]$ by $\{Y_{t_j}\}_{0 \le j \le n}$ satisfies*

$$\mathcal{E}^{-\mu}[U_a^b[Y, n]] \le \frac{\mathcal{E}^{\mu}[(Y_T - a)^-]}{b - a}. \tag{64}$$

Sketch of Proof. We only prove the case $d = 1$. For $j = 1, 2, \cdots n$, we consider the following BSDE

$$y_t^j = Y_{t_j} + \int_t^{t_j} g(s, y_s^j, z_s^j) ds - \int_t^{t_j} z_s^j dB_s, \quad t \in [t_{j-1}, t_j].$$

Then we define, for $s \in [t_{j-1}, t_j]$,

$$a_s^j := \begin{cases} (z_s^j)^{-1} g(s, y_s^j, z_s^j), & \text{if } z_s^j \ne 0; \\ 0, & \text{otherwise.} \end{cases}$$

and then $a_s := \sum_{j=1}^n a_s^j 1_{(t_{j-1}, t_j]}(s)$. Since g is Lipschitz in z and $g(t, y, 0) \equiv 0$, it is clear that $|a_s| \le \mu$. We also have, for each j,

$$g(s, y_s^j, z_s^j) = a_s z_s^j, \quad s \in (t_{j-1}, t_j].$$

We set

$$\frac{dQ}{dP}\Big|_{\mathcal{F}_T} := \exp\{\int_0^T a_s dB_s - \frac{1}{2} \int_0^T |a_s|^2 ds\}.$$

By Girsanov Theorem, Q is a probability measure and

$$\mathbf{E}_Q[Y_{t_j} | \mathcal{F}_{t_{j-1}}] = \mathcal{E}^g[Y_{t_j} | \mathcal{F}_{t_{j-1}}] \le Y_{t_{j-1}}, \quad j = 1, 2, \cdots n,$$

for Y is a g - supermartingale. This implies that $\{Y_{t_j}\}_{j=1}^n$ is a (discrete) Q - supermartingale. We then can apply the classical up crossing theorem ((see e.g., [HWY1992], Theorem 2.14 and 2.42))

$$\mathbf{E}_Q[U_a^b[Y, n]] \le \frac{\mathbf{E}_Q[(Y_T - a)^-]}{b - a}.$$

This with $|a_s| \le \mu$, we then can apply the comparison theorem to prove (64). $\square$

We now consider a more general situation. Let $(Y_t)_{t\in[0,T]}$ be an adapted process. For a given time sequence $t_0, t_1, t_2 \cdots$ in $[0,T]$ with $0 \le t_0 < t_1 < t_2 \cdots$, we denote $\tau_{-1} := t_0$ and

$$\tau_0 := \inf\{t_i \ge t_0; Y_{t_i} \le a\}$$
$$\tau_1 := \inf\{t_i \ge \tau_1; Y_{t_i} \ge b\}$$

$$\cdots\cdots$$

$$\tau_{2i} := \inf\{t_i \ge \tau_{2i-1}; Y_{t_i} \le a\}$$
$$\tau_{2i+1} := \inf\{t_i \ge \tau_{2i+1}; Y_{t_i} \ge b\}$$

$$\cdots\cdots$$

If $\tau_{2j-1} \le T$, sequence $(Y_{\tau_0}, \cdots Y_{\tau_{2i-1}})$ upcrosses the interval $[a,b]$ i times. We denote by $U_a^b(Y,k)$ the number of upcrossing $[a,b]$ of the sequence $(Y_{t_0}, \cdots, Y_{t_k})$. It is clear that

$$\{U_a^b(Y,k) = i\} = \{\tau_{2i-1} \le t_k < \tau_{2i+1}\}$$

We now fix an integer n. We have the following upcrossing inequality

Theorem 3.5. *Let g satisfy (i) and (ii) of (34) and let $(Y_t)_{t\in[0,T]}$ be a g-supermartingale. Then we have*

$$[U_a^b(Y,n)] \le \frac{1}{b-a} e^{2\mu(t_n-t_0)} \{\mathcal{E}^\mu[(Y_{t_n}-a)^- + \mathcal{E}^\mu[\int_{t_0}^{t_n} e^{\mu s}|g_s^0|ds] + a\mu(t_n-t_0)\}$$

$$(65)$$

where $g_s^0 := g(s,0,0)$.

Proof. We set $\tau_i^n := \tau_i \wedge t_n$, for each $i = 0,1,\cdots$, and consider the following BSDE:

$$-dy_t^i = g(t, y_t^i, z_t^i)dt - z_t^i dB_t, \quad t \in [0, \tau_{2i+1}^n],$$
$$y_{\tau_{2i+1}^n}^i = Y_{\tau_{2i+1}^n}.$$

As in the proof of Comparison Theorem, we can write

$$g(t, y_t^i, z_t^i) = \alpha_t^i y_t^i + \beta_t^i \cdot z_t^i + g(t,0,0), \quad t \in [0, \tau_{2i+1}^n],$$

with $|\alpha_s^i| \le \mu$, $|\beta_t^i| \le \mu$. For $t \in [0,T]$, we define

$$\alpha_t := \sum_{i=0}^{n} 1_{(\tau_{2i}^n, \tau_{2i+1}^n)}(t)\alpha_t^i,$$

$$\beta_t := \sum_{i=0}^{n} 1_{(\tau_{2i}^n, \tau_{2i+1}^n)}(t)\beta_t^i.$$

We then introduce a new probability Q by

$$\frac{dQ}{dP}|\mathcal{F}_T := \exp[-\frac{1}{2}\int_0^T |\beta_s|^2 ds + \int_0^T \beta_s dB_s].$$

Since Y is an $\mathcal{E}^g$–supermartingale, we have, for each $i = 0, 1, \cdots$, by Lemma 7.8,

$$Y_{\tau_{2i}^n} \geq \mathcal{E}^g_{\tau_{2i}^n, \tau_{2i+1}^n}[Y_{\tau_{2i+1}^n}]$$

$$= E_Q[Y_{\tau_{2i+1}^n} \exp(\int_{\tau_{2i}^n}^{\tau_{2i+1}^n} \alpha_s ds) + \int_{\tau_{2i}^n}^{\tau_{2i+1}^n} \exp(\int_{\tau_{2i}^n}^s \alpha_r dr) g_s^0 ds)|\mathcal{F}_{\tau_{2i}^n}] \quad (66)$$

We now estimate the term $u_i := E_Q[\exp(\int_0^{\tau_{2i+1}^n} \alpha_s ds) 1_{\{\tau_{2i+1} \leq t_n\}}]$. Since $(Y_{\tau_{2i+1}^n} - a) \geq b - a$ on $\{\tau_{2i+1} \leq t_n\}$ and $\{\tau_{2i} < t_n\} = \{\tau_{2i+1} \leq t_n\} + \{\tau_{2i} < t_n < \tau_{2i+1}\}$, we have

$$u_i \leq \frac{1}{b-a} E_Q[(Y_{\tau_{2i+1}^n} - a) \exp(\int_0^{\tau_{2i+1}^n} \alpha_s ds) I_{\{\tau_{2i+1} \leq t_n\}}]$$

$$\leq \frac{1}{b-a} E_Q[(Y_{\tau_{2i+1}^n} - a) \exp(\int_0^{\tau_{2i+1}^n} \alpha_s ds) I_{\{\tau_{2i} < t_n\}}]$$

$$+ \frac{1}{b-a} E_Q[(Y_{t_n} - a)^- \exp(\int_0^{t_n} \alpha_s ds) I_{\{\tau_{2i} < t_n < \tau_{2i+1}\}}]$$

With $\{\tau_{2i} < t_n\} \in \mathcal{F}_{\tau_{2i}}$, we apply (66) to the first term of the right side:

$$u_i \leq \frac{1}{b-a} E_Q[\{(Y_{\tau_{2i}^n} - a) I_{\{\tau_{2i} < t_n\}} + \int_{\tau_{2i}^n}^{\tau_{2i+1}^n} e^{\mu s} |g_s^0| ds\} \exp(\int_0^{\tau_{2i}^n} \alpha_s ds)]$$

$$+ \frac{a}{b-a} E_Q[|\exp(\int_0^{\tau_{2i}^n} \alpha_s ds) - \exp(\int_0^{\tau_{2i+1}^n} \alpha_s ds)|]$$

$$+ \frac{1}{b-a} e^{\mu t_n} E_Q[(Y_{t_n} - a)^- I_{\{\tau_{2i} < t_n < \tau_{2i+1}\}}]$$

Since $I_{\{\tau_{2i} < t_n\}}(Y_{\tau_{2i}^n} - a) = I_{\{\tau_{2i} < t_n\}}(Y_{\tau_{2i}} - a) \leq 0$, and

$$|\exp(\int_0^{\tau_{2i}^n} \alpha_s ds) - \exp(\int_0^{\tau_{2i+1}^n} \alpha_s ds)|$$

$$= |\int_{\tau_{2i}^n}^{\tau_{2i+1}^n} \alpha_s \exp(\int_{\tau_{2i}^n}^s \alpha_r dr) ds| \leq \mu e^{\mu(t_n - t_0)}(\tau_{2i+1}^n - \tau_{2i-1}^n),$$

we thus have

$$u_i \leq \frac{1}{b-a} e^{\mu(t_n - t_0)} E_Q[a\mu(\tau_{2i+1}^n - \tau_{2i-1}^n)$$

$$+ \int_{\tau_{2i-1}^n}^{\tau_{2i+1}^n} e^{\mu s} |g_s^0| ds + (Y_{t_n} - a)^- I_{\{\tau_{2i} < t_n < \tau_{2i+1}\}}]$$

We observe that $I_{\{\tau_{2i} < t_n < \tau_{2i+1}\}} \leq I_{\{U_a^b(Y,n)=i\}}$ and, in the expression of u_i, $\{\tau_{2i+1} \leq t_n\} = \{U_a^b(Y,n) > i\}$. Thus

$$e^{-\mu(t_n-t_0)} E_Q[I_{\{U_a^b(Y,n)>i\}}] \leq \frac{1}{b-a} e^{\mu(t_n-t_0)} \{E_Q[(Y_{t_n} - a)^- I_{\{U_a^b(Y,n)=i\}}]$$
$$+ E_Q[\int_{\tau_{2i-1}^n}^{\tau_{2i+1}^n} e^{\mu s} |g_s^0| ds] + a\mu E_Q[\tau_{2i+1}^n - \tau_{2i-1}^n]\}.$$

Summering both sides for all i yields

$$e^{-\mu(t_n-t_0)} E_Q[U_a^b(Y,n)] \leq \frac{1}{b-a} e^{\mu(t_n-t_0)} \{E_Q[(Y_{t_n} - a)^-]$$
$$+ \frac{1}{b-a} E_Q[\int_{t_0}^{t_n} e^{\mu s} |g_s^0| ds] + a\mu(t_n - t_0)\}.$$

This with $\mathcal{E}^{-\mu}[\cdot] \leq E_Q[\cdot] \leq \mathcal{E}^\mu[\cdot]$ derives the upcrossing inequality. $\square$

Remark 3.7. Since, $\mathcal{E}^{-\mu}[\cdot]_{\mu=0} = \mathcal{E}^\mu[\cdot]_{\mu=0} = E[\cdot]$, thus in the case where $\mu = 0$ and $g_s^0 \equiv 0$, the about upcrossing inequality becomes a classical one:

$$(b-a)E[U_a^b(Y,n)] \leq E[(Y_{t_n} - a)^-].$$

To extend the above upcrossing inequality to denumerable sets, following (Peng, 1997 [Peng1997b]), we now extend the domain of $\mathcal{E}^g[\cdot]$ from $L^2(\mathcal{F}_T)$ to a larger space . We consider

$$L_2^0(\mathcal{F}_T) := \{X^+ \in L^0(\mathcal{F}_T), \; X^- \in L^2(\mathcal{F}_T)\}.$$

We need the following result:

Lemma 3.4. *Let $X \in L_2^0(\mathcal{F}_T)$ and let $\{X_i\}_{i=1}^\infty$ and $\{X_i'\}_{i=1}^\infty$ be two non decreasing sequences in $L^2(\mathcal{F}_T)$ such that $X_i \nearrow X$, a.s $X_i' \nearrow X$ a.s.. Then we have*

$$\lim_{i\to\infty} \mathcal{E}_g[X_i] = \lim_{i\to\infty} \mathcal{E}_g[X_i'].$$

Proof. We only need to consider the case where $X_i \geq X_i'$, a.s., for all $i = 1, 2, \cdots$. In this case

$$\lim_{i\to\infty} \mathcal{E}_g[X_i] \geq \lim_{i\to\infty} \mathcal{E}_g[X_i'].$$

On the other hand, for each fixed integer i_0, we have $X_{i_0} \wedge X_i' \nearrow X_{i_0}$ in $L^2(\mathcal{F}_T)$. It follows from the continuity of $\mathcal{E}_g[\cdot]$ in L^2 that $\lim_{i\to\infty} \mathcal{E}_g[X_i'] \geq \lim_{i\to\infty} \mathcal{E}_g[X_{i_0} \wedge X_i'] = \mathcal{E}_g[X_{i_0}]$. Thus $\lim_{i\to\infty} \mathcal{E}_g[X_i'] \geq \lim_{i\to\infty} \mathcal{E}_g[X_i]$. $\square$

Definition 3.6. *For each $X \in L_2^0(\mathcal{F}_T)$, we define*

$$\mathcal{E}_g[X] = \lim_{i\to\infty} \mathcal{E}_g[X_i],$$

where $\{X_i\}_{i=1}^\infty$ is a non decreasing sequence in $L^2(\mathcal{F}_T)$ such that $X_i \nearrow X$, a.s.

From the above lemma, the functional $\mathcal{E}_g[\cdot] : L_2^0(\mathcal{F}_T) \to R \cup \{+\infty\}$ is clearly defined. We are interested in the situation where $g = g_{-\mu}(z) = -\mu|z|$.

Lemma 3.5. *For each nonnegative* $X \in L_2^0(\mathcal{F}_T)$, *if* $\mathcal{E}^{-\mu}[X] = \mathcal{E}_{g_{-\mu}}[X] < +\infty$, *then* $X < +\infty$, dP-*a.s.*

Proof. We set $A := \{\omega \in \Omega : X(\omega) = +\infty\}$. It is clear that $\lambda 1_A \leq X$, a.s, for each $\lambda \in [0, \infty)$. Thus, by comparison theorem,

$$\mathcal{E}^{-\mu}[\lambda 1_A] \leq \mathcal{E}^{-\mu}[X], \quad \forall \lambda \in [0, \infty).$$

But we have $\mathcal{E}^{-\mu}[\lambda 1_A] = \lambda \mathcal{E}^{-\mu}[1_A]$ and, by strict comparison theorem, $\mathcal{E}^{-\mu}[1_A] > 0 \Leftrightarrow P(A) > 0$. It follows that A must be a P–zero subset. The proof is complete.   □

Let $Y = (Y_t)_{t\in[0,T]}$ be an $\mathcal{F}_t$–adapted process, $u = \{t_1, t_2, \cdots, t_n\} \subset [0, T]$ with $t_1 < \cdots < t_n$. We denote by $U_a^b(Y, u)$ the upcrossing number of $\{Y_{t_1}, \cdots, Y_{t_n}\}$. For any subset D of $[0, T]$, define

$$U_a^b(Y, D) := \sup\{U_a^b(Y, u): u \text{ is a finite subset of } D\}.$$

If D is a denumerable dense subset of $[0, T]$. Let $\{u_n\}_{n=1}^\infty$ be a sequence of finite subsets in D such that $u_n \subset u_{n+1}$ for each n with $\cup_n u_n = D$. It is clear that

$$U_a^b(Y, D) = \lim_{n \to \infty} U_a^b(Y, u_n).$$

Theorem 3.6. *We assume that* g *satisfies* (i) *and* (ii) *of* (34). *Let* $Y = (Y_t)_{t\in[0,T]}$ *be a* $\mathcal{E}^g$–*supermartingale,* D *be a denumerable dense subset of* $[0, T]$. *Then for each* $a, b \in R$, $r, s \in [0, T]$ *such that* $a < b$ *and* $r < s$, *we have*

$$\mathcal{E}^{-\mu}[U_a^b(Y, D\cap[r, s])] \leq \frac{e^{2\mu(s-r)}}{b-a}\{\mathcal{E}^\mu[(Y_s - a)^-] + \mathcal{E}^\mu[\int_r^s e^{\mu t}|g_t^0|dt] + a\mu(s-r)\},$$
(67)

where μ *is the Lipschitz constant of* g *and* $g_s^0 = g(s, 0, 0)$. *In particular*

$$\mathcal{E}^{-\mu}[U_a^b(Y, D)] \leq \frac{e^{2\mu T}}{b-a}\{\mathcal{E}^\mu[(Y_T - a)^-] + \mathcal{E}^\mu[\int_0^T e^{\mu t}|g_t^0|dt] + a\mu T\}. \quad (68)$$

Moreover, $U_a^b(Y, D) < \infty$, *a.s.*

Proof. Let $u_n = \{t_0, t_1, t_2, \cdots, t_n\}$ be defined as the above with $t_0 = r$ and $t_n = s$. Since $\{U_a^b(Y, u_n)\}_{n=1}^\infty$ is an increasing and positive sequence such that $U_a^b(Y, u_n) \in L^2(\mathcal{F}_T)$ for each n, it follows that

$$\mathcal{E}^{-\mu}[U_a^b(Y, D \cap [r, s])] = \lim_{n \to \infty} \mathcal{E}^{-\mu}[U_a^b(Y, D \cap u_n)].$$

The sequence $\{\mathcal{E}^{-\mu}[U_a^b(Y, D \cap u_n)]\}_{n=1}^\infty$ is increasing and uniformly bounded by the left hand of (65). It follows that $\mathcal{E}^{-\mu}[U_a^b(Y, D\cap[r, s])]$ and $\mathcal{E}^{-\mu}[U_a^b(Y, D)]$ are well–defined and bounded. By Lemma 3.5, $U_a^b(Y, D) < \infty$, a.s. .   □

Remark 3.8. From the above upcrossing inequality we can deduce a down-crossing inequality of a $\mathcal{E}^g$–submartingale Y. In fact, from the relation $D_a^b(Y,n) = U_{-b}^{-a}(-Y,n)$, one can directly obtain the downcrossing inequality of $D_a^b(Y,n)$ of a $\mathcal{E}^g$–submartingale Y from the corresponding upcrossing inequality of $U_{-b}^{-a}(-Y,n)$ of $\mathcal{E}^{\bar{g}}$–supermartingale $-Y$, where $\bar{g}(s,y,z) := -g(s,-y,-z)$.

From the above result, and combine the condition $E[\sup_{t\in[0,T]} |Y_t|^2] < \infty$, we have the following classical result.

Theorem 3.7. *We assume that g satisfies (i) and (ii) of (34). Let $Y = (Y_t)_{t\in[0,T]}$ be a $\mathcal{E}^g$-supermartingale, D be a denumerable dense subset of $[0,T]$. Then for almost all ω and for any $t \in [0,T]$, $\lim_{s\in D, s\searrow t} Y_s$ and $\lim_{s\in D, s\nearrow t} Y_s$ exist and are finite. Furthermore the process $(\bar{Y}_t)_{t\in[0,T]}$ defined by*

$$\bar{Y}_t := \lim_{s\in D, s\searrow t} Y_s, \, t \in [0,T)$$

is an $\mathcal{F}_t$-adapted process with $E[\sup_{0\leq t\leq T} |\bar{Y}_t|^2] < \infty$. If g also satisfies (iii) of (34), then $\bar{Y}$ is an $\mathcal{E}^g$-supermartingale.

Proof. We only need to prove that $\bar{Y}$ is an $\mathcal{E}^g$–supermartingale. The rest of the proofs can be find in, e.g., [HWY1992]. Let $s < t$, $s,t \in [0,T]$ and $s_n \in D$, $s_n < t$, $s_n \downarrow\downarrow s$, $t_n \in D$, $t_n \downarrow\downarrow t$ and $s_n \leq t_n$. Then, for $m \geq n$,

$$\mathcal{E}_{s_m,t_n}^g[Y_{t_n}] \leq Y_{s_m}.$$

We fix n and let $m \to \infty$. We have $Y_{s_m} \to \bar{Y}_s$ and, by $\mathcal{E}_{s,t_n}^g[Y_{t_n}])_{s\in[0,t_n]} \in S_{\mathcal{F}}^2(0,t_n)$, we also have $\mathcal{E}_{s_m,t_n}^g[Y_{t_n}] \to \mathcal{E}_{s,t_n}^g[Y_{t_n}]$, we derive

$$\mathcal{E}_{s,t_n}^g[Y_{t_n}] \leq \bar{Y}_s, \text{ a.s.}$$

Now let $n \to \infty$. We have $Y_{t_n} \to \bar{Y}_t$, in $L^2(\mathcal{F}_T)$. It follows that

$$|\mathcal{E}_{s,t_n}^g[Y_{t_n}] - \mathcal{E}_{s,t}^g[\bar{Y}_t]| \leq |\mathcal{E}_{s,t_n}^g[Y_{t_n}] - \mathcal{E}_{s,t_n}^g[\bar{Y}_t]| + |\mathcal{E}_{s,t_n}^g[\bar{Y}_t] - \mathcal{E}_{s,t}^g[\bar{Y}_t]|.$$

We then can apply a technique used in the estimate of (141) to prove that $|\mathcal{E}_{s,t_n}^g[Y_{t_n}] - \mathcal{E}_{s,t}^g[\bar{Y}_t]| \to 0$. Thus

$$\mathcal{E}_{s,t}^g[\bar{Y}_t] \leq \bar{Y}_s.$$

$\square$

Remark 3.9. By this proposition we can prove that, in many typical cases a g - supermartingale Y admits a RCLL modification. More details on this topic will be given in Lemma 4.8, for a more general situation. We will always take its RCLL version.

Lemma 3.6. *Let Y be an RCLL g-supermartingale on $[0,T]$ and let σ and τ be two $\mathcal{F}_t$-stopping times. Then we have*

$$\mathcal{E}_g[Y_\tau | \mathcal{F}_\sigma] \leq Y_{\tau\wedge\sigma}.$$

Proof. See Theorem 7.4.

3.3 A Monotonic Limit Theorem of BSDE

For a given stopping time $\tau \leq T < \infty$, we consider a process (y_t) the solution of the following BSDE

$$y_t = \xi + \int_{t\wedge\tau}^{\tau} g(y_s, z_s, s)ds + (A_\tau - A_{t\wedge\tau}) - \int_{t\wedge\tau}^{\tau} z_s dB_s \qquad (69)$$

where $\xi \in L^2(\mathcal{F}_\tau)$, A is a given RCLL increasing process with $\mathbf{E}[(A_\tau)^2] < \infty$. The following terms will be frequently used.

Definition 3.7. *If (y, z) is a solution of BSDE (69) then we call (y_t) a g-supersolution on $[0, \tau]$. If $A_t \equiv 0$ on $[0, \tau]$, then we call y a g-solution on $[0, \tau]$.*

We recall that a g-solution y on $[0, \tau]$ is uniquely determined if its terminal condition $y_\tau = \xi$ is given, a g-supersolution y on $[0, \tau]$ is uniquely determined if y_τ and $(A_t)_{0 \leq t \leq \tau}$ are given. If y is a g-solution and y' is a g-supersolution on $[0, \tau]$ such that $y_\tau \leq y'_\tau$ a.s., then for all stopping time $\sigma \leq \tau$ we have also $y_\sigma \leq y'_\sigma$.

Proposition 3.12. *Let y be a g-supersolution defined on an interval $[0, \tau]$. Then there is a unique $z \in L^2(0, \tau; \mathbf{R}^d)$ and a unique increasing RCLL process A on $[0, \tau]$ with $\mathbf{E}[(A_\tau)^2] < \infty$ such that the triple (y_t, z_t, A_t) satisfies (69).*

Proof. If both (y, z, A) and (y, z', A') satisfy (69), then we apply Itô's formula to $(y_t - y_t)^2 (\equiv 0)$ on $[0, \tau]$ and take expectation:

$$\mathbf{E} \int_0^{\tau} |z_s - z'_s|^2 ds + \mathbf{E}\left[\sum_{t \in (0, \tau]} (\Delta(A_t - A'_t))^2 \right] = 0.$$

Thus $z_t \equiv z'_t$. From this it follows that $A_t \equiv A'_t$. $\qquad\square$

Thus we can define

Definition 3.8. *Let y be a g–supersolution on $[0, \tau]$ and let (y, A, z) be the related unique triple in the sense of BSDE (69). Then we call (A, z) the (unique) decomposition of (y_t).*

Let us now consider the following sequence of g-supersolution $\{y^i\}_{i=1}^{\infty}$ on $[0, T]$, i.e.,

$$y_t^i = y_T^i + \int_t^T g(y_s^i, z_s^i, s)ds + (A_T^i - A_t^i) - \int_t^T z_s^i dB_s, \qquad i = 1, 2, \cdots. \quad (70)$$

Here A^i are RCLL increasing processes with $A_0^i = 0$ and $\mathbf{E}[(A_T^i)^2] < \infty$.

The following theorem shows that the limit of $\{y^i\}_{i=1}^{\infty}$ is still a g–supersolution.

Theorem 3.8. *We assume that g satisfies (i) and (ii) of Assumptions (34). For each $i = 1, 2, \cdots$, let A^i be a continuous and increasing processes with $A_0^i = 0$ and $\mathbf{E}[(A_T^i)^2] < \infty$ and (y^i, z^i) be the solution of BSDE (70). If, as $i \to \infty$, $\{y^i\}_{i=1}^{\infty}$ converges monotonically up to a process y with $\mathbf{E}[esssup_{0 \le t \le T} |y_t|^2] < \infty$. Then this limit y is still a g-supersolution, i.e., there exists $z \in L_\mathcal{F}^2(0, T; \mathbf{R}^d)$ and an RCLL increasing process A with $\mathbf{E}[(A_T)^2] < \infty$ such that*

$$y_t = y_T + \int_t^T g(y_s, z_s, s)ds + (A_T - A_t) - \int_t^T z_s dB_s, \quad t \in [0, T]. \quad (71)$$

To prove this theorem, we need the following lemma. This lemma says that both $\{z^i\}$ and $\{(A_T^i)^2\}$ are uniformly bounded in L^2:

Lemma 3.7. *Under the assumptions of Theorem 3.8, there exists a constant C that is independent of i such that*

$$\begin{array}{ll} (i) & \mathbf{E}\int_0^T |z_s^i|^2 ds \le C, \\ (ii) & \mathbf{E}[(A_T^i)^2] \le C. \end{array} \quad (72)$$

Proof. From BSDE (70), we have

$$A_T^i = y_0^i - y_T^i - \int_0^T g(y_s^i, z_s^i, s)ds + \int_0^T z_s^i dB_s$$

$$\le |y_0^i| + |y_T^i| + \int_0^T [\nu|y_s^i| + \mu|z_s^i| + |g(0, 0, s)|]ds + |\int_0^T z_s^i dB_s|.$$

We observe that $|y_t^i|$ is dominated by $|y_t^1| + |y_t|$. Thus there exists a constant, independent of i, such that

$$\mathbf{E}[\sup_{0 \le t \le T} |y_t^i|^2] \le C. \quad (73)$$

It follows that, there exists a constant C_1, independent of i, such that

$$\mathbf{E}|A_T^i|^2 \le C_1 + 2(1 + \mu^2 T)\mathbf{E}\int_0^T |z_s^i|^2 ds. \quad (74)$$

On the other hand, we use Itô's formula applied to $|y_t^i|^2$:

$$|y_0^i|^2 + \mathbf{E}\int_0^T |z_s^i|^2 ds = \mathbf{E}|y_T^i|^2 + 2\mathbf{E}\int_0^T y_s^i g(y_s^i, z_s^i, s)ds + 2\mathbf{E}\int_0^T y_s^i dA_s^i$$

The last two terms are bounded by

$$2y_s^i g(y_s^i, z_s^i, s) \le 2|y_s^i|(\nu|y_s^i| + \mu|z_s^i| + |g(0, 0, s)|)$$

$$\le 2(\nu + \mu^2)|y_s^i|^2 + \frac{1}{2}|z_s^i|^2 + |g(0, 0, s)|$$

and $2\mathbf{E}\int_0^T |y_s^i| dA_s^i \le 2[\mathbf{E}\sup_{0\le s\le T} |y_s^i|^2]^{1/2}[\mathbf{E}|A_T^i|^2]^{1/2}$. Thus

$$\mathbf{E}\int_0^T |z_s^i|^2 ds \le C + 4[\mathbf{E}\sup_{0\le s\le T} |y_s^i|^2]^{1/2}[\mathbf{E}|A_T^i|^2]^{1/2}$$

$$\le C + 16(1+\mu^2 T)\mathbf{E}[\sup_{0\le s\le T} |y_s^i|^2] + \frac{1}{4(1+\mu^2 T)}\mathbf{E}|A_T^i|^2$$

$$= C_1 + \frac{1}{4(1+\mu^2 T)}\mathbf{E}|A_T^i|^2,$$

where, from (73), the constants C and C_1 are all independent of i. This with (74) it follows that (72)–(i) and then (72)–(ii) holds true. The proof is complete. □

Combining this Lemma with Theorem 7.2 in Appendix, we can easily prove Theorem 3.8.

Proof of Theorem 3.8. In (70), we set $g_t^i := -g(y_t^i, z_t^i, t)$; Since $\{z^i\}$ is bounded in $L_{\mathcal{F}}^2(0,T;\mathbf{R}^d)$, thanks to the monotonic limit theorem of Itô processes (see Appendix: Theorem 7.2), there exists a $z \in L_{\mathcal{F}}^2(0,T;\mathbf{R}^d)$ such that, for each $p \in [0,2)$, $\{z^i\}_{i=1}^\infty$ strongly converges to z in $L_{\mathcal{F}}^p(0,T;\mathbf{R}^d)$.

As result, $\{g^i\} = \{-g(y^i, z^i, \cdot)\}$ also strongly converges in $L_{\mathcal{F}}^p(0,T;\mathbf{R}^d)$ to g^0 and

$$g^0(s) = -g(y_s, z_s, s), \quad \text{a.s., a.e.}$$

From this it follows immediately that (y,z) is the solution of the BSDE (71). The proof is complete. □

3.4 g–Martingales and (Nonlinear) g–Supermartingale Decomposition Theorem

More general than the martingales under g–expectations, we now introduce the notion of g–martingales under g–evaluations. Under this general framework, we will prove a general g–supermartingale decomposition theorem of Doob–Meyer's type.

Definition 3.9. An $\mathcal{F}_t$-progressively measurable real-valued process Y with

$$\mathbf{E}[ess\sup_{0\le t\le T} |Y_t|^2] < \infty, \quad \forall T < \infty$$

is called a g–martingale (resp. g–supermartingale, g–submartingale) on $[0,T]$ if for each $0 \le s \le t \le T$,

$$\mathcal{E}_{s,t}^g[Y_t] = Y_s, \quad (\text{resp. } \le Y_s, \quad \ge Y_s) \text{ a.s.}$$

In this subsection we will consider g–supermartingales. By Comparison Theorem of BSDE, it is easy to prove the following result

Proposition 3.13. *We assume that g satisfies (i) and (ii) of (34). Let $(A_t)_{0 \leq t < \infty}$ be an RCLL increasing (resp. decreasing) process with $\mathbf{E}[(A_T)^2] < \infty$ for each $T > 0$. Let (y, z) be the solution of the following BSDE, for each $T > 0$,*

$$y_t = y_T + \int_t^T g(y_s, z_s, s)ds + (A_T - A_t) - \int_t^T z_s dB_s, \quad t \in [0, T], \quad (75)$$

Then $(y_t)_{0 \leq t \leq T}$ is a g-supermartingale (resp. g-submartingale).

In this section we are concerned with the inverse problem: can we say that a right-continuous $\mathcal{E}^g$-supermartingale is also a $\mathcal{E}^g$-supersolution? This problem is more difficult since it is in fact a nonlinear version of Doob-Meyer Decomposition Theorem. We claim

Theorem 3.9. *We assume that g satisfies (i) and (ii) of (34). Let (Y_t) be a right-continuous g-supermartingale on $[0, T]$. Then (Y_t) is an g-supersolution: there exists a unique RCLL increasing process (A_t) with $\mathbf{E}[(A_T)^2] < \infty$, for each $T > 0$, such that (Y_t) coincides with the unique solution (y_t) of the BSDE. For each $T > 0$,*

$$y_t = Y_T + \int_t^T g(y_s, z_s, s)ds + (A_T - A_t) - \int_t^T z_s dB_s, \quad t \in [0, T], \quad (76)$$

In order to prove this theorem, we consider the following family of BSDE parameterized by $i = 1, 2, \cdots$.

$$y_t^i = Y_T + \int_t^T g(y_s^i, z_s^i, s)ds + i \int_t^T (Y_s - y_s^i)ds - \int_t^T z_s^i dB_s. \quad (77)$$

An important observation is that, for each i, y_t^i is always bounded from above by Y_t. Thus y^i is a g-supersolution on $[0, T]$:

Lemma 3.8. *We have, for each $i = 1, 2, \cdots$,*

$$Y_t \geq y_t^i, \ \forall t \in [0, T], \ a.s..$$

Proof. For a $\delta > 0$ and a given integer $i > 0$, we define

$$\sigma^{i, \delta} := \inf\{t; \ y_t^i \geq Y_t + \delta\} \wedge T.$$

If $P(\sigma^{i, \delta} < T) = 0$, for all i and δ, then the proof is done. If it is not the case, then there exist $\delta > 0$ and a positive integer i such that $P(\sigma^{i, \delta} < T) > 0$. We can then define the following stopping times

$$\tau := \inf\{t \geq \sigma^{i, \delta}; \ y_t^i \leq Y_t\}$$

It is clear that $\sigma^{i, \delta} \leq \tau \leq T$. Since $Y. - y^i$ is RCLL, we have

$$y_\tau^i \le Y_\tau.$$

But since $(Y(s) - y^i(s)) \le 0$ on $[\sigma^{i,\delta}, \tau]$, by monotonicity of $\mathcal{E}^g[\cdot]$,

$$
\begin{aligned}
y_{\sigma^{i,\delta}}^i &\le \mathcal{E}_{\sigma^{i,\delta},\tau}^g [y_\tau^i | \mathcal{F}_{\sigma^{i,\delta}}] \\
&\le \mathcal{E}_{\sigma^{i,\delta},\tau}^g [Y_\tau | \mathcal{F}_{\sigma^{i,\delta}}] \\
&\le Y_{\sigma^{i,\delta}}. \text{ a.s.}
\end{aligned}
$$

The last step is due to Theorem 7.3. But on the other hand, we have $P(\sigma^{i,\delta} < T) > 0$ and, by the definition of $\sigma^{i,\delta}$, $y_{\sigma^{i,\delta}}^i \ge Y_{\sigma^{i,\delta}} + \delta$ on $\{\sigma^{i,\delta} < T\}$. This induces a contradiction. The proof is complete.   □

Remark 3.10. From the above result, the term $i(Y_s - y_s^i)$ in (77) equals to $i(Y_s - y_s^i)^+$. By Comparison Theorem y_t^i are pushed up to be above the supermartingale Y_t. But in fact they can never surpass Y_t. We will see that this effect will force y^i to converge to the supermartingale Y itself. Thus, by Limit Theorem 3.8 Y itself is also a form of (76). Specifically, we have:

Proof of Theorem 3.9. The uniqueness is due to the uniqueness of g-supersolution i.e. Proposition 3.12. We now prove the existence. We rewrite BSDE (77) as

$$y_t^i = Y_T + \int_t^T g(y_s^i, z_s^i, s)ds + A_T^i - A_t^i - \int_t^T z_s^i dB_s,$$

where we denote

$$A_t^i := i \int_0^t (Y_s - y_s^i)ds.$$

From Lemma 3.8, $Y_t - y_t^i = |Y_t - y_t^i|$. It follows from the Comparison Theorem that $y_t^i \le y_t^{i+1}$. Thus $\{y^i\}$ is a sequence of continuous $\mathcal{E}^g$-supersolutions that is monotonically converges up to a process (y_t). Moreover (y_t) is bounded from above by Y_t. It is then easy to check that all conditions in Theorem 3.8 are satisfied. (y_t) is a $\mathcal{E}^g$-supersolution on $[0, T]$ of the following form.

$$y_t = Y_T + \int_t^T g(y_s, z_s, s)ds + (A_T - A_t) - \int_t^T z_s dB_s, \quad t \in [0, T],$$

where (A_t) is a RCLL increasing process. It then remains to prove that $y = Y$. From Lemma 3.7–(ii) we have

$$\mathbf{E}[|A_T^i|^2] = i^2 \mathbf{E}\left[\int_0^T |Y_t - y_t^i| dt\right]^2 \le C.$$

It then follows that $Y_t \equiv y_t$. The proof is complete   □ .

4 Finding the Mechanism: Is an $\mathcal{F}$–Expectation a g–Expectation?

4.1 $\mathcal{E}^\mu$-Dominated $\mathcal{F}$-Expectations

Now we will study $\mathcal{F}$-expectations dominated by $\mathcal{E}^\mu = \mathcal{E}^{g_\mu}$, with $g_\mu(z) := \mu|z|$, for some large enough $\mu > 0$, according to the following

Definition 4.1. ($\mathcal{E}^\mu$-domination) *Given $\mu > 0$, we say that an $\mathcal{F}$ - expectation $\mathcal{E}$ is dominated by $\mathcal{E}^\mu$ if*

$$\mathcal{E}[X + Y] - \mathcal{E}[X] \le \mathcal{E}^\mu[Y], \ \forall X, Y \in L^2(\mathcal{F}_T) \tag{78}$$

By Proposition 3.6, for any g satisfying (i), (ii) (iii") of (34), the associated g-expectation is dominated by $\mathcal{E}^\mu$, where μ is the Lipschitz constant in (34).

Lemma 4.1. *If $\mathcal{E}$ is dominated by $\mathcal{E}^\mu$ for some $\mu > 0$, then*

$$\mathcal{E}^{-\mu}[Y] \le \mathcal{E}[X + Y] - \mathcal{E}[X] \le \mathcal{E}^\mu[Y]. \tag{79}$$

Proof. It is a simple consequence of

$$\mathcal{E}^{-\mu}[Y|\mathcal{F}_t] = -\mathcal{E}^\mu[-Y|\mathcal{F}_t].$$

$\square$

Lemma 4.2. *If $\mathcal{E}$ is dominated by $\mathcal{E}^\mu$ for some $\mu > 0$, then $\mathcal{E}[\cdot]$ is a continuous operator on $L^2(\mathcal{F}_T)$ in the following sense:*

$$\exists C > 0, \quad |\mathcal{E}[\xi_1] - \mathcal{E}[\xi_2]| \le C \|\xi_1 - \xi_2\|_{L^2}, \quad \forall \xi_1, \xi_2 \in L^2(\mathcal{F}_T). \tag{80}$$

Proof. The claim follows easily from Lemma 4.1 above and Lemma 3.3. $\square$

From now on we will deal with $\mathcal{F}$-expectations $\mathcal{E}[\cdot]$ also satisfying the following condition:

$$\mathcal{E}[X + Y|\mathcal{F}_t] = \mathcal{E}[X|\mathcal{F}_t] + Y, \quad \forall X \in L^2(\mathcal{F}_T) \quad \text{and} \quad Y \in L^2(\mathcal{F}_t) \tag{81}$$

Recall that, when $\mathcal{E}[\cdot]$ is a g-expectation, (81) means that g satisfies (34)–(iii") (see (57)). We observe that an expectation $E_Q[\cdot]$ under a Girsanov transformation $\dfrac{dQ}{dP}$ satisfies this assumption.

We need to introduce a new notation: for a given $\zeta \in L^2(\mathcal{F}_T)$, we consider the mapping $\mathcal{E}_\zeta[\cdot]$ defined by

$$\mathcal{E}_\zeta[X] := \mathcal{E}[X + \zeta] - \mathcal{E}[\zeta] : L^2(\mathcal{F}_T) \longmapsto R. \tag{82}$$

Lemma 4.3. *If $\mathcal{E}[\cdot]$ is an $\mathcal{F}$-expectation satisfying (78) and (81), then the mapping $\mathcal{E}_\zeta[\cdot]$ is also an $\mathcal{F}$-expectation satisfying (78) and (81). Its conditional expectation under $\mathcal{F}_t$ is*

$$\mathcal{E}_\zeta[X|\mathcal{F}_t] = \mathcal{E}[X + \zeta|\mathcal{F}_t] - \mathcal{E}[\zeta|\mathcal{F}_t]. \tag{83}$$

Proof. It is easily seen that $\mathcal{E}_\zeta[\cdot]$ is a nonlinear expectation.

We now prove that the notion $\mathcal{E}_\zeta[X|\mathcal{F}_t]$ defined in (83) is actually the conditional expectation induced by $\mathcal{E}_\zeta[\cdot]$ under $\mathcal{F}_t$.

Indeed, put $G(X, \zeta, \mathcal{F}_t) = \mathcal{E}[X + \zeta|\mathcal{F}_t] - \mathcal{E}[\zeta|\mathcal{F}_t]$. We want to show that, for all $A \in \mathcal{F}_t$, $\mathcal{E}_\zeta(G(X, \zeta, \mathcal{F}_t)1_A) = \mathcal{E}_\zeta(X1_A)$. Computations give:

$$\begin{aligned}
\mathcal{E}_\zeta[G(X, \zeta, \mathcal{F}_t)] &= \mathcal{E}[\mathcal{E}[X + \zeta|\mathcal{F}_t] - \mathcal{E}[\zeta|\mathcal{F}_t] + \zeta] - \mathcal{E}[\zeta] \quad \text{(by (9))} \\
&= \mathcal{E}[\mathcal{E}[X + \zeta|\mathcal{F}_t] - \mathcal{E}[\zeta|\mathcal{F}_t] + \mathcal{E}[\zeta|\mathcal{F}_t]] - \mathcal{E}[\zeta] \quad \text{(by (81))} \\
&= \mathcal{E}[\mathcal{E}[X + \zeta|\mathcal{F}_t]] - \mathcal{E}[\zeta] \\
&= \mathcal{E}[X + \zeta] - \mathcal{E}[\zeta].
\end{aligned}$$

Thus we have

$$\mathcal{E}_\zeta[G(X, \zeta, \mathcal{F}_t)] = \mathcal{E}_\zeta[X], \quad \forall X. \tag{84}$$

Now for each $A \in \mathcal{F}_t$, we have,

$$\begin{aligned}
G(X1_A, \zeta, \mathcal{F}_t) &= \mathcal{E}[X1_A + \zeta 1_A + \zeta 1_{A^c}|\mathcal{F}_t] - \mathcal{E}[\zeta|\mathcal{F}_t] \\
&= \mathcal{E}[(X + \zeta)1_A + \zeta 1_{A^c}|\mathcal{F}_t] - \mathcal{E}[\zeta|\mathcal{F}_t] \\
&= \mathcal{E}[X + \zeta|\mathcal{F}_t]1_A + \mathcal{E}[\zeta|\mathcal{F}_t]1_{A^c} - \mathcal{E}[\zeta|\mathcal{F}_t] \\
&= (\mathcal{E}[X + \zeta|\mathcal{F}_t] - \mathcal{E}[\zeta|\mathcal{F}_t])1_A \\
&= G(X, \zeta, \mathcal{F}_t)1_A.
\end{aligned}$$

From this with (84) it follows that $\mathcal{E}_\zeta[X|\mathcal{F}_t]$ satisfies (7):

$$\mathcal{E}_\zeta[G(X, \zeta, \mathcal{F}_t)1_A] = \mathcal{E}_\zeta[G(X1_A, \zeta, \mathcal{F}_t)] = \mathcal{E}_\zeta[X1_A], \quad \forall A \in \mathcal{F}_t.$$

Thus $\mathcal{E}_\zeta[\cdot]$ is an $\mathcal{F}$-expectation with $\mathcal{E}_\zeta[\cdot|\mathcal{F}_t]$ given by (83).

We now check that (78) is satisfied. For each $X, Y \in L^2(\mathcal{F}_T)$,

$$\begin{aligned}
\mathcal{E}_\zeta[X + Y] - \mathcal{E}_\zeta[X] &= (\mathcal{E}[X + Y + \zeta] - \mathcal{E}[\zeta]) - (\mathcal{E}[X + \zeta] - \mathcal{E}[\zeta]) \\
&= \mathcal{E}[X + Y + \zeta] - \mathcal{E}[X + \zeta].
\end{aligned}$$

Since $\mathcal{E}[\cdot]$ satisfies (78), $\mathcal{E}_\zeta[\cdot]$ satisfies

$$\mathcal{E}_\zeta[X + Y] - \mathcal{E}_\zeta[X] \le \mathcal{E}^\mu[Y].$$

Finally, let $Y \in L^2(\mathcal{F}_t)$; since $\mathcal{E}[\cdot]$ satisfies property (81), thus

$$\begin{aligned}
\mathcal{E}_\zeta[X + Y|\mathcal{F}_t] &= \mathcal{E}[X + \zeta|\mathcal{F}_t] - \mathcal{E}[\zeta|\mathcal{F}_t] + Y \\
&= \mathcal{E}_\zeta[X|\mathcal{F}_t] + Y.
\end{aligned}$$

Thus $\mathcal{E}_\zeta[\cdot]$ also satisfies property (81). The proof is complete. $\quad\square$

Lemma 4.4. *Let $\mathcal{E}[\cdot]$ be an $\mathcal{F}$-expectation satisfying (78) and (81). Then, for each $t \leq T$, we have a.s.*

$$\mathcal{E}^{-\mu}[X|\mathcal{F}_t] \leq \mathcal{E}_\zeta[X|\mathcal{F}_t] \leq \mathcal{E}^\mu[X|\mathcal{F}_t], \ \forall X, \ \zeta \in L^2(\mathcal{F}_T).$$

This lemma is a simple consequence of the following one, whose proof is inspired by [BCHMP2000].

Lemma 4.5. *Let $\mathcal{E}_1[\cdot]$ and $\mathcal{E}_2[\cdot]$ be two $\mathcal{F}$-expectations satisfying (78) and (81). If*

$$\mathcal{E}_1[X] \leq \mathcal{E}_2[X], \quad \forall X \in L^2(\mathcal{F}_T),$$

then a.s. and for all t,

$$\mathcal{E}_1[X|\mathcal{F}_t] \leq \mathcal{E}_2[X|\mathcal{F}_t], \quad \forall X \in L^2(\mathcal{F}_T).$$

Proof. Indeed, for all $Y \in L^2(\mathcal{F}_T)$, we have by (81)

$$
\begin{aligned}
\mathcal{E}_1[Y - \mathcal{E}_1[Y|\mathcal{F}_t]] &= \mathcal{E}_1[\mathcal{E}_1[Y - \mathcal{E}_1[Y|\mathcal{F}_t]|\mathcal{F}_t]] \\
&= \mathcal{E}_1[\mathcal{E}_1[Y|\mathcal{F}_t] - \mathcal{E}_1[Y|\mathcal{F}_t]] \\
&= \mathcal{E}_1[0] = 0.
\end{aligned}
$$

On the other hand,

$$
\begin{aligned}
\mathcal{E}_1[Y - \mathcal{E}_1[Y|\mathcal{F}_t]] &\leq \mathcal{E}_2[Y - \mathcal{E}_1[Y|\mathcal{F}_t]] \\
&= \mathcal{E}_2[\mathcal{E}_2[Y - \mathcal{E}_1[Y|\mathcal{F}_t]|\mathcal{F}_t]].
\end{aligned}
$$

Thus

$$\mathcal{E}_2[\mathcal{E}_2[Y|\mathcal{F}_t] - \mathcal{E}_1[Y|\mathcal{F}_t]] \geq 0, \quad \forall Y \in L^2(\mathcal{F}_T).$$

Now, for a fixed $X \in L^2(\mathcal{F}_T)$, we set $\eta = \mathcal{E}_2[X|\mathcal{F}_t] - \mathcal{E}_1[X|\mathcal{F}_t]$. Since

$$
\begin{aligned}
\eta 1_{\{\eta<0\}} &= 1_{\{\eta<0\}}\mathcal{E}_2[X|\mathcal{F}_t] - 1_{\{\eta<0\}}\mathcal{E}_1[X|\mathcal{F}_t] \\
&= \mathcal{E}_2[X1_{\{\eta<0\}}|\mathcal{F}_t] - \mathcal{E}_1[X1_{\{\eta<0\}}|\mathcal{F}_t],
\end{aligned}
$$

we have then

$$\mathcal{E}_2[\eta 1_{\{\eta<0\}}] = 0.$$

But since $\eta 1_{\{\eta<0\}} \leq 0$, it follows from the strict monotonicity of $\mathcal{E}_2[\cdot]$ that $\eta 1_{\{\eta<0\}} = 0$ a.s.. Thus

$$\mathcal{E}_2[X|\mathcal{F}_t] - \mathcal{E}_1[X|\mathcal{F}_t] \geq 0 \quad \text{a.s.}$$

The proof is complete. □

Lemma 4.6. *If $\mathcal{E}$ meets (78) and (81), there exists a positive constant C such that, for all X and Y in $L^2(\mathcal{F}_T)$, and for all $t \geq 0$,*

$$\mathcal{E}[\mathcal{E}[X + Y|\mathcal{F}_t] - \mathcal{E}[X|\mathcal{F}_t]] \leq C\|Y\|_{L^2}.$$

Proof. Indeed, Lemmas 4.3 and 4.4 above imply that

$$\mathcal{E}[\mathcal{E}[X + Y|\mathcal{F}_t] - \mathcal{E}[X|\mathcal{F}_t]] = \mathcal{E}[\mathcal{E}_X[Y|\mathcal{F}_t]]$$
$$\leq \mathcal{E}[\mathcal{E}^\mu[Y|\mathcal{F}_t]]$$
$$\leq \mathcal{E}^\mu[\mathcal{E}^\mu[Y|\mathcal{F}_t]] = \mathcal{E}^\mu[Y] \leq C\|Y\|_{L^2}.$$

The last inequality is from Lemma 4.2.

$\square$

4.2 $\mathcal{F}_t$-Consistent Martingales

In this subsection we assume that $\mathcal{E}$ is an $\mathcal{F}$-expectation satisfying (78) for some $\mu > 0$, and (81) as well.

Definition 4.2. *A process* $(X_t)_{t \in [0,T]} \in L^2_\mathcal{F}(0,T)$ *is called an $\mathcal{E}$-martingale (resp. $\mathcal{E}$-supermartingale, -submartingale) if for each* $0 \leq s \leq t \leq T$

$$X_s = \mathcal{E}[X_t|\mathcal{F}_s], \ (resp. \ \geq \mathcal{E}[X_t|\mathcal{F}_s], \ \leq \mathcal{E}[X_t|\mathcal{F}_s]).$$

Lemma 4.7. *An $\mathcal{E}^\mu$-supermartingale (ξ_t) is both an $\mathcal{E}$ - supermartingale and $\mathcal{E}^{-\mu}$ - supermartingale. An $\mathcal{E}^{-\mu}$ - submartingale (ξ_t) is both an $\mathcal{E}$ - and $\mathcal{E}^\mu$ - submartingale. An $\mathcal{E}$ - martingale (ξ_t) is an $\mathcal{E}^{-\mu}$ - supermartingale and an $\mathcal{E}^\mu$-submartingale.*

Proof. It comes simply from the fact that, for each $0 \leq s \leq t \leq T$,

$$\mathcal{E}^{-\mu}[\xi_t|\mathcal{F}_s] \leq \mathcal{E}[\xi_t|\mathcal{F}_s] \leq \mathcal{E}^\mu[\xi_t|\mathcal{F}_s].$$

$\square$

The next result is the first step in a procedure that will eventually prove that every $\mathcal{E}$-martingale admits continuous paths.

Lemma 4.8. *For each $X \in L^2(\mathcal{F}_T)$ the process $\mathcal{E}[X|\mathcal{F}_t]$, $t \in [0,T]$ admits a unique modification with a.s. RCLL paths.*

Proof. We can deduce from Lemma 4.7 that the process $\mathcal{E}[X|\mathcal{F}_t]$, $t \in [0,T]$, is an $\mathcal{E}^{-\mu}$-supermartingale. Hence we can apply the downcrossing inequality of Proposition 3.11.

This downcrossing inequality tells us that $\mathcal{E}[X|\mathcal{F}_t]$, $t \in [0,T]$ has P-a.s. finitely many downcrossings of every interval $[a,b]$ with rational $a < b$. By classical methods, this imply the almost sure existence of left and right limits for the paths of $\mathcal{E}[X|\mathcal{F}.]$.

We thus can define $Y_t = \lim\limits_{\substack{s \searrow t \\ s \in \mathbf{Q} \cap [0,T]}} \mathcal{E}[X|\mathcal{F}_s]$. For each $A \in \mathcal{F}_t$, we have that

$$Y_t 1_A = \lim_{\substack{s \searrow t \\ s \in \mathbf{Q} \cap [0,T]}} \mathcal{E}[X|\mathcal{F}_s]1_A, \text{ in } L^2(\mathcal{F}_T).$$

From Lemma 4.2, it follows that

$$\mathcal{E}[Y_t 1_A] = \lim_{\substack{s \searrow t \\ s \in \mathbf{Q} \cap [0,T]}} \mathcal{E}[\mathcal{E}[X|\mathcal{F}_s]1_A].$$

But

$$\mathcal{E}[\mathcal{E}[X|\mathcal{F}_s]1_A] = \mathcal{E}[1_A \mathcal{E}[X|\mathcal{F}_t]].$$

It follows that a.s. $Y_t = \mathcal{E}[X|\mathcal{F}_t]$.

Now it's again classical to prove, using the existence of left and right limits, that the process Y defined above is a RCLL modification of $\mathcal{E}[X|\mathcal{F}_t]$, $t \in [0,T]$, and the lemma is proved.   □

Henceforth, and without needing to recall it, we will always consider the RCLL modifications of the $\mathcal{E}$-martingales we have to deal with.

Lemma 4.8 has an immediate consequence as follows :

Lemma 4.9. *Let $\mathcal{E}[\cdot]$ be an $\mathcal{F}$-expectation satisfying (78) and (81). Then for each $X \in L^2(\mathcal{F}_T)$ and $g \in L^2_{\mathcal{F}}(0,T)$ the process $\mathcal{E}[X + \int_t^T g_s ds|\mathcal{F}_t]$, $t \in [0,T]$ is RCLL a.s.*

Proof. Indeed, we can write

$$\mathcal{E}[X + \int_t^T g_s ds|\mathcal{F}_t] = \mathcal{E}[X + \int_0^T g_s ds - \int_0^t g_s ds|\mathcal{F}_t]$$

$$= \mathcal{E}[X + \int_0^T g_s ds|\mathcal{F}_t] - \int_0^t g_s ds$$

because of (81). The claim follows then easily from Lemma 4.8.   □

Lemma 4.10. *For each $X \in L^2(\mathcal{F}_T)$, let*

$$y_t = \mathcal{E}[X|\mathcal{F}_t].$$

Then there exists a pair $(g(\cdot), z(\cdot)) \in L^2_{\mathcal{F}}(0,T; R \times R^d)$ with

$$|g_t| \leq \mu|z_t| \tag{85}$$

such that

$$y_t = X + \int_t^T g_s ds - \int_t^T z_s dB_s. \tag{86}$$

Furthermore, take $X' \in L^2(\mathcal{F}_T)$, put $y'_t = \mathcal{E}[X'|\mathcal{F}_t]$, and let $(g'(\cdot), z'(\cdot)) \in L^2_{\mathcal{F}}(0,T; R \times R^d)$ be the corresponding pair. Then we have

$$|g_t - g'_t| \leq \mu|z_t - z'_t| \tag{87}$$

Proof. Since

$$y_t = \mathcal{E}[X|\mathcal{F}_t], \quad 0 \le t \le T,$$

is an $\mathcal{E}$ - martingale, and since it is RCLL, it is a right-continuous $\mathcal{E}^\mu$-submartingale (resp. $\mathcal{E}^{-\mu}$ - supermartingale). By the domination $\mathcal{E}^{-\mu}[X|\mathcal{F}_t] \le \mathcal{E}[X|\mathcal{F}_t] \le \mathcal{E}^\mu[X|\mathcal{F}_t]$, we also have $\mathbf{E}[\sup_{t\in[0,T]} |y_t|^2] < \infty$. Thus, from the g - supermartingale decomposition theorem (Theorem 3.9) that there exist (z^μ, A^μ) and $(z^{-\mu}, A^{-\mu})$ in $L^2_{\mathcal{F}}([0,T]; R \times R^d)$ with A^μ and $A^{-\mu}$ RCLL and increasing such that $A^\mu(0) = 0$, $A^{-\mu}(0) = 0$ and such that

$$y_t = y_T + \int_t^T \mu|z^\mu_s|ds - A^\mu_T + A^\mu_t - \int_t^T z^\mu_s dB_s$$

and

$$y_t = y_T - \int_t^T \mu|z^{-\mu}_s|ds + A^{-\mu}_T - A^{-\mu}_t - \int_t^T z^{-\mu}_s dB_s.$$

Hence, the martingale parts and the bounded variation parts of the above two processes must coincide:

$$z^\mu_t \equiv z^{-\mu}_t,$$
$$-\mu|z^\mu_t|dt + dA^\mu_t \equiv \mu|z^\mu_t|dt - dA^{-\mu}_t,$$

whence

$$2\mu|z^\mu_t|dt \equiv dA^\mu_t + dA^{-\mu}_t.$$

It follows that A^μ and $A^{-\mu}$ are both absolutely continuous and we can write:

$$dA^\mu_t = a^\mu_t dt, \quad dA^{-\mu}_t = a^{-\mu}_t dt$$

with

$$0 \le a^\mu_t, \quad 0 \le a^{-\mu}_t.$$

We also have

$$a^\mu_t + a^{-\mu}_t \equiv 2\mu|z^\mu_t|,$$

so, if we define

$$z_t = z^\mu_t$$
$$g_t = \mu|z_t| - a^\mu_t,$$

we get (86) and (85).

Now, we prove (87). We have

$$\begin{aligned} y_t - y'_t &= \mathcal{E}[X|\mathcal{F}_t] - \mathcal{E}[X'|\mathcal{F}_t] \\ &= \mathcal{E}[X - X' + X'|\mathcal{F}_t] - \mathcal{E}[X'|\mathcal{F}_t] \\ &= \mathcal{E}_{X'}[X - X'|\mathcal{F}_t] \end{aligned}$$

Recall (Lemma 4.3) that $\mathcal{E}_{X'}[\cdot]$ is another $\mathcal{F}$-expectation satisfying (78) and (81). Thus there also exists a pair $(\tilde{g}(\cdot), \tilde{z}(\cdot)) \in L^2_{\mathcal{F}}(0,T; R \times R^d)$ with

$$|\tilde{g}_t| \le \mu |\tilde{z}_t| \tag{88}$$

such that the $\mathcal{E}_{X'}$-martingale $y_t - y_t'$ satisfies

$$y_t - y_t' = X - X' + \int_t^T \tilde{g}_s ds - \int_t^T \tilde{z}_s dB_s.$$

On the other hand, we have

$$y_t - y_t' = X - X' + \int_t^T [g_s - g_s'] ds - \int_t^T [z_s - z_s'] dB_s.$$

It follows then that

$$\tilde{g}_t \equiv g_t - g_t', \quad \text{and} \quad \tilde{z}_t \equiv z_t - z_t'.$$

This with (88) yields (87). The proof is complete. □

Remark 4.1. From the above lemma, the result of Lemma 4.9 can be improved to: for each $X \in L^2(\mathcal{F}_T)$ and $g \in L_\mathcal{F}^2(0,T)$, the process $\mathcal{E}[X + \int_t^T g_s ds | \mathcal{F}_t]$, $t \in [0,T]$ is continuous a.s..

4.3 BSDE under $\mathcal{F}_t$–Consistent Nonlinear Expectations

Here again, $\mathcal{E}$ denotes an $\mathcal{F}$-expectation satisfying (78) for some $\mu > 0$, and (81) as well. Let a function f be given

$$f(\omega, t, y) : \Omega \times [0,T] \times R \longmapsto R$$

satisfying, for some constant $C_1 > 0$,

$$\begin{cases} \text{(i)} \;\; f(\cdot, y) \in L_\mathcal{F}^2(0,T), \quad \text{for each } y \in R; \\ \text{(ii)} \;\; |f(t,y_1) - f(t,y_2)| \le C_1 |y_1 - y_2|, \quad \forall y_1, \, y_2 \in R. \end{cases} \tag{89}$$

For a given terminal data $X \in L^2(\mathcal{F}_T)$, we consider the following type of equation:

$$Y_t = \mathcal{E}[X + \int_t^T f(s, Y_s) ds | \mathcal{F}_t] \tag{90}$$

Theorem 4.1. *We assume (89). Then there exists a unique process $Y(\cdot)$ solution of (90). Moreover, $Y(\cdot)$ admits continuous paths.*

Proof. Define a mapping $\Lambda(y(\cdot)) : L_\mathcal{F}^2(0,T) \longmapsto L_\mathcal{F}^2(0,T)$ by

$$\Lambda_t(y(\cdot)) := \mathcal{E}[X + \int_t^T f(s, y_s) ds | \mathcal{F}_t].$$

Using Lemma 78,

$$\Lambda_t(y_1(\cdot)) - \Lambda_t(y_2(\cdot)) \leq \mathcal{E}^\mu[\int_t^T (f(s, y_1(s)) - f(s, y_2(s)))ds|\mathcal{F}_t].$$

Thus

$$|\Lambda_t(y_1(\cdot)) - \Lambda_t(y_2(\cdot))| \leq \mathcal{E}^\mu[\int_t^T |f(s, y_1(s)) - f(s, y_2(s))|ds|\mathcal{F}_t]$$

$$\leq C_1\mathcal{E}^\mu[\int_t^T |y_1(s) - y_2(s)|ds|\mathcal{F}_t], \text{ by (89).}$$

Using Lemma 3.3, it follows that

$$E[|\Lambda_t(y_1(\cdot)) - \Lambda_t(y_2(\cdot))|^2] \leq C_1^2 \mathbf{E}[\mathcal{E}^\mu[\int_t^T |y_1(s) - y_2(s)|ds|\mathcal{F}_t]^2]$$

$$\leq C_1^2 e^{\mu^2(T-t)} \mathbf{E}[\int_t^T |y_1(s) - y_2(s)|ds]^2$$

$$\leq C_2 \mathbf{E}[\int_t^T |y_1(s) - y_2(s)|^2 ds].$$

where $C_2 := TC_1^2 e^{\mu^2 T}$.

We observe that, for any finite number β, the following two norms are equivalent in $L_\mathcal{F}^2(0, T)$

$$\mathbf{E}\int_0^T |\phi_s|^2 ds \sim \mathbf{E}\int_0^T |\phi_s|^2 e^{\beta s} ds.$$

Thus we multiply $e^{2C_2 t}$ on both sides of the above inequality and then integrate them on $[0, T]$. It follows that

$$\mathbf{E}\int_0^T |\Lambda_t(y_\cdot) - \Lambda_t(y_\cdot')|^2 e^{2C_2 t} dt \leq C_2 \mathbf{E}\int_0^T e^{2C_2 t} \int_t^T |y_s - y_s'|^2 ds dt$$

$$= C_2 \mathbf{E}\int_0^T \int_0^s e^{2C_2 t} dt |y_s - y_s'|^2 ds$$

$$= (2C_2)^{-1} C_2 \mathbf{E}\int_0^T (e^{2C_2 s} - 1)|y_s - y_s'|^2 ds.$$

We then have

$$\mathbf{E}\int_0^T |\Lambda_t(y_\cdot) - \Lambda_t(y_\cdot')|^2 e^{2C_2 t} dt \leq \frac{1}{2}\mathbf{E}\int_0^T |y_t - y_t'|^2 e^{2C_2 t} dt.$$

Namely, Λ is a contraction mapping on $L_\mathcal{F}^2(0, T)$. It follows that this mapping has a unique fixed point Y:

$$Y_t = \mathcal{E}[X + \int_t^T f(s, Y_s)ds|\mathcal{F}_t].$$

Finally, Lemma 4.9 and Remark 4.1 proves that the solution of (90) admits continuous paths, and the proof is complete.   □

Theorem 4.2. (Comparison Theorem). *Let Y be the solution of (90) and let Y' be the solution of*

$$Y'_t = \mathcal{E}[X' + \int_t^T [f(s, Y'_s) + \phi_s]ds|\mathcal{F}_t]$$

where $X' \in L^2(\mathcal{F}_T)$ and $\phi \in L^2_{\mathcal{F}}(0, T)$. If

$$X' \geq X, \quad \phi_t \geq 0, \quad dP \times dt\text{-}a.e., \tag{91}$$

then we have

$$Y'_t \geq Y_t, \quad dP \times dt\text{-}a.e. \tag{92}$$

(92) becomes equality if and only if (91) become equalities.

Proof. We begin with the case $\phi_t \equiv 0$. For each $\delta > 0$, we define

$$\tau_1^\delta = \inf\{t \geq 0; Y'_t \leq Y_t - \delta\} \wedge T.$$

It is clear that if, for all $\delta > 0$, $\tau_1^\delta = T$ a.s., then (92) holds. Now if for some $\delta > 0$ we have

$$P(A) > 0, \quad \text{with } A = \{\tau_1^\delta < T\} \in \mathcal{F}_{\tau_1^\delta}$$

we then can define

$$\tau_2 = \inf\{t \geq \tau_1^\delta; Y'_t \geq Y_t\}.$$

Since $Y'_T = X' \geq X = Y_T$, thus $\tau_2 \leq T$ and $1_A Y'(\tau_2) = 1_A Y(\tau_2)$. It follows that, for $\tau \in [\tau_1^\delta, \tau_2]$,

$$1_A Y_\tau = \mathcal{E}[1_A Y_{\tau_2} + \int_\tau^{\tau_2} 1_A f(s, 1_A Y_s)ds|\mathcal{F}_\tau],$$

$$1_A Y'_\tau = \mathcal{E}[1_A Y_{\tau_2} + \int_\tau^{\tau_2} 1_A f(s, 1_A Y'_s)ds|\mathcal{F}_\tau].$$

By the uniqueness result of Theorem 4.1, the solutions of the above two equations must coincide with each other. Thus $Y'_{\tau_1^\delta} 1_A = Y_{\tau_1^\delta} 1_A$. This contradicts $P(A) > 0$.

In order to prove the general case when $\phi_s \geq 0$, we define for $n = 1, 2, 3, \cdots$, $Y^n(\cdot)$ to be the solution of

$$Y^n_t = \mathcal{E}\left[[X' + \int_{\frac{iT}{n}}^T \phi_s ds] + \int_t^T f(s, Y^n_s)ds|\mathcal{F}_t \right],$$

$$\text{for } t \in [t^n_i, t^n_{i+1}), \ t^n_i := \frac{iT}{n}, \ i = 0, 1, \cdots, n-1..$$

This equation can be written, piece by piece, as

$$Y_t^n = \mathcal{E}\left[[Y_{t_{i+1}^n}^n + \int_{t_t^n}^{t_{i+1}^n} \phi_s ds] + \int_t^{t_{i+1}^n} f(s, Y_s^n)ds \big| \mathcal{F}_t\right],$$

$$t \in [t_i^n, t_{i+1}^n),\ Y_T^n = Y_{t_n^n}^n = X'.$$

From the first part of the proof. We have, for $i = n-1$, $Y_t^n \geq Y_t$, $t \in [t_{n-1}^n, T)$. In particular, $Y_{t_{n-1}^n}^n \geq Y_{t_{n-1}^n}$. An obvious iteration of this algorithm gives

$$Y_t^n \geq Y_t,\ t \in [t_i^n, t_{i+1}^n),\quad i = 0, \cdots, n-2.$$

Thus $Y_t^n \geq Y_t$, $t \in [0, T]$.

In order to prove that $Y_t' \geq Y_t$, It suffices to show the convergence of the sequence (Y^n) to Y'. A computation analogous to the proof of Theorem 4.1 shows that, for fixed $t \in [t_i^n, t_{i+1}^n)$ and an appropriate constant C,

$$E[|Y_t^n - Y_t'|^2] \leq C\mathbf{E}[(\int_{\frac{iT}{n}}^t |\phi_s|ds + C_1 \int_t^T |Y_s^n - Y_s'|ds)^2]$$

Using Schwartz inequality, one has for all $t \in [0, T]$

$$\mathbf{E}[|Y_t^n - Y_t'|^2] \leq 2C\frac{T}{n}\mathbf{E}\int_0^T |\phi_s|^2 ds + 2CC_1^2 TE\int_t^T |Y_s^n - Y_s'|^2 ds. \qquad (93)$$

Gronwall's Lemma applied to the above inequality shows that

$$\mathbf{E}[|Y_t^n - Y_t'|^2] \to 0,$$

and finally $Y_t' \geq Y_t$.

Finally, we investigate possible equality in (92). From $Y_t' \equiv Y_t$, one has

$$\mathcal{E}[X + \int_0^T f(s, Y_s)ds] = \mathcal{E}[X' + \int_0^T f(s, Y_s)ds + \int_0^T \Phi_s ds]$$

Since $X' \geq X$ and $\int_0^T \Phi_s ds \geq 0$, it follows from the strict monotonicity of $\mathcal{E}$ that $X' = X$ a.s., and $\int_0^T \Phi_s ds = 0$, whence $\Phi = 0$ $dt \times dP$ a.e. and the end of the proof. $\qquad\square$

4.4 Decomposition Theorem for $\mathcal{E}$-Supermartingales

Our next result generalizes the decomposition theorem for g-supermartingales proved in Theorem. 3.9 to continuous $\mathcal{E}$-supermartingales. The proof is very similar. It also uses mainly arguments from Theorem 3.9.

Theorem 4.3. (Decomposition theorem for $\mathcal{E}$-supermartingales) Let $\mathcal{E}[\cdot]$ be an $\mathcal{F}$-expectation satisfying (78) and (81), and let $Y \in S_{\mathcal{F}}^2(0,T)$ be a $\mathcal{E}$-supermartingale. Then there exists an $A(\cdot) \in S_{\mathcal{F}}^2(0,T)$ with $A(0) = 0$ such that $Y + A$ is an $\mathcal{E}$-martingale.

Proof. For $n \geq 1$, we define $y^n(\cdot)$, solution of the following BSDE:

$$y_t^n = \mathcal{E}[Y_T + \int_t^T n(Y_s - y_s^n)ds|\mathcal{F}_t]$$

We have then the following

Lemma 4.11. We have, for each t and $n \geq 1$,

$$Y_t \geq y_t^n, \ a.s.$$

Proof. For a $\delta > 0$ and a given integer $n > 0$, we define

$$\sigma^{n,\delta} := \inf\{t; \ y_t^n \geq Y_t + \delta\} \wedge T.$$

If $P(\sigma^{n,\delta} < T) = 0$, for all n and δ, then the proof is done. If it is not the case, then there exist $\delta > 0$ and a positive integer n such that $P(\sigma^{n,\delta} < T) > 0$. We can then define the following stopping times

$$\tau := \inf\{t \geq \sigma^{n,\delta}; \ y_t^n \leq Y_t\}$$

It is clear that $\sigma^{n,\delta} \leq \tau \leq T$. Because of Theorem 4.1, $Y_t - y_t^n$ is continuous. Hence we have

$$y_\tau^n \leq Y_\tau \tag{94}$$

But since $(Y_s - y_s^n) \leq 0$ in $[\sigma^{n,\delta}, \tau]$, by monotonicity of $\mathcal{E}[\cdot]$,

$$\begin{aligned} y_{\sigma^{n,\delta}}^n &= \mathcal{E}[y_\tau^n + \int_{\sigma^{n,\delta}}^\tau n(Y_s - y_s^n)ds|\mathcal{F}_{\sigma^{n,\delta}}] \\ &\leq \mathcal{E}[y_\tau^n|\mathcal{F}_{\sigma^{n,\delta}}] \\ &\leq \mathcal{E}[Y_\tau|\mathcal{F}_{\sigma^{n,\delta}}] \end{aligned}$$

Finally, since Y is an $\mathcal{E}$-supermartingale, by (optional stopping theorem) Theorem 7.4, we have

$$Y_{\sigma^{n,\delta}} \geq y_{\sigma^{n,\delta}}^n.$$

But on the other hand, we have $P(\sigma^{n,\delta} < T) > 0$ and, by the definition of $\sigma^{n,\delta}$, $y_{\sigma^{n,\delta}}^n \geq Y_{\sigma^{n,\delta}} + \delta$ on $\{\sigma^{n,\delta} < T\}$. This induces a contradiction. The proof is complete. $\qquad \square$

Lemma 4.11 with Theorem 4.2 above imply that $y^n(\cdot)$ monotonically converges to some $Y^0(\cdot) \leq Y(\cdot)$. Indeed, writing $\phi_t = Y_t - y_t^{(n+1)} \geq 0$ shows that $(y^n(\cdot))$ is an increasing sequence of functions.

Observe then that $y_t^n + \int_0^t n(Y_s - y_s^n)ds$ is an $\mathcal{E}$-martingale. By Lemma 4.10, there exists $(g^n, z^n) \in L_{\mathcal{F}}^2(0, T; R \times R^d)$ with

$$|g_s^n| \leq \mu|z_s^n|, \quad n = 1, 2, \cdots, \tag{95}$$

such that

$$y_t^n + \int_0^t n(Y_s - y_s^n)ds = y_T^n + \int_0^T n(Y_s - y_s^n)ds$$
$$+ \int_t^T g_s^n ds - \int_t^T z_s^n dB_s,$$

hence, as $y_T^n = Y_T$,

$$y_t^n = Y_T + \int_t^T [g_s^n + n(Y_s - y_s^n)]ds - \int_t^T z_s^n dB_s. \tag{96}$$

(87) also tells us that

$$|g_s^n - g_s^m| \leq \mu|z_s^n - z_s^m|, \quad n, m = 1, 2, \cdots \tag{97}$$

Let us denote, for each $n = 1, 2, \cdots$,

$$A_t^n = n \int_0^t (Y_s - y_s^n)ds$$

A^n is a continuous increasing process such that $A^n(0) = 0$.

We are now going to identify the limit of $y^n(\cdot)$. To this end, we shall use the following lemma :

Lemma 4.12. *There exists a constant C which is independent of n such that*

$$(i) \quad \mathbf{E} \int_0^T |z_s^n|^2 ds \leq C; \quad (ii) \quad \mathbf{E}[(A_T^n)^2] \leq C. \tag{98}$$

Proof. By $y_t^1 \leq y_t^n \leq y_t^{n+1} \leq Y_t$, $n = 1, 2, \cdots$ with $E[\sup_{t \in [0,T]} |Y_t|^2] < \infty$, we have $|y_t^n| \leq |y_t^1| + |Y_t|$. Thus there exists a constant C, independent of n, such that

$$\mathbf{E}[\sup_{0 \leq t \leq T} |y_t^n|^2] \leq C. \tag{99}$$

We then can apply (28) and (95) to prove (98) step by step as the proof of Lemma 3.7. $\qquad\square$

With the help of Lemma 4.12 we can now end the proof of the Decomposition Theorem.

Note first that (98)–(i) with (95) also implies

$$\mathbf{E} \int_0^T |g_s^n|^2 ds \leq \mu^2 C$$

(98)–(ii) implies that

$$y^n(\cdot) \nearrow Y(\cdot).$$

From by the monotonic limit Theorem 7.2 (in Appendix), it follows that we can write Y under the form

$$Y_t = Y_T + \int_t^T g_s ds + A_T - A_t - \int_t^T z_s dB_s \qquad (100)$$

for some $(g, z) \in L_{\mathcal{F}}^2(0, T; R \times R^d)$ and an increasing process A with $A_0 = 0$ and $\mathbf{E}[A_T^2] < \infty$. Observe that $Y(\cdot)$ and then $A(\cdot)$ is continuous. It follows from Theorem 7.2 that

$$z^n(\cdot) \to z(\cdot), \quad \text{strongly in } L_{\mathcal{F}}^2(0, T; R^d).$$

It follows from (97) that

$$g^n(\cdot) \to g(\cdot), \quad \text{strongly in } L_{\mathcal{F}}^2(0, T).$$

And finally, (28) gives

$$A_t^n \longmapsto A_t, \quad \text{strongly in } L^2(\mathcal{F}_T).$$

Thanks to Lemma 4.6, we can pass to the L^2-limit in both sides of

$$y_t^n = \mathcal{E}[Y_T + A_T^n - A_t^n | \mathcal{F}_t].$$

It follows that

$$Y_t = \mathcal{E}[Y_T + A_T - A_t | \mathcal{F}_t].$$

Thus $Y_t + A_t = \mathcal{E}[Y_T + A_T | \mathcal{F}_t]$ is an $\mathcal{E}$-martingale (because of (81)). Since A is increasing, the Theorem is proved. $\square$

4.5 Representation Theorem
of an $\mathcal{F}$–Expectation by a g–Expectation

In this subsection, we will prove an important result: an $\mathcal{F}_t$–consistent nonlinear expectation can be identified as a g-expectation, provided that (78) and (81) hold.

Theorem 4.4. *We assume that an $\mathcal{F}$-expectation $\mathcal{E}[\cdot]$ satisfies (78) and (81) for some $\mu > 0$. Then there exists a function $g = g(t, z) : \Omega \times [0, T] \times R^d$ satisfying (i), (ii) and (iii") of (34) such that*

$$\mathcal{E}[X] = \mathcal{E}_g[X], \quad \forall X \in L^2(\mathcal{F}_T).$$

In particular, every $\mathcal{E}$-martingale is continuous a.s.
Moreover, we have $|g(t, z)| \le \mu|z|$ for all $t \in [0, T]$.

Proof. For each given $z \in R^d$, we consider the following forward equation

$$\begin{cases} dY_t^z = -\mu|z|dt + z dB_t, \\ Y^z(0) = 0. \end{cases}$$

We have $E[\sup_{t \in [0,T]} |Y_t^z|^2] < \infty$. It is also clear that Y^z is an $\mathcal{E}^\mu$-martingale, thus an $\mathcal{E}[\cdot]$-supermartingale. Indeed, we can write $Y_t^z = \mathcal{E}^\mu[Y_T^z|\mathcal{F}_t]$. From Theorem 4.3, there exists an increasing process $A^z(\cdot)$ with $A^z(0) = 0$ and $E[A_T^{z2}] < \infty$ such that

$$Y_t^z = \mathcal{E}[Y_T^z + A_T^z - A_t^z|\mathcal{F}_t].$$

Or

$$Y_t^z + A_t^z = \mathcal{E}[Y_T^z + A_T^z|\mathcal{F}_t], \quad t \in [0, T].$$

Then, from Lemma 4.10. there exists $(g(z, \cdot), Z^z(\cdot)) \in L_{\mathcal{F}}^2(0, T; R \times R^d)$ with $|g(z, t)| \le \mu|Z_t^z|$ such that

$$Y_t^z + A_t^z = Y_T^z + A_T^z + \int_t^T g(z, s)ds - \int_t^T Z_s^z dB_s. \tag{101}$$

We also have

$$|g(z, t) - g(z', t)| \le \mu|Z_t^z - Z_t^{z'}|. \tag{102}$$

But on the other hand, since

$$Y_t^z = Y_T^z + \int_t^T \mu|z|ds - \int_t^T z dB_s,$$

it follows that

$$A_t^z \equiv \mu|z|t - \int_0^t g(z, s)ds,$$
$$Z_t^z \equiv z.$$

In particular, (102) becomes

$$|g(z, t) - g(z', t)| \le \mu|z - z'|. \tag{103}$$

Moreover,

$$Y_t^z + A_t^z = Y^z(r) + A^z(r) - \int_r^t g(z,s)ds + \int_r^t zdB_s, \quad 0 \le r \le t \le T,$$

and $Y_t^z + A_t^z$ is an $\mathcal{E}$-martingale. But with the assumption (81) one has, for each $z \in R^d$ and $r \le t$

$$\mathcal{E}[-\int_r^t g(z,s)ds + \int_r^t zdB_s|\mathcal{F}_r] = \mathcal{E}[Y_t^z + A_t^z - (Y^z(r) + A^z(r))|\mathcal{F}_r],$$

i.e.

$$\mathcal{E}[-\int_r^t g(z,s)ds + \int_r^t zdB_s|\mathcal{F}_r] = 0 \quad 0 \le r \le t \le T \qquad (104)$$

Now let $\{A_i\}_{i=1}^N$ be a $\mathcal{F}_r$-measurable partition of Ω (i.e., A_i are disjoint, $\mathcal{F}_r$-measurable and $\cup A_i = \Omega$) and let $z_i \in R^d$, $i = 1, 2, \cdots, N$. From (11), it follows that

$$\mathcal{E}[-\int_r^t g(\sum_{i=1}^N z_i 1_{A_i}, s)ds + \int_r^t \sum_{i=1}^N z_i 1_{A_i} dB_s|\mathcal{F}_r]$$

$$= \mathcal{E}[\sum_{i=1}^N 1_{A_i} \left(-\int_r^t g(z_i,s)ds + \int_r^t z_i dB_s\right)|\mathcal{F}_r]$$

$$= \sum_{i=1}^N 1_{A_i}\mathcal{E}[-\int_r^t g(z_i,s)ds + \int_r^t z_i dB_s|\mathcal{F}_r]$$

$$= 0$$

(because of (104)). In other words, for each simple function $\eta \in L^2(\Omega, \mathcal{F}_r, P)$,

$$\mathcal{E}[-\int_r^t g(\eta,s)ds + \int_r^t \eta dB_s|\mathcal{F}_r] = 0.$$

From this, the continuity of $\mathcal{E}[\cdot]$ in L^2 given by (80) and the fact that g is Lipschitz in z, it follows that the above equality holds for $\eta(\cdot) \in L^2_{\mathcal{F}}(0, T; R^d)$:

$$\mathcal{E}[-\int_r^t g(\eta_s,s)ds + \int_r^t \eta_s dB_s|\mathcal{F}_r] = 0. \qquad (105)$$

We just have to prove now that

$$\mathcal{E}_g[X] = \mathcal{E}[X], \quad \forall X \in L^2(\mathcal{F}_T).$$

To this end we first solve the following BSDE

$$-dy_s = g(s, z_s)ds - z_s dB_s,$$
$$y_T = X.$$

Since g is Lipschitz in z, there exists a unique solution $(y(\cdot), z(\cdot)) \in L^2_{\mathcal{F}}(0, T; R \times R^d)$. By the definition of g-expectation,

$$\mathcal{E}_g[X] = y(0).$$

On the other hand, using (105), one finds

$$\mathcal{E}[X] = \mathcal{E}[y(0) - \int_0^T g(z_s, s)ds + \int_0^T z_s dB_s]$$

$$= y(0) + \mathcal{E}[-\int_0^T g(z_s, s)ds + \int_0^T z_s dB_s]$$

$$= y(0) = \mathcal{E}_g[X].$$

It follows that this g-expectation $\mathcal{E}_g[\cdot]$ coincides with $\mathcal{E}[\cdot]$ and we are finished.
□

4.6 How to Test and Find g?

Let $g(s,z)$ be the generator of the investigated agent. An very important problem is how to find this function g. We will treat this problem for the case where g is a deterministic function: $g(t,z) : [0,\infty) \times \mathbf{R}^d \to \mathbf{R}$. We assume that

$$|g(t,z) - g(t,z')| \le \mu|z - z'|, \ \forall t \ge 0, \ \forall z, z' \in \mathbf{R}^d,$$
$$g(t,0) \equiv 0, \ \forall t \ge 0,. \tag{106}$$

In this case we can find such g by the following testing method.

Proposition 4.1. *We assume (106). Let $\bar{z} \in \mathbf{R}^d$ be given, then*

$$\int_t^T g(s, \bar{z})ds = \mathcal{E}_g[\bar{z}B_T|\mathcal{F}_t] - \bar{z}B_t \tag{107}$$

In particular

$$\int_0^T g(s, \bar{z})ds = \mathcal{E}_g[\bar{z}B_T] \tag{108}$$

Proof. We denote $Y_t := \mathcal{E}_g[\bar{z}B_T|\mathcal{F}_t]$, it is the solution of the following BSDE

$$Y_t = \bar{z}B_T + \int_t^T g(s, Z_s)ds - \int_t^T Z_s dB_s$$

Or

$$Y_t - \bar{z}B_t = \int_t^T g(s, Z_s - \bar{z} + \bar{z})ds - \int_t^T (Z_s - \bar{z})dB_s.$$

It follows that $(\bar{Y}_t, \bar{Z}_t) := (Y_t - \bar{z}B_t, Z_t - \bar{z})$ solves the BSDE

$$\bar{Y}_t = \int_t^T g(s, \bar{Z}_s + \bar{z})ds - \int_t^T \bar{Z}_s dB_s.$$

This BSDE has a unique solution $(\bar{Y}_t, \bar{Z}_t) \equiv (\int_t^T g(s, \bar{z})ds, 0)$. We thus have (107).

Remark 4.2. It is meaningful to test the generator g of an agent: at a time $t \leq T$, we let the agent evaluate $\bar{z}B_T$ and result $\mathcal{E}_g[\bar{z}B_T|\mathcal{F}_t]$. Then the deterministic data $\int_t^T g(s, \bar{z})ds$ is obtained by $\bar{Y}_t = \mathcal{E}_g[\bar{z}B_T|\mathcal{F}_t] - \bar{z}B_t$, where B_t is a known value at the time t.

Example 4.1. If g is time–invariant: $g = g(z)$, then we have

$$g(\bar{z})(T - t) = \mathcal{E}_g[\bar{z}B_T|\mathcal{F}_t] - \bar{z}B_t$$

and

$$g(\bar{z})T = \mathcal{E}_g[\bar{z}B_T], \quad \bar{z} \in \mathbf{R}^d.$$

Example 4.2. If we already know that $g = g_0(\theta, z)$, where $g_0 : [a, b] \times \mathbf{R}^d \to \mathbf{R}$ is a given function but we have to find the parameter $\theta \in [a, b]$, assume that for some $\bar{z} \in \mathbf{R}^d$, $g_0(\theta, z)$ is a strictly increasing function of θ in $[a, b]$. Then we can only test the agent once at the time, say $t = 0$. Using the formula

$$g_0(\theta, \bar{z})T = \mathcal{E}_g[\bar{z}B_T],$$

we can uniquely determine θ.

4.7 A General Situation: $\mathcal{F}_t$–Evaluation Representation Theorem

Theorem 4.4 is only valid for a part of $\mathcal{F}_t$–consistent nonlinear expectations. For a general situation we have the following result [Peng2003b]. By the limitation of the size of this lecture, we will only state the result without given the proof. We are given an $\mathcal{F}_t$–consistent nonlinear evaluation defined on $L^2(\mathcal{F}_T)$:

$$\mathcal{E}_{s,t}[\cdot] : L^2(\mathcal{F}_t) \to L^2(\mathcal{F}_s), \quad 0 \leq s \leq t \leq T.$$

It satisfies the axiomatic assumptions (A1)–(A4), with the following additional $\mathcal{E}^{g_{\mu,\mu}}$–dominated assumption $(g_{\mu,\mu}(y, z) := \mu(|y| + |z|))$, weaker than (A5):

(A5') There a sufficiently large number $\mu > 0$ such that, for each $0 \leq s \leq t \leq T$,
$$\mathcal{E}_{s,t}[X] - \mathcal{E}_{s,t}[X'] \leq \mathcal{E}_{s,t}^{g_{\mu,\mu}}[X - X'], \quad \forall X, X' \in L^2(\mathcal{F}_t).$$

The g–evaluation representation theorem is as follows:

Theorem 4.5. *Let $\mathcal{E}_{s,t}[\cdot] : L^2(\mathcal{F}_t) \to L^2(\mathcal{F}_s)$, $0 \leq s \leq t \leq T$, satisfy (A1)–(A4) and (A5'). Then there exists a function $g(\omega, t, y, z)$ satisfying (34)–(i), (ii) and (iii), such that, for each $0 \leq s \leq t \leq T$,*

$$\mathcal{E}_{s,t}[X] = \mathcal{E}_{s,t}^g[X], \quad \forall X \in L^2(\mathcal{F}_t).$$

Remark 4.3. In this result we do not need the assumption (81). Thus g may depend on (y, z).

Remark 4.4. In [Peng2003b] we also consider the situation where (A4) is weakened by (A4'): $1_A \mathcal{E}_{s,t}[X] = 1_A \mathcal{E}_{s,t}[1_A X]$, for each $A \in \mathcal{F}_s$. In this case the corresponding g satisfies only (34)–(i) and (ii) without the condition $g(s, 0, 0) \equiv 0$.

Remark 4.5. From the above g–evaluation reprentation theorems, we see that the dominating term, such as $\mathcal{E}^{g_{\mu,\mu}}[\cdot]$, plays an important role. A general formulation is:

$$\mathcal{E}_{s,t}[X] - \mathcal{E}_{s,t}[X'] \leq \mathcal{E}^*_{s,t}[X - X'],$$

where $\mathcal{E}^*_{s,t}[X - X']$ is a given self–dominated nonlinear evaluation: i.e., it is a concrete evaluation satisfying (A1)–(A4) and

$$\mathcal{E}^*_{s,t}[X] - \mathcal{E}^*_{s,t}[X'] \leq \mathcal{E}^*_{s,t}[X - X'], \quad \forall X, X' \in L^2(\mathcal{F}_t).$$

5 Dynamic Risk Measures

Recently Rosazza Gianin [Roazza2003] considered a type of dynamic risk measures induced from g–expectations. We consider a more general situation. Let $\mathcal{E}_{s,t}[\cdot]$ be an $\mathcal{F}_t$–consistent nonlinear evaluation defined on $L^2(\mathcal{F}_T)$. It satisfies (A1)–(A4). We set, for each $0 \leq s \leq t \leq T$, and $X \in L^2(\mathcal{F}_t)$, $\rho_{s,t}[X] := \mathcal{E}_{s,t}[-X]$. $\{\rho_{s,t}[\cdot]\}_{0 \leq s \leq t \leq T}$ is called a dynamic risk measure defined on $L^2(\mathcal{F}_T)$. We consider an $\mathcal{F}$–consistent evaluation $\{\mathcal{E}_{s,t}[\cdot]\}_{0 \leq s \leq tT}$ satisfying some of the following axiomatic conditions: for each $0 \leq s \leq t \leq T$ and X, $Y \in L^2(\mathcal{F}_t)$, it satisfies

(e1) subadditivity: $\mathcal{E}_{s,t}[X + Y] \leq \mathcal{E}_{s,t}[X] + \mathcal{E}_{s,t}[Y]$;
(e2) positively homogeneity: $\mathcal{E}_{s,t}[\alpha X] = \alpha \mathcal{E}_{s,t}[X]$;
(e3) constant translability: $\mathcal{E}_{s,t}[X + \eta] = \mathcal{E}_{s,t}[X] + \eta$, $\forall \eta \in L^2(\mathcal{F}_s)$
(e4) convexity:

$$\mathcal{E}_{s,t}[\alpha X + (1 - \alpha)Y] \leq \alpha \mathcal{E}_{s,t}[X] + (1 - \alpha)\mathcal{E}_{s,t}[Y], \quad \forall \alpha \in [0, 1].$$

Similar to [ADEH1999] and [FoSc2002] for static situations, we can define the following type of dynamic risk measures.

Definition 5.1. *A dynamic risk measure $\{\rho_{s,t}[\cdot]\}_{0 \leq s \leq t \leq T}$ is said to be coherent if the corresponding nonlinear evaluation $\{\mathcal{E}_{s,t}[\cdot]\}_{0 \leq s \leq t \leq T}$ satisfies (e1)–(e3). It is said to be convex and constant translable if $\mathcal{E}_{s,t}[\cdot]$ satisfies (e3) and (e4).*

For the situation of $\mathcal{E}^g$–evaluation, we have the corresponding ρ^g–risk measure defined by $\rho^g_{s,t}[X] := \mathcal{E}^g_{s,t}[-X]$. A very interesting point is that the concrete function g perfectly reflexes the attitude of an investor towards risks. In fact we have the following properties:

Proposition 5.1. *We assume that g satisfies (34)–(i), (ii). Then $\mathcal{E}^g[\cdot]$ is subadditive (resp. superadditive) if g is subadditive (resp. superadditive) in $(y, z) \in R^{1+d}$. It is positively homogegeous if g is positively homogegeous in $(y, z) \in R^{1+d}$. It is convex (resp. concave) if g is convex (resp. concave) in $(y, z) \in R^{1+d}$. It has constant translability if g is independent of y. Moreover, if, for each (y, z) and P-a.s., $g(\cdot, y, z) \in D_\mathcal{F}^2(0, T)$, then all the above "if" can be replaced by "if and only if".*

For the proof of this proposition we refer to [EPQ1997], [BCHMP2000], [Roazza2003] and [Peng2003c]

6 Numerical Solution of BSDEs: Euler's Approximation

Let $(\epsilon_i^n)_{i=1,2,\cdots,n}$ be a Bernouil sequence, i.e., an i.i.d. sequence such that with

$$P\{\epsilon_i^n = 1\} = P\{\epsilon_i^n = -1\} = \frac{1}{2}.$$

We set

$$B_k^n := \sqrt{n} \sum_{i=1}^k \epsilon_i^n, \quad \mathcal{F}_k^n := \sigma\{B_k^n; 1 \leq k \leq n\}$$

$$\Delta B_{k+1}^n := B_{k+1}^n - B_k^n = \sqrt{n}\epsilon_k^n,$$

Let ξ be $\mathcal{F}_k^n$-measurable. This implies that there exists a function: Φ : $\{1, -1\}^k \to \mathbf{R}$, such that

$$\xi^n = \Phi_n(\epsilon_1^n, \cdots, \epsilon_k^n).$$

All processes are assumed to be $\mathcal{F}_k^n$-adapted. We make the following assumption

(H1) B^n converges to B in $\mathcal{S}^2$

(H2) ξ^n converges to ξ in $L^2(P)$.

f and $f^n : [0, 1] \times \Omega \times \mathbf{R} \times \mathbf{R} \longrightarrow \mathbf{R}$ such that for each $(y, z) \in \mathbf{R} \times \mathbf{R}$, $\{f^n(t, y, z)\}_{0 \leq t \leq 1}$ (resp. $\{f(t, y, z)\}_{0 \leq t \leq 1}$) are progressively measurable with respect to $\bar{\mathcal{F}}_t^n$ (resp. to $\mathcal{F}_t$) such that

(H3)–(i):

$$|f^n(t, y, z) - f^n(t, y', z')| \leq C(|y - y'| + |z - z'|)$$
$$|f(t, y, z) - f(t, y', z')| \leq C(|y - y'| + |z - z'|)$$

(ii) For each (y, z) paths $\{f^n(t, y, z)\}_{0 \leq t \leq 1}$ have RCLL paths and converges to $\{f(t, y, z)\}_{0 \leq t \leq 1}$ in $\mathcal{S}^2(R)$ with

$$|Y|_{\mathcal{S}^2} := \{E[\sup_{0 \leq t \leq 1} |Y_t|^2]\}^{1/2}.$$

We set

$$f^n(t, y, z) \equiv g_k^n(y, z), \ \ t \in [\frac{k}{n}, \frac{k+1}{n}), \ k = 0, 1, \cdots, n.$$

and

$y_n^n = \xi^n$: a given $\mathcal{F}_n^n$-measurable random variable. Then we solve backwardly

$$y_k^n = y_{k+1}^n + g_k^n(y_k^n, z_k^n)\frac{1}{n} - z_k^n \Delta B_{k+1}^n, \ k = n-1, \cdots, 3, 2, 1.$$

Or $y_t^n \equiv y_k^n$, $z_t^n \equiv z_k^n$, $t \in [\frac{k}{n}, \frac{k+1}{n})$. We call (y^n, z^n) the solution to (g, ξ).

$$dy_t^n = f^n(t, y_t^n, z_t^n)d\langle B^n \rangle_t - z_t^n dB_t^n,$$
$$y_T^n = \xi^n.$$

Theorem 6.1. *(Existence and Uniqueness and Comparison) Let*

$$g_k^n(\omega, y, z) : \Omega \times \mathbf{R} \times \mathbf{R} \to \mathbf{R}, \ k = 1, \cdots, n-1$$

be $\mathcal{F}_k^n$-measurable and C-Lipschitz with respect to y with $n > C$. Then there exists a unique $\mathcal{F}_k^n$-adapted pair (y^n, z^n), solution to (g, ξ). Moreover, if $(y^{n\prime}, z^{n\prime})$ is the solution corresponding to (g', ξ'), and if

$$g_k^{n\prime}(\omega, y, z) \ge g_k^n(\omega, y, z), \ \xi^{n\prime} \ge \xi^n,$$

then the corresponding solution $(y^{n\prime}, z^{n\prime})$ satisfies

$$y_k^{n\prime} \ge y_k^n.$$

Corollary. If $A_1(\cdot)$ and $A_2(\cdot)$ satisfies the above conditions with $A_1(y) \ge A_2(y)$, for all $y \in R$. Then $A_1^{-1}(x) \le A_2^{-1}(x)$, for all $x \in R$.

Proof of the theorem. Assume that y_{k+1}^n are solved, we then solve (y_k^n, z_k^n).

$$y_k^n = y_{k+1}^n + g_k^n(y_k^n, z_k^n)\frac{1}{n} - z_k^n \Delta B_{k+1}^n \tag{109}$$

Since y_{k+1}^n has the form: $y_{k+1}^n = \Phi_{k+1}(\epsilon_1, \cdots, \epsilon_{k+1})$. We set

$$y_{k+1}^{(+)} := \Phi_{k+1}(\epsilon_1, \cdots, 1),$$
$$y_{k+1}^{(-)} := \Phi_{k+1}(\epsilon_1, \cdots, -1).$$

y_{k+1}^+ and y_{k+1}^- are $\mathcal{F}_k^n$-measurable. We set $\epsilon_{k+1} = \pm 1$, in (109):

$$y_k^n = y_{k+1}^+ + g_k^n(y_k^n, z_k^n)\frac{1}{n} + - z_k^n n^{-1/2}$$

$$y_k^n = y_{k+1}^- + g_k^n(y_k^n, z_k^n)\frac{1}{n} + + z_k^n n^{-1/2}$$

z_k^n can be uniquely solved by $z_k^n = \frac{y_{k+1}^{(+)} - y_{k+1}^{(-)}}{2}$. The equation for y_k^n is

$$y_k^n - g_k^n(y_k^n, z_k^n)\frac{1}{n} = \frac{y_{k+1}^{(+)} + y_{k+1}^{(-)}}{2} \tag{110}$$

When $n > C$, the mapping $A(y) := y - g_k^n(y, z_k^n)\frac{1}{n}$ is strictly monotonic function of y with $A(y) \to +\infty$ (resp.$-\infty$) as $y \to +\infty$ (resp. $-\infty$). Thus the solution y_k^n of (3) exists and is unique. By the Corollary, the comparison theorem also holds. □

We consider
(a) $y_t = \xi + \int_t^1 f(s, y_s, z_s)ds - \int_t^1 z_s dB_s$
(b)$_n$ $y_t^n = \xi^n + \int_t^1 f_n(s, y_s^n, z_s^n)d\langle B^n\rangle_t - \int_t^1 z_s^n dB_s^n$

Theorem 6.2. *(Briand, Delyon & Memin, 2001) We assume (H1), (H2) and (H3). Let (y^n, z^n) be the solution of (b)$_n$ and (y, z) be the solution of (a). Then, in $\mathcal{S}^2 \times \mathcal{S}^2$,*

$$\left(y^n, \int_0^{\cdot} z_s^n dB_s^n\right) \to \left(y, \int_0^{\cdot} z_s dB_s\right), \text{ as } n \to \infty$$

and in $\mathcal{S}^2 \times \mathcal{S}^2$

$$\left(\int_0^{\cdot} z_s^n d\langle B^n\rangle_s, \int_0^{\cdot} |z_s^n|^2 d\langle B^n\rangle_s\right) \to \left(\int_0^{\cdot} z_s^n d\langle B^n\rangle_s, \int_0^{\cdot} |z_s^n|^2 d\langle B^n\rangle_s\right) \text{ as } n \to \infty.$$

7 Appendix

7.1 Martingale Representation Theorem

The existence theorem of BSDE requires the following result: any element $\xi \in L^2(\mathcal{F}_T)$ can be represent by

$$\xi = \mathbf{E}[\xi] + \int_0^T \phi_s dB_s.$$

For notational simplification, we assume that B is 1–dimensional, i.e., $d = 1$. We need the following lemma.

Lemma 7.1. *Let $\eta \in L^2(\mathcal{F}_T)$ be given such that*

$$\mathbf{E}[\eta(1 + \int_0^T \phi_s dB_s)] = 0, \quad \forall \phi \in L_{\mathcal{F}}^2(0, T).$$

Then $\eta = 0$, a.s..

Proof. For each deterministic $\mu(\cdot) \in L^\infty(0,T;\mathbb{C})$, we denote by X^μ, the solution of the following SDE

$$dX_t^\mu = \mu(t)X_t^\mu dB_t, \quad X_0^\mu = 1.$$

It suffices to prove that if, for each $\mu(\cdot) \in L^\infty(0,T;\mathbb{C})$ we have $\mathbf{E}[\eta X_T^\mu] = 0$, then $\eta = 0$, a.s.

For each $N \in \mathbb{Z}$, $x = (x_1, \cdots, x_N) \in \mathbf{R}^N$ and $0 \le t_1 < \cdots < t_N \le T$, we set $\mu(t) = i\sum_{j=1}^N x_j 1_{[0,t_j]}(t)$. It is easy to check that

$$X_t^\mu = \exp\{i\int_0^t \mu(s)dB_s - \frac{1}{2}\int_0^t |\mu(s)|^2 ds\}$$
$$= e^{i\sum_{j=1}^N x_j B_{t_j \wedge t}} \exp\{-\frac{1}{2}\int_0^t |\mu(s)|^2 ds\}$$

Thus the condition $\mathbf{E}[\eta X_T^\mu] = 0$ implies

$$\Phi_\mu(x) := \mathbf{E}[\eta e^{i\sum_{j=1}^N x_j B_{t_j}}] = 0.$$

Now for an arbitrary $g \in C_0^\infty(\mathbf{R}^N)$, let $\hat{g}$ be its Fourier transform. We then have

$$\mathbf{E}[g(B_{t_1}, \cdots, B_{t_N})\eta]$$
$$= \mathbf{E}[(2\pi)^{-\frac{N}{2}}\int_{\mathbf{R}^N} \hat{g}(x_1, \cdots, x_N)e^{i\sum_{j=1}^N x_j B_{t_j}} dx\eta]$$
$$= (2\pi)^{-\frac{N}{2}}\int_{\mathbf{R}^N} \hat{g}(x)\, \Phi_\mu(x)dx = 0.$$

Since the subset

$$\{g(B(t_1), \cdots, B(t_N)); 0 \le t_1, \cdots, t_N \le T,\ g \in C_0^\infty(\mathbf{R}^N), N \in \mathbb{Z}\}$$

is dense in $L^2(\mathcal{F}_T)$, it follows that $\eta = 0$.

We now can prove the representation theorem.

Theorem 7.1. (*Representation theorem of an element of $L^2(\mathcal{F}_T)$ by Itô's integral*) For each $\xi \in L^2(\mathcal{F}_T)$ there exists a unique $z \in L_{\mathcal{F}}^2(0,T)$ such that

$$\xi = \mathbf{E}[\xi] + \int_0^T z_s dB_s, \quad a.s. \tag{111}$$

Proof. Let $\xi \in L^2(\mathcal{F}_T)$ be given. We define the following functional

$$f(\phi) := \mathbf{E}[\xi \int_0^T \phi_s dB_s], \quad \phi \in L_{\mathcal{F}}^2(0,T;\mathbf{R}^d).$$

By Schwards inequality $|f(\phi)| \le \mathbf{E}[|\xi|^2]^{1/2} \cdot \mathbf{E}[\int_0^T |\phi_s|^2 ds]^{1/2}$. Thus f is a bounded linear functional defined on $L_{\mathcal{F}}^2(0,T)$. It follows from the well–known

Riesz representation theorem (see for example [Yosida1980] p90) that, there exists a unique process $z \in L^2_{\mathcal{F}}(0, T)$, such that

$$f(\phi) = \mathbf{E}[\int_0^T \phi_s z_s ds], \ \forall \phi \in L^2_{\mathcal{F}}(0, T),$$

or

$$\mathbf{E}[\int_0^T \phi_s dB_s(\xi - \int_0^T z_s dB_s)] = 0, \ \forall \phi \in L^2_{\mathcal{F}}(0, T).$$

Thus we have

$$\mathbf{E}[(1 + \int_0^T \phi_s dB_s)(\xi - \mathbf{E}[\xi] - \int_0^T z_s dB_s)] = 0, \ \forall \phi \in L^2_{\mathcal{F}}(0, T).$$

But by Lemma 7.1, this implies (111). □

7.2 A Monotonic Limit Theorem of Itô's Processes

We present a convergence result of a sequence of Itô processes, called "monotonic limit theorem". In this lecture we use this result to prove nonlinear supermartingale decomposition theorems. We consider the following sequence of Itô processes:

$$y_t^i = y_0^i + \int_0^t g_s^i ds - A_t^i + \int_0^t z_s^i dB_s, \ i = 1, 2, \cdots. \tag{112}$$

for each i, the adapted process $g^i \in L^2_{\mathcal{F}}(0, T)$ are given, we also assume that, for each i,

$$A^i \in S^2_{\mathcal{F}}(0, T) \text{ is increasing with } A_0^i = 0, \tag{113}$$

and

    (i) (g_t^i) and (z_t^i) are bounded in $L^2_{\mathcal{F}}(0, T)$: $\mathbf{E} \int_0^T [|g_s^i|^2 + |z_s^i|^2] ds \leq C$;
    (ii) (y_t^i) increasingly converges to (y_t) with $\mathbf{E}[\sup_{0 \leq t \leq T} |y_t|^2] < \infty$, (114)

where the constant C is independent of i. It is clear that

    (i) $\mathbf{E}[\sup_{0 \leq t \leq T} |y_t^i|^2] \leq C$;
    (ii) $\mathbf{E} \int_0^T |y_t^i - y_t|^2 dt \to 0$, (115)

where the constant C is independent of i.

Remark 7.1. It is not hard to check that the limit y has the following form

$$y_t = y_0 + \int_0^t g_s^0 ds - A_t + \int_0^t z_s dB_s, \tag{116}$$

where g^0 and z are respectively the weak limit of $\{g^i\}_{i=1}^\infty$ and $\{z^i\}_{i=1}^\infty$ in $L_{\mathcal{F}}^2(0,T)$, $(A_t)_{t\in[0,T]}$ is an increasing process. In general, we can not prove the strong convergence of $\left\{\int_0^T z_s^i dB_s\right\}_{i=1}^\infty$. Our new observation is: for each $p \in [1,2)$, $\{z^i\}$ converges strongly in $L_{\mathcal{F}}^p(0,T;\mathbf{R}^d)$. This observation is crucially important, since we will treat nonlinear cases.

The limit theorem is as follows.

Theorem 7.2. *We assume (113) and (114). Then the limit y_t of $\{y^i\}_{i=1}^\infty$ has a form (116), where $g^0 \in L_{\mathcal{F}}^2(0,T)$ and $z \in L_{\mathcal{F}}^2(0,T;\mathbf{R}^d)$ are respectively the weak limit of $\{g^i\}_{i=1}^\infty$ and $\{z^i\}_{i=1}^\infty$ in $L_{\mathcal{F}}^2(0,T)$ and $L_{\mathcal{F}}^2(0,T;\mathbf{R}^d)$. For each $t \in [0,T]$, A_t is a weak limit of $\{A_t^i\}_{i=1}^\infty$ in $L^2(\mathcal{F}_T)$. $(A_t)_{t\in[0,T]}$ is an RCLL square–integrable increasing process. Furthermore, for any $p \in [0,2)$, $\{z^i\}_{i=1}^\infty$ strongly converges to z in $L_{\mathcal{F}}^p(0,T,\mathbf{R}^d)$, i.e.,*

$$\lim_{i\to\infty} \mathbf{E} \int_0^T |z_s^i - z_s|^p ds = 0, \quad p \in [0,2). \tag{117}$$

If moreover $(y)_{t\in[0,T]}$ is continuous, then we have

$$\lim_{i\to\infty} \mathbf{E} \int_0^T |z_s^i - z_s|^2 ds = 0. \tag{118}$$

Remark 7.2. An interesting open problem is: does (118) hold without the additional continuous assumption for y?

In order to prove this theorem, we need the several Lemmas. The following lemma will be applied to prove that the limit processes y is RCLL.

Lemma 7.2. *Let $\{x^i(\cdot)\}_{i=1}^\infty$ be a sequence of (deterministic) RCLL processes defined on $[0,T]$ that increasingly converges to $x(\cdot)$ such that, for each $t \in [0,T]$, and $i = 1, 2, \cdots$, $x^i(t) \leq x^{i+1}(t)$, with $x(t) = b(t) - a(t)$, where $b(\cdot)$ is an RCLL process and $a(\cdot)$ is an increasing process with $a(0) = 0$ and $a(T) < \infty$. Then $x(\cdot)$ and $a(\cdot)$ are also RCLL processes.*

Proof. Since $b(\cdot)$, $a(\cdot)$ and thus $x(\cdot)$ have left and right limits, thus we only need to check that $x(\cdot)$ is right continuous. For each $t \in [0,T)$, since $a(t+) \geq a(t)$, thus

$$x(t+) = b(t) - a(t+) \leq x(t). \tag{119}$$

On the other hand, for any $\delta > 0$, there exists a positive integer $j = j(\delta, t)$ such that $x(t) \leq x^j(t) + \delta$. Since $x^j(\cdot)$ is RCLL, thus there exists a positive number $\epsilon_0 = \epsilon_0(j, t, \delta)$ such that $x^j(t) \leq x^j(t+\epsilon)+\delta, \forall\epsilon \in (0, \epsilon_0]$. These imply that, for any $\epsilon \in (0, \epsilon_0]$,

$$x(t) \leq x^j(t + \epsilon) + 2\delta \leq x^{i+j}(t + \epsilon) + 2\delta \uparrow\uparrow x(t + \epsilon) + 2\delta.$$

Particularly, we have $x(t) \leq x(t+) + 2\delta$ and thus $x(t) \leq x(t+)$. This with (119) implies the right continuity of $x(\cdot)$. □

We need some estimates for the jumps of A. We first have

Lemma 7.3. *Let A be an increasing RCLL process defined on $[0, T]$ with $A_0 = 0$ and $\mathbf{E}(A_T)^2 < \infty$. Then, for any $\epsilon > 0$, there exists a finite number of stopping times σ_k, $k = 0, 1, 2, \cdots N + 1$ with $\sigma_0 = 0 < \sigma_1 \leq \cdots \leq \sigma_N \leq T = \sigma_{N+1}$ and with disjoint graphs on $(0, T)$ such that*

$$\sum_{k=0}^{N} \mathbf{E} \sum_{t \in (\sigma_k, \sigma_{k+1})} (\Delta A_t)^2 \leq \epsilon. \tag{120}$$

Proof. For each $\nu > 0$, we denote

$$A_t(\nu) := A_t - \sum_{s \leq t} \Delta A_s 1_{\{\Delta A_s > \nu\}}.$$

$A_{\cdot}(\nu)$ has jumps of $A_{\cdot}$ smaller than ν. Thus there is a sufficiently small $\nu > 0$ such that

$$\mathbf{E}[\sum_{s \leq T} (\Delta A_s(\nu))^2] \leq \frac{\epsilon}{2}.$$

Now let τ_k, $k = 1, 2, \cdots$ be the successive times of jumps of A with size bigger than ν; they are stopping times, and there is N such that

$$\mathbf{E} \left(\sum_{s \in (\tau_N, T)} (\Delta A_s)^2 \right) \leq \frac{\epsilon}{2}.$$

We then set $\sigma_k := \tau_k \wedge T$ for $k \leq N$, and $\sigma_{N+1} = T$. It is clear that $\{\sigma_k\}_{k=0}^{N+1}$ satisfies (120).   □

For applying the formula of the integral by part to the limit process y (with jumps), the above open intervals (σ_k, σ_{k+1}) is not so convenient. Thus we will cut a sufficiently small interval (σ_k, τ_k) and only work on the remaining subintervals $(\sigma_k, \tau_k]$. This is possible since our filtration is continuous. In fact we have:

Lemma 7.4. *Let $0 < \sigma \leq T$ be a stopping time. Then there exists a sequence of $\mathcal{F}_t$–stopping times $\{\tau^i\}$ with $0 < \tau^i < \sigma$, a.s. for each $i = 1, 2, \cdots$, such that $\tau^i \uparrow \sigma$.*

For the continuous filtration $\mathcal{F}_t$, this lemma is quite classical. The proof is omitted.

The following lemma tells that, for any given RCLL increasing process, the contribution of the jumps of A is mainly concentrated within a finite number of left–open right–closed intervals with "sufficiently small total length". Specifically, we have

Lemma 7.5. *Let A be an increasing RCLL process defined on $[0, T]$ with $A_0 = 0$ and $\mathbf{E}A_T^2 < \infty$. Then, for any $\delta, \epsilon > 0$, there exists a finite number of pairs of stopping times $\{\sigma_k, \tau_k\}$, $k = 0, 1, 2, \cdots N$ with $0 < \sigma_k \leq \tau_k \leq T$, such that*

(i) $(\sigma_j, \tau_j] \cap (\sigma_k, \tau_k] = \emptyset$ for each $j \neq k$;

(ii) $\mathbf{E} \sum_{k=0}^{N} (\tau_k - \sigma_k) \geq T - \epsilon$

(iii) $\sum_{k=0}^{N} \mathbf{E} \sum_{\sigma_k < t \leq \tau_k} (\Delta A_t)^2 \leq \delta$

Proof. We first apply Lemma 7.3 to construct a sequence of non-decreasing stopping times $\{\sigma_k\}_{k=0}^{N+1}$ with $\sigma_0 = 0$ and $\sigma_{N+1} = T$ such that, $\sigma_k < \sigma_{k+1}$ whenever $\sigma_k < T$ and that

$$\sum_{k=0}^{N} \mathbf{E} \sum_{t \in (\sigma_k, \sigma_{k+1})} (\Delta A_t)^2 \leq \delta.$$

Then for each $0 \leq k \leq N$, we apply Lemma 7.4 to construct a stopping time $0 < \tau_k' < \sigma_{k+1}$, such that

$$\mathbf{E} \sum_{k=0}^{N} (\sigma_{k+1} - \tau_k') \leq \epsilon.$$

Finally we set

$$\tau_0 = \tau_0', \ \tau_1 = \sigma_1 \vee \tau_1', \ \cdots, \ \tau_N = \sigma_N \vee \tau_N'.$$

It is clear that $\tau_k \in [\sigma_k, \sigma_{k+1}) \cap [\tau_{k+1}', \sigma_{k+1}]$. We have also $\tau_k < \sigma_{k+1}$ whenever $\sigma_k < T$. Thus $(\sigma_k, \tau_k] \in (\sigma_k, \sigma_{k+1})$. It follows that

$$\mathbf{E} \sum_{k=0}^{N} (\sigma_{k+1} - \tau_k) \leq \epsilon,$$

or

$$\mathbf{E} \sum_{k=0}^{N} (\tau_k - \sigma_k) \geq T - \epsilon,$$

and

$$\sum_{k=0}^{N} \mathbf{E} \sum_{t \in (\sigma_k, \tau_k]} (\Delta A_t)^2 \leq \sum_{k=0}^{N} \mathbf{E} \sum_{t \in (\sigma_k, \sigma_{k+1})} (\Delta A_t)^2 \leq \delta.$$

Thus the above conditions (i)-(iii) are satisfied.   □

We now give the

Proof of Theorem 7.2. Since (g^i) (resp. (z^i)) is weakly compact in $L_{\mathcal{F}}^2(0, T)$ (resp. $L_{\mathcal{F}}^2(0, T; \mathbf{R}^d)$), there is a subsequence, still denoted by (g^i) (resp. (z^i)) which converges weakly to (g_t^0) (resp. (z_t)).

Thus, for each stopping time $\tau \leq T$, the following weak convergence holds in $L^2(\mathcal{F}_\tau)$.

$$\int_0^\tau z_s^i dB_s \rightharpoonup \int_0^\tau z_s dB_s, \qquad \int_0^\tau g_s^i ds \rightharpoonup \int_0^\tau g_s^0 ds.$$

Since

$$A^i_\tau = -y^i_\tau + y^i_0 + \int_0^\tau g^i_s ds + \int_0^\tau z^i_s dB_s$$

thus we also have the weak convergence

$$A^i_\tau \rightharpoonup A_\tau := -y_\tau + y_0 + \int_0^\tau g^0_s ds + \int_0^\tau z_s dB_s.$$

Obviously, $\mathbf{E}[A^2_T] < \infty$. For any two stopping times $\sigma \leq \tau \leq T$, we have $A_\sigma \leq A_\tau$ since $A^i_\sigma \leq A^i_\tau$. From this it follows that A is an increasing process. Moreover, from Lemma 7.2, both A and y are RCLL. Thus y has a form of (116). Since y is given, it is clear that z is uniquely determined. Thus not only the subsequence of $\{z^i\}_{i=1}^\infty$ but also the sequence itself converges weakly to z. Our key point is to show that $\{z^i\}_{i=1}^\infty$ converges to z in the strong sense of (117). In order to prove this we use Itô's formula applied to $(y^i_t - y_t)^2$ on a given subinterval $(\sigma, \tau]$. Here $0 \leq \sigma \leq \tau \leq T$ are two stopping times. Observe that $\Delta y_t \equiv \Delta A_t$ and the fact that y^i and then A^i are continuous. We have

$$\mathbf{E}|y^i_\sigma - y_\sigma|^2 + \mathbf{E} \int_\sigma^\tau |z^i_s - z_s|^2 ds$$

$$= \mathbf{E}|y^i_\tau - y_\tau|^2 - \mathbf{E} \sum_{t \in (\sigma, \tau]} (\Delta A_t)^2 - 2\mathbf{E} \int_\sigma^\tau (y^i_s - y_s)(g^i_s - g^0_s) ds$$

$$+ 2\mathbf{E} \int_{(\sigma, \tau]} (y^i_s - y_s) dA^i_s - 2\mathbf{E} \int_{(\sigma, \tau]} (y^i_s - y_{s-}) dA_s$$

$$= \mathbf{E}|y^i_\tau - y_\tau|^2 + \mathbf{E} \sum_{t \in (\sigma, \tau]} (\Delta A_t)^2 - 2\mathbf{E} \int_\sigma^\tau (y^i_s - y_s)(g^i_s - g^0_s) ds$$

$$+ 2\mathbf{E} \int_{(\sigma, \tau]} (y^i_s - y_s) dA^i_s - 2\mathbf{E} \int_{(\sigma, \tau]} (y^i_s - y_{s-}) dA_s$$

Since $\int_{(\sigma, \tau]}(y^i_s - y_s) dA^i_s \leq 0$, we then have

$$\mathbf{E} \int_\sigma^\tau |z^i_s - z_s|^2 ds \leq \mathbf{E}|y^i_\tau - y_\tau|^2 + \mathbf{E} \sum_{t \in (\sigma, \tau]} (\Delta A_t)^2 \qquad (121)$$

$$+ 2\mathbf{E} \int_\sigma^\tau |y^i_s - y_s||g^i_s - g^0_s| ds + 2\mathbf{E} \int_{(\sigma, \tau]} |y^i_s - y_s| dA_s.$$

The third term on the right side tends to zero since

$$\mathbf{E} \int_0^T |y^i_s - y_s||g^i_s - g^0_s| ds \leq C \left[\mathbf{E} \int_0^T |y^i_s - y_s|^2 ds \right]^{\frac{1}{2}} \to 0. \qquad (122)$$

For the last term, we have, P–almost surely,

$$|y_s^1 - y_s| \geq |y_s^i - y_s| \to 0, \qquad \forall s \in [0, T].$$

Since

$$\mathbf{E} \int_0^T |y_s^1 - y_s| dA_s \leq (\mathbf{E}[\sup_s (|y_s^1 - y_s|^2)])^{\frac{1}{2}} (\mathbf{E}(A_T)^2)^{\frac{1}{2}} < \infty.$$

It then follows from Lebesgue's dominated convergence theorem that

$$\mathbf{E} \int_{(0,T]} |y_s^i - y_s| dA_s \to 0. \tag{123}$$

By convergence of (122) and (123), it is clear from the estimate (121) that, once A is continuous (thus $\Delta A_t \equiv 0$) on $[0, T]$, then z^i tends to z strongly in $L_{\mathcal{F}}^2(0, T; \mathbf{R}^d)$. Thus the second assertion of the theorem, i.e., (118) follows.

But for the general case, the situation becomes complicated. Thanks to Lemma 7.5, for any positive δ and ϵ, there exist a finite number of disjoint intervals $(\sigma_k, \tau_k]$, $k = 0, 1, \cdots, N$, such that $\sigma_k \leq \tau_k \leq T$ are all stopping times satisfying

$$\begin{array}{l} \text{(i)} \quad \mathbf{E} \sum_{k=0}^N [\tau_k - \sigma_k](\omega) \geq T - \frac{\epsilon}{2}; \\ \text{(ii)} \quad \sum_{k=0}^N \sum_{\sigma_k < t \leq \tau_k} \mathbf{E}(\Delta A_t)^2 \leq \frac{\delta \epsilon}{3}. \end{array} \tag{124}$$

Now, for each $\sigma = \sigma_k$ and $\tau = \tau_k$, we apply estimate (121) and then take the sum. It follows that

$$\sum_{k=0}^N \mathbf{E} \int_{\sigma_k}^{\tau_k} |z_s^i - z_s|^2 ds \leq \sum_{k=0}^N \mathbf{E} |y_{\tau_k}^i - y_{\tau_k}|^2 + \sum_{k=0}^N \mathbf{E} \sum_{t \in (\sigma_k, \tau_k]} (\Delta A_t)^2$$

$$+ 2\mathbf{E} \int_0^T |y_s^i - y_s| |g_s^i - g_s^0| ds + 2\mathbf{E} \int_{(0,T]} |y_s^i - y_s| dA_s.$$

By using the convergence results (122) and (123) and taking in consideration of (124)-(ii), it follows that

$$\overline{\lim_{i \to \infty}} \sum_{k=0}^N \mathbf{E} \int_{\sigma_k}^{\tau_k} |z_s^i - z_s|^2 ds \leq \sum_{k=0}^N \mathbf{E} \sum_{t \in (\sigma_k, \tau_k]} (\Delta A_t)^2 \leq \frac{\epsilon \delta}{3}$$

Thus there exists an integer $l_{\epsilon \delta} > 0$ such that, whenever $i \geq l_{\epsilon \delta}$, we have

$$\sum_{k=0}^N \mathbf{E} \int_{\sigma_k}^{\tau_k} |z_s^i - z_s|^2 ds \leq \frac{\epsilon \delta}{2}$$

Thus, in the product space $([0, T] \times \Omega, \mathcal{B}([0, T]) \times \mathcal{F}, m \times P)$ (here m stands for the Lebesgue measure on $[0, T]$), we have

$$m \times P \left\{ (s, \omega) \in \bigcup_{k=0}^N (\sigma_k(\omega), \tau_k(\omega)] \times \Omega; \ |z_s^i(\omega) - z_s(\omega)|^2 \geq \delta \right\} \leq \frac{\epsilon}{2}$$

This with (124)-(i) implies

$$m \times P\left\{(s,\omega) \in [0,T] \times \Omega; \quad |z_s^i(\omega) - z_s(\omega)|^2 \geq \delta\right\} \leq \epsilon, \qquad \forall \; i \geq l_{\epsilon\delta}.$$

From this it follows that, for any $\delta > 0$,

$$\lim_{i \to \infty} m \times P\left\{(s,\omega) \in [0,T] \times \Omega; \quad |z_s^i(\omega) - z_s(\omega)|^2 \geq \delta\right\} = 0.$$

Thus, on $[0,T] \times \Omega$, the sequence $\{z^i\}_{i=1}^\infty$ converges in measure to z. Since $\{z^i\}_{i=1}^\infty$ is also bounded in $L_{\mathcal{F}}^2(0,T;\mathbf{R}^d)$, then for each $p \in [1,2)$, it converges strongly in $L_{\mathcal{F}}^p(0,T;\mathbf{R}^d)$. □

7.3 Optional Stopping Theorem for $\mathcal{E}^g$–Supermartingale

In this subsection the function g satisfies (i), (ii) of (34). We will discuss $\mathcal{E}_{\sigma,\tau}^g[\cdot]$ for stopping times $\sigma, \tau \in \mathcal{S}_T$. A BSDE with a given terminal condition $X \in \mathcal{F}_\tau$ at a given terminal time $\tau \in \mathcal{S}_T$ is formulated as

$$Y_s = X + \int_s^\tau g(r, Y_r, Z_r)dr - \int_s^\tau Z_r dB_r, \; s \in [0,\tau], \tag{125}$$

or equivalently, on $s \in [0,T]$,

$$Y_s = X + \int_s^T 1_{[0,\tau]}(r)g(r, Y_r, Z_r)dr - \int_s^\tau 1_{[0,\tau]}(r)Z_r dB_r. \tag{126}$$

We define

$$\mathcal{E}_{\sigma,\tau}^g[X] := Y_\sigma. \tag{127}$$

It is clear that, when $\sigma = s$ and $\tau = t$ for deterministic time parameters $s \leq t$, then $\mathcal{E}_{\sigma,\tau}^g[\cdot] = \mathcal{E}_{s,t}^g[\cdot]$. We have

Proposition 7.1. *The system of operators*

$$\mathcal{E}_{\sigma,\tau}^g[\cdot] : L^2(\mathcal{F}_\tau) \to L^2(\mathcal{F}_\sigma), \; \sigma \leq \tau, \; \sigma, \tau \in \mathcal{S}_T,$$

is an $\mathcal{F}_t$-consistent nonlinear evaluation, i.e., it satisfies (A1)–(A5) in the following sense: for each $X, X' \in L^2(\mathcal{F}_\tau)$,
(a1) $\mathcal{E}_{\sigma,\tau}^g[X] \geq \mathcal{E}_{\sigma,\tau}^g[X']$, a.s., if $X \geq X'$, a.s.
(a2) $\mathcal{E}_{\tau,\tau}^g[X] = X$;
(a3) $\mathcal{E}_{\rho,\sigma}^g[\mathcal{E}_{\sigma,\tau}^g[X]] = \mathcal{E}_{\rho,\tau}^g[X]$, $\forall 0 \leq \rho \leq \sigma \leq \tau$;
(a4) $1_A \mathcal{E}_{\sigma,\tau}^g[X] = 1_A \mathcal{E}_{\sigma,\tau}^g[1_A X]$, $\forall A \in \mathcal{F}_\tau$;
(a5) for each $0 \leq \sigma \leq \tau \leq T$,

$$\mathcal{E}_{\sigma,\tau}^g[X] - \mathcal{E}_{\sigma,\tau}^g[X'] \leq \mathcal{E}_{\sigma,\tau}^{g_\mu}[X - X'], \; \forall X, X' \in L^2(\mathcal{F}_\tau). \tag{128}$$

The proof is similar as in the case where ρ, σ and τ are deterministic. We omit it.

Another easy property is that $\mathcal{E}_{\cdot \wedge \tau, \tau}[X]$ has continuous paths:

$$(\mathcal{E}^g_{t \wedge \tau, \tau}[X])_{0 \leq t \leq T} \in S^2_{\mathcal{F}}(0, T). \tag{129}$$

By (32) and (33) with $1_{[\sigma, \tau]}(s)g(s, y, z)$ in the place of g, we also have the following estimates

$$\mathbf{E}[|\mathcal{E}^g_{\sigma, \tau}[X]|^2] \leq c\mathbf{E}[|X|^2] + c\mathbf{E}\int_\sigma^\tau |g(s, 0, 0)|^2)ds, \tag{130}$$

and

$$\mathbf{E}[|\mathcal{E}^g_{\sigma, \tau}[X - X']|^2] \leq c\mathbf{E}[|X - X'|^2]. \tag{131}$$

where the constant c depends only on T and the Lipschitz constant C of the function g w.r.t. (y, z). As a consequence of

We also have the following estimate:

Lemma 7.6. *Let σ, $\tau \in \mathcal{S}_T$, $\sigma \leq \tau$ and $X \in L^2(\mathcal{F}_\tau)$. If $X \in L^2(\mathcal{F}_\sigma)$, then we have*

$$\mathbf{E}[|\mathcal{E}^g_{\sigma, \tau}[X] - X|^2] \leq c\mathbf{E}[\int_\sigma^\tau |g(s, X, 0)|^2 ds].$$

where the constant c depends only on T and the Lipschitz constant C of g.

Proof. Observe that $\mathcal{E}^g_{\sigma, \tau}[X] = y_\sigma$, where $(y_t)_{t \in [0, T]}$ is the solution of the BSDE

$$y_t = X + \int_t^T 1_{[\sigma, \tau]}(s)g(s, y_s, z_s)ds - \int_t^T z_s dB_s.$$

We set $\bar{y}_t \equiv y_t - X$, $\bar{z}_t \equiv z_t$, on $[\sigma, \tau]$. This pair of adapted process is the solution of the BSDE

$$\bar{y}_t = \int_t^T 1_{[\sigma, \tau]}(s)\bar{g}(s, \bar{y}_s, \bar{z}_s)ds - \int_t^T \bar{z}_s dB_s, \, t \in [\sigma, \tau].$$

With $\bar{g}(t, y, z) := g(t, y + X, z)$, we have $\mathcal{E}^g_{\sigma, \tau}[X] - X = \mathcal{E}^{\bar{g}}_{\sigma, \tau}[0]$. From (130),

$$\mathbf{E}[|\mathcal{E}^g_{\sigma, \tau}[X] - X|^2] = \mathbf{E}[|\mathcal{E}^{\bar{g}}_{\sigma, \tau}[0]|^2]$$

$$\leq c\mathbf{E}[\int_\sigma^\tau |\bar{g}(s, 0, 0)|^2 ds]$$

$$= c\mathbf{E}[\int_\sigma^\tau |g(s, X, 0)|^2 ds].$$

$\square$

We will prove the following optional stopping theorem:

Theorem 7.3. *We assume that the function g satisfies (i), (ii) of (34). Let* $Y \in D^2_{\mathcal{F}}(0,T)$ *be an $\mathcal{E}$-supermartingale (resp. $\mathcal{E}$-submartingale). Then for each $\sigma, \tau \in \mathcal{S}_T$ such that $\sigma \leq \tau$, we have*

$$\mathcal{E}^g_{\sigma,\tau}[Y_\tau] \leq Y_\sigma \ (\text{resp. } \geq Y_\sigma), \ \text{a.s.} \ . \tag{132}$$

To prove the above theorem, we need several lemmas.

Lemma 7.7. *Let $\tau \in \mathcal{S}^0_T$ be valued in $\{t_0, \cdots, t_n\}$ with $0 = t_0 \leq t_1 < \cdots < t_n \leq t_{n+1} = T$, and let*

$$t_i \leq s < t \leq t_{i+1}, \ \text{for some } i \in \{1, 2, \cdots, n\}. \tag{133}$$

Then, for each $X \in \mathcal{F}_{t \wedge \tau}$,

$$\begin{cases} (i) \ \mathcal{E}^g_{t \wedge \tau, t \wedge \tau}[X] = X; \\ (ii) \ \mathcal{E}^g_{s \wedge \tau, t \wedge \tau}[X] = 1_{\{t \wedge \tau \leq s\}} X + 1_{\{t \wedge \tau = t\}} \mathcal{E}^g_{s,t}[X]. \end{cases} \tag{134}$$

Proof. (i) is easy. To prove (ii), we first observe that

$$\{t \wedge \tau \leq s\}^C = \{t \wedge \tau = t\} \tag{135}$$

and $\{t \wedge \tau \leq s\} = \{t \wedge \tau \leq t_i\}$. Thus $1_{\{t \wedge \tau \leq s\}} X \in \mathcal{F}_{t_i}$. We also have $1_{\{t \wedge \tau = t\}} X \in \mathcal{F}_t$. We now solve $Y_{s \wedge \tau} = \mathcal{E}^g_{s \wedge \tau, t \wedge \tau}[X]$ by, as in (126),

$$Y_{s \wedge \tau} = X + \int_s^T 1_{[0, t \wedge \tau]}(r) g(r, Y_r, Z_r) dr - \int_s^T 1_{[0, t \wedge \tau]}(r) Z_r dB_r. \tag{136}$$

Since $1_{[0, t \wedge \tau]} = 1_{\{t \wedge \tau \leq t_i\}} 1_{[0, t_i]} + 1_{\{t \wedge \tau = t\}} 1_{[0, t]}$. By respectively multiplying $1_{\{t \wedge \tau \leq t_i\}}$ and $1_{\{t \wedge \tau = t\}}$ on both sides of (136), we have, on $s \in [t_i, t)$,

$$Y_{s \wedge \tau} 1_{\{t \wedge \tau \leq t_i\}} = X 1_{\{t \wedge \tau \leq t_i\}}, \tag{137}$$

and

$$Y_{s \wedge \tau} \, 1_{\{t \wedge \tau = t\}} = 1_{\{t \wedge \tau = t\}} X + \int_s^T 1_{[0,t]}(r) 1_{\{t \wedge \tau = t\}} g(r, Y_r, Z_r) dr$$

$$- \int_s^T 1_{[0,t]} 1_{\{t \wedge \tau = t\}}(r) Z_r dB_r$$

$$= 1_{\{t \wedge \tau = t\}} X + \int_s^t g(r, 1_{\{t \wedge \tau = t\}} Y_r, 1_{\{t \wedge \tau = t\}} Z_r) dr$$

$$- \int_s^t 1_{[0, t_{i+1}]} Z_r dB_r.$$

We observe that, the last relation implies that, on $[t_i, t]$,

$$Y_{s \wedge \tau} \, 1_{\{t \wedge \tau = t\}} = 1_{\{t \wedge \tau = t\}} \mathcal{E}^g_{s,t}[1_{\{t \wedge \tau = t\}} X] = 1_{\{t \wedge \tau = t\}} \mathcal{E}^g_{s,t}[X].$$

This with (137) and (135), we then have (ii). $\quad\square$

We now treat a simple situation of the above optional stopping theorem.

Lemma 7.8. Let $Y \in D^2_{\mathcal{F}}(0,T)$ be an $\mathcal{E}^g$- martingale (respectively $\mathcal{E}^g$ - supermartingale, $\mathcal{E}^g$- submartingale). Then for each $\sigma, \tau \in \mathcal{S}^0_T$ such that $\sigma \leq \tau$, we have

$$\mathcal{E}^g_{\sigma,\tau}[Y_\tau] = Y_\sigma, \ (resp. \ \leq Y_\sigma, \ \geq Y_\sigma) \ a.s. \tag{138}$$

Proof. We only prove the case for $\mathcal{E}^g$–supermartingale. It is clear that, once we have

$$\mathcal{E}^g_{t \wedge \tau, \tau}[Y_\tau] \leq Y_{t \wedge \tau}, \ \forall t \in [0,T], \tag{139}$$

then, (138) hold for each $\sigma \in \mathcal{S}^0_T$ valued in $\{s_1, \cdots, s_m\}$ since

$$\mathcal{E}^g_{\sigma,\tau}[Y_\tau] = \sum_{i=1}^m 1_{\{\sigma=s_i\}} \mathcal{E}^g_{s_i \wedge \tau, \tau}[Y_\tau] \leq \sum_{i=1}^m 1_{\{\sigma=s_i\}} Y_{s_i \wedge \tau} = Y_\sigma.$$

We will prove (139) by deduction. Let $\tau \in \mathcal{S}^0_T$ be valued in $\{t_0, \cdots, t_n\}$ with $0 = t_0 \leq t_1 < \cdots < t_n \leq t_{n+1} = T$. Firstly, when $t \geq t_n$, (139) holds since $\mathcal{E}^g_{t \wedge \tau, \tau}[Y_\tau] = \mathcal{E}^g_{\tau,\tau}[Y_\tau] = Y_\tau$. Now suppose that for a fixed $i \in \{1, \cdots, n\}$, (139) holds for $t \geq t_i$. We shall prove that it also holds for $t \geq t_{i-1}$. We need to check the case $t \in [t_{i-1}, t_i)$.

Since $1_{\{t_i \wedge \tau = t_i\}}$ is $\mathcal{F}_t$–measurable, by (a4) we have

$$\begin{aligned}
1_{\{t_i \wedge \tau = t_i\}} \mathcal{E}^g_{t,t_i}[Y_{t_i \wedge \tau}] &= 1_{\{t_i \wedge \tau = t_i\}} \mathcal{E}^g_{t,t_i}[1_{\{t_i \wedge \tau = t_i\}} Y_{t_i}] \\
&= 1_{\{t_i \wedge \tau = t_i\}} \mathcal{E}^g_{t,t_i}[Y_{t_i}] \\
&\leq 1_{\{t_i \wedge \tau = t_i\}} Y_t.
\end{aligned}$$

It follows from (134)–(ii)

$$\begin{aligned}
\mathcal{E}^g_{t \wedge \tau, t_i \wedge \tau}[Y_{t_i \wedge \tau}] &= 1_{\{t_i \wedge \tau \leq t\}} Y_{t_i \wedge \tau} + 1_{\{t_i \wedge \tau = t_i\}} \mathcal{E}^g_{t,t_i}[Y_{t_i \wedge \tau}] \\
&\leq 1_{\{t_i \wedge \tau \leq t\}} Y_{t_i \wedge \tau} + 1_{\{t_i \wedge \tau = t_i\}} Y_t \\
&= Y_{t \wedge \tau}.
\end{aligned}$$

The last step is from $\{t_i \wedge \tau \leq t\} + \{t_i \wedge \tau = t_i\} = \Omega$ and then $t \wedge \tau = t_i \wedge \tau 1_{\{t_i \wedge \tau \leq t\}} + t 1_{\{t_i \wedge \tau = t_i\}}$. From this result we derive

$$\begin{aligned}
\mathcal{E}^g_{t \wedge \tau, \tau}[Y_\tau] &= \mathcal{E}^g_{t \wedge \tau, t_i \wedge \tau}[\mathcal{E}^g_{t_i \wedge \tau, \tau}[Y_\tau]] \\
&\leq \mathcal{E}^g_{t \wedge \tau, t_i \wedge \tau}[Y_{t_i \wedge \tau}] \\
&\leq Y_{t \wedge \tau}.
\end{aligned}$$

Thus (139) holds for $t \geq t_{i-1}$. It follows by deduction that (139) holds for $t \in [0,T]$. The proof is complete. $\square$

We now give

Proof of Theorem 7.3. We only prove the supermartingale part. For each $n = 1, 2, \cdots$, we set

$$\sigma_n := T \sum_{k=1}^{2^n-1} 2^{-n}k 1_{\{2^{-n}(k-1) \le \sigma < 2^{-n}k\}} + T1_{\{\sigma=T\}},$$

$$\tau_n := T \sum_{k=1}^{2^n-1} 2^{-n}k 1_{\{2^{-n}(k-1) \le \tau < 2^{-n}k\}} + T1_{\{\tau=T\}}.$$

It is clear that $\sigma_n \searrow \sigma$, $\tau_n \searrow \tau$ and $\sigma_n \le \tau_n$. By the above lemma, for each $m \ge n$ we have

$$\mathcal{E}^g_{\sigma_m,\tau_n}[Y_{\tau_n}] \le Y_{\sigma_m}, \text{ a.s.}$$

It follows from (129) and $Y \in D^2_{\mathcal{F}}(0,T)$ that, for each fixed n, $\mathcal{E}^g_{\sigma_m,\tau_n}[Y_{\tau_n}] \to \mathcal{E}^g_{\sigma,\tau_n}[Y_{\tau_n}]$ and $Y_{\sigma_m} \to Y_\sigma$ in $L^2(\mathcal{F}_T)$ as $m \to \infty$. We then have

$$\mathcal{E}^g_{\sigma,\tau_n}[Y_{\tau_n}] \le Y_\sigma, \text{ a.s.} \tag{140}$$

Moreover, we have

$$|\mathcal{E}^g_{\sigma,\tau_n}[Y_{\tau_n}] - \mathcal{E}^g_{\sigma,\tau}[Y_\tau]| \le |\mathcal{E}^g_{\sigma,\tau_n}[Y_{\tau_n}] - \mathcal{E}^g_{\sigma,\tau_n}[Y_\tau]| + |\mathcal{E}^g_{\sigma,\tau_n}[Y_\tau] - \mathcal{E}^g_{\sigma,\tau}[Y_\tau]|. \tag{141}$$

Since $Y_{\tau_n} \to Y_\tau$, in $L^2(\mathcal{F}_T)$, the first term on the right tends to zero in $L^2(\mathcal{F}_T)$ because of (131). For the second one, we still use (131):

$$\mathbf{E}[|\mathcal{E}^g_{\sigma,\tau_n}[Y_\tau] - \mathcal{E}^g_{\sigma,\tau}[Y_\tau]|^2] = \mathbf{E}[|\mathcal{E}^g_{\sigma,\tau}[\mathcal{E}^g_{\tau,\tau_n}[Y_\tau]] - \mathcal{E}^g_{\sigma,\tau}[Y_\tau]|^2]$$
$$\le c\mathbf{E}[|\mathcal{E}^g_{\tau,\tau_n}[Y_\tau] - Y_\tau|].$$

But by Lemma 7.6 this term is bounded by $c^2\mathbf{E}[\int_\tau^{\tau_n} |g(s,Y_\tau,0)|^2 ds]$. It follows that the term on the left side of (140) tends to $\mathcal{E}^g_{\sigma,\tau}[Y_\tau]$ in $L^2(\mathcal{F}_T)$ as $n \to \infty$. The proof is complete. $\square$

We will also prove the following optional stopping theorem:

Theorem 7.4. *We assume that an $\mathcal{F}$-expectation $\mathcal{E}[\cdot]$ satisfies (78) and (81) for some $\mu > 0$. Let $Y \in D^2_{\mathcal{F}}(0,T)$ be an $\mathcal{E}$-supermartingale (resp. $\mathcal{E}$-submartingale). Then for each $\sigma, \tau \in \mathcal{S}_T$ we have*

$$\mathcal{E}[Y_\tau|\mathcal{F}_\sigma] \le Y_{\tau\wedge\sigma}, \text{ (resp. } \ge Y_{t\wedge\tau}), \text{ a.s. } . \tag{142}$$

Proof. We only consider the supermartingale case. We first prove that

$$\mathcal{E}[Y_\tau|\mathcal{F}_t] \le Y_{t\wedge\tau} \text{ (resp. } \ge Y_{t\wedge\tau}), \text{ a.s. } . \tag{143}$$

Let τ be a finite valued: $\tau = \sum_{i=1}^n 1_{\{\tau=t_i\}} t_i$, for some $0 \le t_1 \le \cdots \le t_n \le T$. If $t_n \le t$, then it is clear that

$$\mathcal{E}[Y_\tau|\mathcal{F}_t] = Y_\tau = Y_{t\wedge\tau}.$$

If $t \in [t_{n-1}, t_n]$, then both $\{\tau \le t_{n-1}\}$ and $\{\tau = t_n\}$ are $\mathcal{F}_t$-measurable. By (11) we have

$$\mathcal{E}[Y_\tau|\mathcal{F}_t] = \mathcal{E}[Y_{t_n}1_{\{\tau=t_n\}} + Y_{\tau\wedge t_{n-1}}1_{\{\tau\leq t_{n-1}\}}|\mathcal{F}_t]$$
$$= 1_{\{\tau=t_n\}}\mathcal{E}[Y_{t_n}|\mathcal{F}_t] + 1_{\{\tau\leq t_{n-1}\}}\mathcal{E}[Y_{\tau\wedge t_{n-1}}|\mathcal{F}_t]$$
$$\leq 1_{\{\tau=t_n\}}Y_t + 1_{\{\tau\leq t_{n-1}\}}Y_{\tau\wedge t_{n-1}} = Y_{t\wedge\tau}.$$

If $t \in [t_{n-2}, t_{n-1}]$, then we have $\mathcal{E}[Y_\tau|\mathcal{F}_t] = \mathcal{E}[\mathcal{E}[Y_\tau|\mathcal{F}_{t_{n-1}}]|\mathcal{F}_t] \leq \mathcal{E}[Y_{t_{n-1}\wedge\tau}|\mathcal{F}_t]$ $\leq Y_{t\wedge\tau}$. We thus can prove an arbitrary case $t \in [t_i, t_{i+1}]$ by reduction. Thus (142) holds for all finite valued stopping times.

Now for $\tau \in \mathcal{S}_T$, we take τ_n as in the proof of Theorem 7.3. Since $Y \in D_\mathcal{F}^2(0,T)$, thus $Y_{\tau_n} \to Y_\tau$ in $L^2(\mathcal{F}_T)$. We have

$$\mathcal{E}^{-\mu}[Y_{\tau_n} - Y_\tau|\mathcal{F}_t] \leq \mathcal{E}[Y_{\tau_n}|\mathcal{F}_t] - \mathcal{E}[Y_\tau|\mathcal{F}_t] \leq \mathcal{E}^\mu[Y_{\tau_n} - Y_\tau|\mathcal{F}_t].$$

By Lemma 3.3, the right side tends to zero in $L^2(\mathcal{F}_T)$. So does the right side since

$$\mathcal{E}^{-\mu}[Y_{\tau_n} - Y_\tau|\mathcal{F}_t] = -\mathcal{E}^\mu[Y_\tau - Y_{\tau_n}|\mathcal{F}_t].$$

It follows that $\mathcal{E}[Y_{\tau_n}|\mathcal{F}_t] \to \mathcal{E}[Y_\tau|\mathcal{F}_t]$ in $L^2(\mathcal{F}_T)$. We then can pass two sides of the inequality

$$Y_{\tau_n\wedge t} \geq \mathcal{E}[Y_{\tau_n}|\mathcal{F}_t]$$

to the limit to get (143).

Since both $(\mathcal{E}[Y_\tau|\mathcal{F}_t])_{t\in[0,T]} \in S_\mathcal{F}^2(0,T)$ and $(Y_{t\wedge\tau})_{t\in[0,T]} \in D_\mathcal{F}^2(0,T)$ we can easily derive from (143) that for each $\sigma, \tau \in \mathcal{S}_T$, we have (142). $\qquad\square$

Notes

The expectation $\mathbf{E}[\cdot]$ on the probability space $(\Omega, \mathcal{F}, P)$ with $\mathcal{F}_t \subset \mathcal{F}, t \geq 0$ is clearly $\mathcal{F}_t$–consistent. Another example of linear $\mathcal{F}_t$–consistent expectation is $\mathbf{E}_Q[\cdot]$, the expectation under Girsanov transformation dQ/dP. But it seems that the study of $\mathcal{F}_t$–consistent nonlinear expectations is still a very new subject. In 1997, [Peng1997b] (see also [Peng1997a]) introduced the notion of g–expectations which is nonlinear and $\mathcal{F}_t$–consistent. In the same year, the notion of g–evaluation was introduced in [Peng1997a] under the name "stochastic backward semigroup". See also [30]. The term "$\mathcal{F}_t$–consistent nonlinear expectation" was named in [CHMP2002].

Linear BSDE was first introduced by Bismut in [Bis1973], [Bis1978]. Bensoussan developed this approach in [Ben1981] and [Ben1982]. The existence and uniqueness theorem of a nonlinear BSDE, i.e., Theorem 3.1 was obtained in Pardoux and Peng [PP1990]. The present version of the proof was based on El Karoui, Peng and Quenez [EPQ1997]. [EPQ1997] is also a good survey of BSDE and related fields. Comparison Theorem of BSDE i.e., Theorem 3.3 was obtained in [Peng1992] for the case g is C^1 in (y, z). The present case where g is Lipschitz in (y, z) was obtained in [EPQ1997]. [EPQ1997] also observed and investigated a natural relation between BSDE theory and the problem of pricing financial derivatives. We also refer to Yong and Zhou [YZ1999] for a systematic presentation of BSDE theory. Due to the limitation of the size of this lecture, we can not present many important subjects of BSDE theory.

In 1998, Chen [Chen98] has proved the following interesting property: if $\mathcal{E}_{g^1}[X] = \mathcal{E}_{g^2}[X']$, for all $X \in L^2(\mathcal{F}_T)$, then the two generators g^1 and g^2 also coincide: $g^1(s, y, z) \equiv g^2(s, y, z)$. This result was generalized to an "inverse comparison theorem" by [BCHMP2000] and then [CHMP2001]: if $\mathcal{E}_{g^1}[X] \geq \mathcal{E}_{g^2}[X']$, for all $X \in L^2(\mathcal{F}_T)$, then $g^1 \geq g^2$.

The well - known Doob - Meyer decomposition theorem can be found in most standard text books of stochastic analysis e.g., [DM1978-1982], [HWY1992], [IW1981], [KShr1998] and [RW2000]. Decomposition theorem of g - supermartingale of Doob - Meyer's type, i.e., Theorem 3.9 was obtained by Peng [Peng1999]. A new method, i.e., penalization method, was applied to prove this nonlinear decomposition theorem. This method was firstly introduced in BSDE theory by [EKPPQ1997]. The monotonic limit theorem for Itô's processes (Theorem 7.2) as well as for BSDEs (Theorem 3.8) are also obtained in [Peng1999]. Using this penalization method, Chen and Peng [CP1998] to the L^1 case with the usual filtration, which generalizes the Meyer's result to a nonlinear situation. These penalization method and limit theorem were then applied to prove the nonlinear supermartingale decomposition theorem for an abstract $\mathcal{E}$–expectation, i.e., Theorem 4.3. Theorem 4.3 was proved in [CHMP2002]. This type of decomposition theorem for a more general situation, i.e., the case for $\mathcal{F}_t$–evaluation, was recently obtained in [Peng2003b].

The representation theorem of an $\mathcal{F}_t$–expectation by a g - expectation, i.e., Theorem 4.4 was obtained in [CHMP2002]. The more general case, i.e., Theorem 4.5 was obtained in [Peng2003b].

References

[ADEH1999] Artzner, P., Delbaen, F., Eber, J.M. and Heath, D. Coherent measures of risk, Math. Finance, **9**, 203–228, 1999.

[Bis1973] Bismut, J.M. *Conjugate Convex Functions in Optimal Stochastic Control*, J.Math. Anal. Apl. 44, pp.384-404, 1973.

[Bis1978] Bismut, J.M. *Contrôle des systemes linéaires quadratiques : applications de l'integrale stochastique*, Sémin. Proba. XII., Lect. Notes in Mathematics, 649, pp.180-264, 1978, Springer.

[Ben1981] Bensoussan, A. Lecture on stochastic control, Lecture Notes in Mathematics, Vol.972, Springer–Verlag, 1981.

[Ben1982] Bensoussan, A., Stochastic Control by Functional Analysis Methods, North–Holland, 1982.

[BCHMP2000] P. Briand, F. Coquet, Y. Hu, J. Mémin and S. Peng, *A converse comparison theorem for BSDEs and related properties of g-expectations*, Electron. Comm. Probab, **5** (2000).

[Chen98] Z. Chen, *A property of backward stochastic differential equations*, C.R. Acad. Sci. Paris Sér. I Math. **326** (1998), no 4, 483–488.

[CE2002] Z. Chen and L. Epstein (2002), *Ambiguity, Risk and Asset Returns in Continuous Time*, Econometrica, **70**(4), 1403–1443.

[CHMP2001] P. Briand, F. Coquet, Y. Hu, J. Mémin and S. Peng, (2001) *A general converse comparison theorem for Backward stochastic differential equations,* C.R.Acad. Sci. Paris, **t. 333,** Serie I, 577–581.

[CHMP2002] F. Coquet, Y. Hu J. Memin and S. Peng (2002), *Filtration–consistent nonlinear expectations and related g–expectations,* Probab. Theory Relat. Fields, **123**, 1–27.

[CP1998]   Z. Chen and S. Peng (1998) *A Nonlinear Doob-Meyer type Decomposition and its Application.* SUT Journal of Mathematics (Japan), **34**, No.2, 197–208, 1998.

[CP200]   Z. Chen and S. Peng (2000), A general downcrossing inequality for g-martingales, *Statist. Probab. Lett.* **46**, no. 2, 169–175.

[CQZ2000]   J. Cvitanić, M.C. Quenez and F. Zapatero, (2000) *Incomplete information with recursive preference,* preprint.

[DF1992]   D. Duffie and L. Epstein (1992), Stochastic differential utility, *Econometrica* **60**, no 2, 353–394.

[Duffie]   D. Duffie (2001) *Dynamic Asset Pricing,* Princeton University Press.

[EKPPQ1997] N. El Karoui, C. Kapoudjian, E. Pardoux, S. Peng and M.-C. Quenez (1997), *Reflected Solutions of Backward SDE and Related Obstacle Problems for PDEs, Ann. Probab.* **25**, no 2, 702–737.

[EPQ1997]   N. El Karoui, N., S. Peng and M.-C. Quenez, *Backward stochastic differential equation in finance, Math. Finance* **7** (1997), no 1, 1–71.

[EQ1995]   El Karoui, N., and M.C.Quenez (1995), Dynamic Programming and Pricing of Contingent Claims in Incomplete Marke, SIAM J.of Control and Optimization, 33, n.1.

[FS1992]   Fleming, W.H. and Soner H.M., *Controlled Markov Processes and Viscosity Solutions,* Springer–Verleg, New York, 1992.

[FoSc2002] Föllmer H. and Alexander Schied, Convex measures of risk and trading constraints, preprint, version 2002.

[Frittelli2000] Frittelli, M. (2000) Representing sublinear risk measures and Pricing rules, Working paper no. 10, Universita di Milano Bicocca, Italy.

[FR-G2002] Frittelli, M. and Rosazza Gianin E. (2002) Putting oders in risk measures, J. Banking and Finance, Vol. 26, no.26, 1473–1486.

[HWY1992]   He, S.W., Wang, J.G. and Yan J.-A. (1992) Semimartingale Theory and Stochastic Calculus, CRC Press, Beijing.

[IW1981]   Ikeda, N. and Watanabe, S., *Stochastic Differential Equations and Diffusion Processes,* North–Holland, Amsterdam, 1981.

[KShr1998] Karatzas, I. and Shreve, S. E., *Brownian Motion and Stochastic Calculus,* Springer–Verleg, New York, 1988.

[DM1978-1982] Dellacherie, C. and Meyer, P.A., Probabilities and Potentiel A and B, Chap. North–Holland, 1978 and 1982.

[PP1990]   E. Pardoux and S. Peng (1990), *Adapted solution of a backward stochastic differential equation, Systems and Control Letters* **14**, no 1, 55-61.

[Peng1992] S. Peng (1992), *A generalized dynamic programming principle and Hamilton-Jacobi-Bellman equation, Stochastic stoch. reports* **38**, no 2, 119–134.

[Peng1997a] S. Peng, BSDE and Stochastic Optimizations, in *Topics in Stochastic Analysis,* J. Yan, S. Peng, S. Fang and L.M. Wu, Ch.2, (Chinese vers.), Science Publication 1997.

[Peng1997b] S. Peng (1997), BSDE and related *g*-expectation, *in Pitman Research Notes in Mathematics Series, no.364, "Backward Stochastic Differential Equation", Ed. by N. El Karoui & L. Mazliak,*, 141–159.

[Peng1999] S.Peng (1999), Monotonic limit theorem of BSDE and nonlinear decomposition theorem of Doob-Meyer's type, *Prob. Theory Rel. Fields* **113**, no 4, 473-499.

[Peng2002] S. Peng (2002), *Nonlinear expectations and nonlinear Markov chains*, in The proceedings of the 3rd Colloquium on "Backward Stochastic Differential Equations and Applications", Weihai, 2002.

[Peng2003a] S.Peng, (2003) The mechanism of evaluating risky values and nonlinear expectations, preprings.

[Peng2003b] S.Peng, (2003), Dynamical consistent nonlinear evaluations and expectations, preprint.

[Peng2003c] Peng, S. (2003) Filtration consistent nonlinear expectations and evaluations of contingent Claims, to appear in Acta Math. Appl. Sinica.

[PX2003] Peng, S. and Xu, M. (2003) Numerical calculations to solve BSDE, preprint 2003.

[RW2000] Rogers, L.C.G. & Williams, D., Diffusions, Markov processes and martingales, Cambridge University Press, 2000.

[Roazza2003] Rosazza, E. G., (2003) Some examples of risk measures via *g*–expectations, preprint.

[Yan1985] Yan J.-A. (1985) *On the commutability of essential infimum and conditional expectation operators*, Chinese Science Bulletin, **30**(8), 1013–1018.

[YZ1999] Yong, J. and Zhou, X. Stochastic Controls: Hamiltonian Systems and HJB Equations, Springer, 1999.

[Yosida1980] K. Yosida (1980), Functional Analysis, Springer–Verlag,, 6th edition

References on BSDE and Nonlinear Expectations

1. Fabio Antonelli, Backward-Forward stochastic differential equations, Ann. Appl. Prob. 1993, Vol.3, No.3 777–793

2. Fabio Antonelli, Stability of backward stochastic differential equations STOCH PROC APPL 62 (1): 103-114 MAR 1996.

3. Fabio Antonelli, Emilio Barucci, Mancino, Maria Elvira Asset pricing with a forward–backward stochastic differential utility. Econ. Lett. 72, No.2, 151-157 (2001).

4. Fabio Antonelli, E. Barucci, ME Mancino, A comparison result for FBSDE with applications to decisions theory, Math. Method Oper. Res. 54 (3): 407-423 FEB 2002

5. Fabio Antonelli, Kohatsu-Higa A Filtration stability of backward SDE's STOCH ANAL APPL 18 (1): 11-37 JAN 2000

6. Fabio Antonelli, Jin Ma, Weaksolutions of forward-backward SDE's Stoch. Anal. Appl. 21 (3): 493-514 MAY 2003

7. Khaled Bahlali, Backward-Forward stochastic differential equations with locally Lipschitz coefficient, C.R.A.S. Paris, t.333, I, 481–486, 2001.

8. Khaled Bahlail, Mezerdi Brahim, M.Hassani, Youssef Ouknine, Some generic properties in backward stochastic differential equations with continuous coefficient. Monte Carlo Methods Appl. 7, No.1-2, 15-19 (2001).

9. Khaled Bahlail, E.H.Essaky, M.Hassani, E.Pardoux, Existence, uniqueness and stability of backward stochastic differential equations with locally monotone coefficient. (English. Abridged French version) C. R., Math., Acad. Sci. Paris 335, No.9, 757-762 (2002).

10. Khaled Bahlail, El Essaky, Youssef Ouknine, Reflected backward stochastic differential equation with jumps and locally Lipschitz coefficient. Random Oper. Stoch. Equ. 10, No.4, 335-350 (2002).

11. Vlad Bally, Construction of asymptotically optimal controls for control and game problems, Proba. Theory Relat. Fields 111, 453–467 (1996)

12. Vlad Bally, A. Matoussi, Weak solutions for SPDE's and backward doubly stochastic differential equations. J. Theor. Probab. 14, No.1, 125-164 (2001).

13. Vlad Bally, Pages G Error analysis of the optimal quantization algorithm for obstacle problems, Stoch. Proc. Appl. 106 (1): 1-40 JUL 2003

14. Philippe Briand, A Remark On Generalized Feynman-Kac Formula, CR Acad. Sci. I-Math 321 (10): 1315-1318 Nov 16 1995

15. Philippe Briand, Ying Hu, Stability of BSDEs with Random Terminal Time and homogenization of Semilinear Elliptic PDEs, J. of Functional Analysis 155, 455–494, (1998)

16. Philippe Briand, B.Delyon, J. Memin, On the robustness of backward stochastic differential equations, Stoch. Proc. Appl. 97 (2): 229-253 FEB 2002

17. Philippe Briand, B. Delyon, Y. Hu, E. Pardoux L-p solutions of backward stochastic differential equations, Stoch. Proc. Appl. 108 (1): 109-129, 2003

18. Rainer Buckdahn, Backward stochastic differential equations. Option hedging under additional cost. Bolthausen, Erwin (ed.) et al., Seminar on stochastic analysis, random fields and applications. Proceedings of a seminar held at the Centro Stefano Franscini, Ascona, Switzerland, June 7-12, 1993. Basel: Birkhauser. Prog. Probab. 36, 307-318 (1995).

19. Rainer Buckdahn, Ying Hu, Pricing of American contingent claims with jump stock price and constrained portfolios, Math. Oper. Res. 23 (1): 177-203 Feb 1998.

20. Rainer Buckdahn, Ying Hu, Hedging contingent claims for a large investor in an incomplete market, Adv. Appl. Probab. 30 (1): 239-255 MAR 1998.

21. Rainer Buckdahn, Ying Hu, Probabilistic approach to homogenizations of systems of quasilinear parabolic PDEs with periodic structures, Nonlinear Analysis, Theory and Methods.

22. Rainer Buckdahn, Jin Ma, Stochastic viscosity solutions for nonlinear stochastic partial differential equations. Part I, Stoch. Processes.

23. Rainer Buckdahn, Jin Ma, Stochastic viscosity solutions for nonlinear stochastic partial differential equations. Part II, Stoch. Processes.

24. Rainer Buckdahn, E.Pardoux, Backward stochastic differential equations and integral-partial differential equations. Stochastics Stochastics Rep. 60, No.1-2, 57-83 (1997).

25. Rainer Buckdahn, Mare Quincampolx et Aurel Rascanu, Propriete de viabilite pour des equations diffenentielles stochastiques retrogrades et applications a des equations aux derivees partielles, C.R.A.S. Paris, t.325, I, 1159–1162, 1997.

26. Rainer Buckdahn, Shige Peng, Stationary backward stochastic differential equations and associated partial differential equations. Probab. Theory Relat. Fields 115, No.3, 383-399 (1999).

27. Rainer Buckdahn, M.Quincampoix, A.Rascanu, Viability property for a backward stochastic differential equation and applications to partial differential equations, Probab Theory Rel. Fields 116 (4): 485-504 APR 2000.

28. A. Cadenillas, A stochastic maximum principle for systems with jumps, with applications to finance, Syst Control Lett 47 (5): 433-444 Dec 16 2002

29. Zhigang Cao, Jia-An Yan, A Comparison Theorem for Solutions of Backward Stochastic Differential Equations, Advance in Mathematics, Vol. 28, NO. 4, 1999. 304–308.

30. Z. Chen and S. Peng (2001), Continuous Properties of g-martingales, *Chin. Ann. of Math.* **22B: 1**, 115–128.

31. Shuping Chen, XunJing Li, XunYu Zhou, Stochastic linear quadratic regulators with indefinite control weight costs, SIAM J. Control Optim. Vol. 38, No. 5, 1685–1702, September 1998.

32. Shuping Chen, Jiongmin Yong, Stochastic linear quadratic optimal control problems with random coefficients, Chin. Ann.of Math. 21B: 3(2000), 323–338.

33. Shuping Chen, XunYu Zhou, Stochastic linear quadratic regulators with indefinite control weight costs. II SIAM J Control Optim 39 (4): 1065-1081 Dec. 20 2000

34. Shuping Chen, Jiongmin Yong, Stochastic linear quadratic optimal control problems, Appl. Math. Opt. 43 (1): 21-45 Jan-Feb 2001

35. Zengjing Chen, Existence and uniqueness for BSDE with stopping time, Chinese Sci Bull 43 (2): 96-99 Jan 1998

36. Zengjing Chen, A proberty of backward stochastic differential equations,C.R.A.S. Paris, t.326, I, 483–488, (1998)

37. Zengjing Chen, A new proof of Doob-Meyer decomposition theorem, C.R.A.S. Paris, t.328, I, 919–924, 1999,'

38. Zengjing Chen, Shige Peng, A general Downcrossing Inequality for g-Martingales, *Statistics and Prob. Letters*, **45**, 1999.

39. Zengjing Chen, Bo Wang, Infinite time interval BSDES and the convergence of g-martingales, J Aust Math Soc A 69: 187-211 Part 2 Oct 2000

40. Zengjing Chen, Xiangrong Wang, Comonotonicity of backward stochastic differential equations.

41. Zengjing Chen, ShiGe Peng, Continuous properties of G-martingales, Chinese Ann Math B 22 (1): 115-128 JAN 2001

42. Zengjing Chen, L. Epstein, Ambiguity,risk, and asset returns in continuous time, Econometrica 70 (4):1403-1443 Jul 2002

43. D. Chevance, Discretization of Pardoux-Peng's backward stochastic differential equations, Z Angew Math Mech 76: 323-326 Suppl. 3 1996

44. D. Chevance, Numerical methods for backward stochastic differential equations. Rogers, L. C. G. (ed.) et al., Numerical methods in finance. Session at the Isaac Newton Institute, Cambridge, GB, 1995. Cambridge: Cambridge Univ. Press. 232-244 (1997).

45. Adam Cmiel, Gurgul, Henryk, Stochastic backward-lag-type Leontief model. Cent. Eur. J. Oper. Res. Econ. 5, No.1, 5-22 (1997).

46. Constantin, Adrian A backward stochastic differential equation with non-Lipschitz coefficients. C. R. Math. Acad. Sci., Soc. R. Can. 17, No.6, 280-282 (1995).

47. Francois Coquet, Resolution explicite d'une EDSR conduite par un processus de Poisson avec reflexion a la frontiere. (Explicit solution of a stochastic backward differential equation driven by Poisson process with reflection at the boundary). (French) Fascicule de probabilite. Publications, 1996 – 1997. Rennes: Universite de Rennes I, Institut de Recherche Mathematiques de Rennes, Publ. Inst. Rech. Math. Rennes. 1996, 1-3 (1997).

48. Francois Coquet, Ying Hu, Jean Memin, Shige Peng, A general converse comparison theorem for backward stochastic differential equations. (English. Abridged French version) C. R. Acad. Sci., Paris, Serie. I, Math. 333, No.6, 577-581 (2001).

49. F. Coquet, Y. Hu, J. Memin S. Peng, Filtrition Consistent Nonlinear Expectations and Related g-Expectations, *Probab. Theory Relat. Fields* **123**, 1-27, 2002.

50. Domenico Cuoco, Jaksa Cvitanic, Optimal consumption choices for a 'large' investor, J. of Economic Dynamics and Control 22(1998) 401–436.

51. Jaksa Cvitanic, Ioannis Karatzas, Hedging contingent claims with constrained portfolios, The Annals of Proba. 1993. Vol. 3, No. 4, 652–681.

52. Jaksa Cvitanic, Ioannis Karatzas, Backward Stochastic Differential Equations with Reflection and Dynkin games, The Annals of Prob. 1996, Vol. 24, No. 4 2024–2056.

53. Jaksa Cvitanic, Karatzas, Ioannis, Soner, H.Mete Backward stochastic differential equations with constraints on the gains-process. Ann. Probab. 26, No.4, 1522-1551 (1998).

54. R.W.R.Darling, Constructing gamma-martingales with prescribed limit, using backward sde, The Annals of Proba. 1995, Vol. 23, No.3, 1234–1261

55. F. Delarue On the existence and uniqueness of solutions to FBSDEs in a non-degenerate case STOCH PROC APPL 99 (2): 209-286 JUN 2002

56. A. Dermoune, S. Hamadene and Y. Ouknine, Backward Stochastic Differential Equation with Local time, Stochastic and Stochastic Rep. 66, No.1-2, 103-119 (1999).

57. D. Ding A note on probabilistic interpretation for quasilinear mixed boundary problems, Appl Math Mech-Engl 18 (9): 857-864 SEP 1997

58. N. Doluchaev, Xunyu Zhou Stochastic controls with terminal contingent conditions, J Math Anal Appl 238 (1): 143-165 Oct 1 1999

59. Jim Douglas, Jin Ma, Protter, Philip Numerical methods for forward-backward stochastic differential equations. Ann. Appl. Probab. 6, No.3, 940-968 (1996).

60. Darrell Duffie and Larry G.Epstein Appendix C with Costis Skiadas, Stochastic Differential Utility, Econometrica, Vol.60, No.2 (March, 1992) 353–394.

61. Es-Saky EH, Ouknine Y Convergence of backward stochastic differential equations and homogenization of semilinear variational inequalities in a convex set, B Sci Math 126 (5): 413-431 Jun 2002

62. El Karoui, Backward stochastic differential equations. A general introduction. El Karoui, Nicole (ed.) et al., Backward stochastic differential equations. Harlow: Longman. Pitman Res. Notes Math. Ser. 364, 7-26 (1997).

63. El-Karoui, S.Hamadene, BSDEs and risk-sensitive control, zero-sum and nonzero-sum game problems of stochastic functional differential equations, Stoch Proc Appl 107 (1): 145-169 Sep 2003.

64. El Karoui, S.J.Huang, A general result of existence and uniqueness of backward stochastic differential equations.

65. EL Karoui, M.C. Quenez, Dynamic programming and pricing of contingent claims in an incomplete market (c) SIAM J.Control and Optimization Vol.33.No.1.pp.29-66. 1995

66. El Karoui, M. C. Quenez, Nonlinear pricing theory and backward stochastic differential equations. Biais, B. (ed.) et al., Financial mathematics. Lectures given at the 3rd session of the Centro Internazionale Matematico Estivo (CIME), held in Bressanone, Italy, July 8–13, 1996. Berlin: Springer. Lect. Notes Math. 1656, 191-246 (1997)

67. El-Karoui, M.C.Quenez, Imperfect markets and backward stochastic differential equations. Rogers, L. C. G. (ed.) et al., Numerical methods in finance. Session at the Isaac Newton Institute, Cambridge, GB, 1995. Cambridge: Cambridge Univ. Press. 181-214 (1997)

68. El Karoui, C.Kapoudjian, E.Pardoux S. Peng, M.C. Quenez, Reflected solutions of backward SDE's, and related obstacle problems for PDE's, Ann Probab 25 (2): 702-737 Apr 1997

69. El Karoui, Shige Peng, M.C.Quenez, Backward stochastic differential equations in finance, Math Financ 7 (1): 1-71 Jan 1997

70. El Karoui, Shige Peng, M.C.Quenez, A dynamic maximum principle for the optimization of recursive utilities under constraints Ann. Appl. Prob. 11 (3): 664-693 AUG 2001.

71. El Karoui, L.Pardoux and M.C.quenez, Reflected Backward SDEs and America Options

72. K.D. Elworthy, Stochastic Differential Geometry, Bull. Sc. Math., 2c serie, 117. 1993. 7–27

73. M.Erraoui, Ouknine, Youssef, A.Sbi, Backward stochastic differential equations with distribution as terminal condition. Random Oper. Stoch. Equ. 5, No.4, 349-356 (1997).

74. M. Erraoui, Y. Ouknine, A. Sbi, Reflected solutions of backward stochastic differential equations with distribution as terminal condition. Random Oper. Stoch. Equ. 6, No.1, 1-16 (1998).

75. Anne Estrade, Monique Pontier, Backward stochastic differential equations in a Lie group. Azema, Jacques (ed.) et al., Seminaire de Probabilite XXXV. Berlin: Springer. Lect. Notes Math. 1755, 241-259 (2001).

76. Marco Fuhrman, Gianmario Tessitore, Nonlinear Kolmogorov equations in infinite dimensional spaces: the backward stochastic differential equations approach and applications to optimal control. Ann. Probab. 30, No.3, 1397-1465 (2002).

77. G. Gaudron, E.Pardoux, Backward stochastic differential equations (BSDE), weak convergence and homogenization of semilinear parabolic differential equations (PDE), Ann I H Poincare-Pr 37 (1): 1-42 Jan-Feb 2001

78. G. Gaudron, Convergence of BSDEs and homogenization of elliptic semilinear PDEs, Stoch Anal Appl 20 (4): 791-813 Jul 2002

79. Gegout-Petit, E.Pardoux, Equations diffirentielles stochastiques retrogrades refrichies dans un convexe. (Backward stochastic differential equations reflected in a convex domain). (French) Stochastics Stochastics Rep. 57, No.1-2, 111-128 (1996).

80. S. Hamadene, Nonzero sum Linear-quadratic Stochastic Differential Games and Backward-Forward Equations, Stochastic Analysis and Applications, 17(1), 117–130 (1990).

81. S. Hamadene, Euations diffrentielles stochastiques retrogrades: Les cas locale-ment lipschitzien. (Backward stochastic differential equations: The locally Lip-schitz case). (French) Ann. Inst. Henri Poincare Probab. Stat. 32, No.5, 645-659 (1997).

82. S. Hamadene, Backward-forward SDEs and stochastic differential games, Stochastic process and their applications 77 (1998) 1–15.

83. S. Hamadene, Nonzero sum linear-quadratic stochastic differential games and backward-forward equations. Stochastic Anal. Appl. 17, No.1, 117-130 (1999).

84. S. Hamadene, Multidimensional backward stochastic differential equations with uniformly continuous coefficients Bernoulli 9 (3): 517-534 Jun 2003.

85. S. Hamadene, J.P. Lepeltier, Reflected BSDEs and mixed game problem, Stochastic Process and their Applications 85 (2000) 177–188

86. S. Hamadene, J.P. Lepeltier, Zero-sum stochastic differential games and back-ward equations. Syst. Control Lett. 24, No.4, 259-263 (1995)

87. S. Hamadene, J.P. Lepeltier, Backward equations, stochastic control and zero-sum stochastic differential games. Stochastics Stochastics Rep. 54, No.3-4, 221-231 (1995).

88. S. Hamadene, J.P. Lepeltier and Zhen Wu, Infinite a horizon Reflected Back-ward Stochastic Differential Equations and Applications in Mixed control and Game problems, Probability and Mathematical Statistics, Math. Stat. 19, No.2, 211-234 (1999).

89. S. Hamadene, Y. Ouknine, Reflected backward stochastic differential equation with jumps and random obstacle. Electron. J. Probab. 8, Paper No.2, 20 p., electronic only (2003).

90. M. Hassani, Y. Ouknine, On a general result for backward stochastic differential equations. Stochastics Stochastics Rep. 73, No.3-4, 219-240 (2002).

91. M. Hassani, Y. Ouknine, Infinite dimensional BSDE with jumps, Stoch Anal Appl 20 (3): 519-565 May 2002.

92. Zhiyuan Huang, qingquan Lin, The Weak Solutions for Stochastic Differential Equations with Terminal Conditions, Mathematica Applicata 1997,10(4): 60-64

93. Ying Hu, Shige Peng, Adapted solution of a backward semilinear stochastic evolution equation. Stochastic Anal. Appl. 9, No.4, 445-459 (1991).

94. Ying Hu, Probabilistic interpretations of a system of quasilinear elliptic partial differential equations under Neumann boundary conditions, Stochastic Process and Their Applications 48 (1993) 107-121

95. Ying Hu, Shige Peng, Solution Of Forward-Backward Stochastic Differential-Equations, Probab. Theory Rel. Fiel. 103 (2): 273-283 OCT 1995

96. Ying Hu, Jiongmin Yong, Forward-backward stochastic differential equations with nonsmooth coefficients, Stoch Proc Appl 87 (1): 93-106 May

97. Ying Hu Shige Peng, Solution of forward-backward stochastic differential equa-tions. Probab. Theory Relat. Fields 103, No.2, 273-283 (1995).

98. Ying Hu, Shige Peng, A stability theorem of backward stochastic differential equations and its application. (English. Abridged French version) C. R. Acad. Sci., Paris, Serie. I 324, No.9, 1059-1064 (1997).

99. Ying Hu, On the existence of solution to one-dimensional forward-backward SDEs, Stoch Anal Appl 18 (1): 101-111 Jan 2000

100. Ying Hu Potential Kernels associated with a filtration and Forward CBackward SDEs, Potential Analysis 10:103-118,1999

101. Ying Hu, Jin Ma, Jiongmin Yong, On semi-linear degenerate backward stochastic partial differential equations. Probab. Theory Relat. Fields 123, No.3, 381-411 (2002).

102. Ying Hu, Xunyu Zhou, Indefinite stochastic Riccati equations SIAM J CONTROL OPTIM 42 (1): 123-137 2003

103. Ying Hu, On the solution of forward-backward SDEs with monotone and continuous coefficients, Nonlinear Analysis 42 1-12

104. Ying Hu, Jiongmin Yong, Forward-Backward stochastic differential equations with nonsmooth coefficients

105. Wilfrid S. Kendall, Probability, convexity, and Harmonic Maps II. Smoothness via probabilistic gradient inequalities, J. of Functional Analysis 126.228-257(1994)

106. Magdalena Kobylanski, Existence and uniqueness results for backward stochastic differential equations when the generator has a quadratic growth, CR Acad Sci I-Math 324 (1): 81-86 Jan 1997

107. Magdalena Kobylanski, Backward stochastic differential equations and partial differential equations with quadratic growth. Ann. Probab. 28, No.2, 558-602 (2000).

108. Michael Kohlmann, Reflected forward backward stochastic differential equations and contingent claims. Chen, Shuping (ed.) et al., Control of distributed parameter and stochastic systems. Proceedings of the international conference (IFIP WG 7.2), Hangzhou, China, June 19-22, 1998. Boston, MA: Kluwer Academic Publishers. 223-230 (1999)

109. Michael Kohlmann, Shanjian Tang, New developments in backward stochastic Riccati equations and their applications. Kohlmann, Michael (ed.) et al., Mathematical finance. Workshop of the mathematical finance research project, Konstanz, Germany, October 5-7, 2000. Basel: Birkhauser. 194-214 (2001).

110. Michael Kohlmann, Shanjian Tang, Global adapted solution of one-dimensional backward stochastic Riccati equations, with application to the mean-variance hedging, Stoch Proc Appl 97 (2): 255-288 Feb 2002.

111. Michael Kohlmann, Shanjian Tang, Multidimensional backward stochastic Riccati equations and applications, SIAM J CONTROL OPTIM 41 (6): 1696-1721. 2003

112. Michael Kohlmann, Shanjian Tang, Minimization of risk and linear quadratic optimal control theory, SIAM J Control Optim 42 (3): 1118-1142 2003.

113. Michael Kohlmann, Xunyu Zhou, Relationship between backward stochastic differential equations and stochastic controls: A linear-quadratic approach. SIAM J. Control Optimization 38, No.5, 1392-1407 (2000).

114. Zai Lanjri, A class of two-parameter backward stochastic differential equations driven by a Brownian sheet. Stochastic Anal. Appl. 20, No.4, 883-899 (2002).

115. A.Lazrak, M.C.Quenez MC, A generalized stochastic differential utility, Math Oper Res 28 (1): 154-180 Feb 2003

116. A. Lejay, BSDE driven by Dirichlet process and semi-linear parabolic PDE. Application to homogenization, Stoch Proc Appl 97 (1): 1-39 Jan 2002

117. J.P.Lepeltier, San Martin, Jaime, Backward stochastic differential equations with continuous coefficient. Stat. Probab. Lett. 32, No.4, 425-430 (1997).

118. J.P.Lepeltier, San Martin, Jaime, On the existence or non-existence of solutions for certain backward stochastic differential equations BERNOULLI 8 (1): 123-137 FEB 2002

119. J.P.Lepeltier, Jean-Pierre, San Martin, Jaime, On the existence or non-existence of solutions for certain backward stochastic differential equations. Bernoulli 8, No.1, 123-137 (2002).

120. Juan Li, Zhen Wu, Fully coupled forward-backward stochastic differential equations with Brownian motion and Poisson processes under local Lipschitz condition. (Chinese. English summary) Math. Appl. 15, No.2, 40-47 (2002).

121. Xunjing Li, Shanjian Tang, General necessary conditions for partially observed optimal stochastic controls, J. Appl. Prob.32. 1118-1137(1995)

122. AEB Lim, Xunyu Zhou, Linear-quadratic control of backward stochastic differential equations. SIAM J. Control Optimization 40, No.2, 450-474 (2001).

123. AEB Lim, Xunyu Zhou, Optimal control of linear backward stochastic differential equations with a quadratic cost criterion LECT NOTES CONTR INF 280: 301-317 2002

124. AEB Lim, Xunyu Zhou, Mean-variance portfolio selection with random parameters in a complete market, Math Oper Res 27 (1): 101-120 Feb 2002

125. Jianzhong Lin, Adapted solution of a backward stochastic nonlinear Volterra integral equation. Stochastic Anal. Appl. 20, No.1, 165-183 (2002).

126. Qinquan Lin, Solution of backward stochastic differential equations with jumps and quasi-continuous generator. (Chinese. English summary) J. Shandong Univ., Nat. Sci. Ed. 35, No.2, 121-125 (2000).

127. Qinquan Lin, Shige Peng, Smallest g-supersolution for BSDE with continuous drift coefficients CHINESE ANN MATH B 21 (3): 359-366 JUL 2000

128. Linear, degenerate backward stochastic partial differential equations, Lect Notes Math 1702: 103-136 1999

129. Jicheng Liu, Ren, Jiagang Comparison theorem for solutions of backward stochastic differential equations with continuous coefficient. Stat. Probab. Lett. 56, No.1, 93-100 (2002).

130. Yazeng Liu, Shige Peng, Infinite horizon backward stochastic differential equation and exponential convergence index assignment of stochastic control systems. Automatica 38, No.8, 1417-1423 (2002).

131. Chenghu Ma, An existence theorem of intertemporal recursive utility in the presence of levy jumps, J. of Mathematical Economics 34 (2000) 509-526.

132. Jin Ma, Philip Protter, San Martin Jaime, Torres, Soledad Numerical method for backward stochastic differential equations.Ann. Appl. Probab. 12, No.1, 302-316 (2002).

133. Jin Ma, Philip Protter and Jiongmin Yong, Solving forward –backward stochastic differential equations explicitly, a four step scheme, Probab.Theory Relat.Fields 98.339-359(1994)

134. Jin Ma, Jiongmin Yong, Solvability Of Forward-Backward Sdes And The Nodal Set Of Hamilton-Jacobi-Bellman Equations, Chinese Ann Math B 16 (3): 279-298 Jul 1995

135. Jin Ma, Jiongmin Yong,, Adapted solution of a degenerate backward spde, with applications, Stochastic Process and their Application 70 (1997) 59-84

136. Jin Ma, Jiongmin Yong, Forward-backward stochastic differential equations and their applications - Introduction, Lect Notes Math 1702: 1-24 1999

137. Jin Ma, Jiongmin Yong, On linear, degenerate backward stochastic partial differential equations. Probab. Theory Relat. Fields 113, No.2, 135-170 (1999)

138. Jin Ma, Jiongmin Yong, Approximate solvability of forward-backward stochastic differential equations. Appl. Math. Optimization 45, No.1, 1-22 (2002).

139. Jin Ma, Zajic, Tim Rough asymptotics of forward-backward stochastic differential equations. Chen, Shuping (ed.) et al., Control of distributed parameter and stochastic systems. Proceedings of the international conference (IFIP WG 7.2), Hangzhou, China, June 19-22, 1998. Boston, MA: Kluwer Academic Publishers. 239-246 (1999).

140. Jin Ma, Jianfeng Zhang, Representation theorems for backward stochastic differential equations. Ann. Appl. Probab. 12, No.4, 1390-1418 (2002).

141. Jin Ma, Jianfeng Zhang, Path regularity for solutions of backward stochastic differential equations. Probab. Theory Relat. Fields 122, No.2, 163-190 (2002).

142. Anis Matoussi, A Reflected solutions of backward stochastic differential equations with continuous coefficient, Stat Probabil Lett 34 (4): 347-354 Jul 16 1997

143. Anis Matoussi, Scheutzow, Michael Stochastic PDEs driven by nonlinear noise and backward doubly SDEs. J. Theor. Probab. 15, No.1, 1-39 (2002).

144. L. Mazliak, The maximum principle in stochastic control and backward equations, El Karoui, Nicole (ed.) et al., Backward stochastic differential equations. Harlow: Longman. Pitman Res. Notes Math. Ser. 364, 101-113 (1997).

145. Yuliya S. Mishura, Ol'tsik, Yanina A, Optimal financial strategy with wealth process governed by backward stochastic differential equation. Theory Stoch. Process. 4(20), No.1-2, 222-237 (1998).

146. Modeste N'Zi, Multivalued backward stochastic differential equations with local Lipschitz drift. Stochastics Stochastics Rep. 60, No.3-4, 205-218 (1997).

147. Modeste N'Zi, Multivalued backward stochastic differential equations with local Lipschitz drift. Random Oper. Stoch. Equ. 5, No.2, 163-172 (1997).

148. Modeste N'Zi, Ouknine, Youssef Multivalued backward stochastic differential equations with continuous coefficients. Random Oper. Stoch. Equ. 5, No.1, 59-68 (1997).

149. Modeste N'zi, Ouknine, Youssef Equations diffrentielles stochastiques retrogrades multivoques. (Multidimensional backward stochastic differential equations). (French) Probab. Math. Stat. 17, No.2, 259-275 (1997).

150. Modeste N'zi, Ouknine, Y. Backward stochastic differential equations with jumps involving a subdifferential operator. Random Oper. Stoch. Equ. 8, No.4, 319-338 (2000).

151. Xuerong Mao, Adapted solutions of backward stochastic differential equations with non-Lipschitz coefficients, Stochastic Process and their Application 58(1995) 281-292.

152. David Nualart, Schoutens, Wim Backward stochastic differential equations and Feynman-Kac formula for Levy processes, with applications in finance. Bernoulli 7, No.5, 761-776 (2001).

153. Y.Ouknine, Reflected backward stochastic differential equations with jumps. Stochastics Stochastics Rep. 65, No.1-2, 111-125 (1998).

154. E.Pardoux, Backward stochastic differential equations and applications. Chatterji, S. D. (ed.), Proceedings of the international congress of mathematicians, ICM '94, August 3-11, 1994, Zurich, Switzerland. Vol. II. Basel: Birkhauser. 1502-1510 (1995).

155. E. Pardoux, Backward Stochastic Differential Equations and Viscosity Solutions, 79–128, in Stochastic Analysis and Related Topics, VI, Birkhauser, 1996.

156. E. Pardoux, Generalized discontinuous backward stochastic differential equations. El Karoui, Nicole (ed.) et al., Backward stochastic differential equations. Harlow: Longman. Pitman Res. Notes Math. Ser. 364, 207-219 (1997).

157. E. Pardoux, Homogenization of Linear and semilinear second order parabolic PDEs with periodic coefficients: A probabilistic Approach, J. of Functional Analysis 167, 498–520 (1999)

158. E.Pardoux, Shige Peng, Adapted solution of a backward stochastic differential equation. Syst. Control Lett. 14, No.1, 55-61 (1990).

159. E.Pardoux, Shige Peng, Backward stochastic differential equations and quasilinear parabolic partial differential equations. Stochastic partial differential equations and their applications, Proc. IFIP Int. Conf., Charlotte/NC (USA) 1991, Lect. Notes Control Inf. Sci. 176, 200-217 (1992).

160. E.Pardoux, Shige Peng, Backward doubly stochastic differential equations and systems of quasilinear SPDEs. Probab. Theory Relat. Fields 98, No.2, 209-227 (1994).

161. E. Pardoux, F.Pradeilles, Zusheng Rao, Probabilistic interpretation of a system of semi-linear parabolic partial differential equations Ann.inst.Henri Poincar, Vol.33, no 4,1997, p.467-490.

162. E. Pardoux, A.Rascanu, Backward stochastic differential equations with subdifferential operator and related variation inequalities Stochastic Processes and their Applications 76 (1998) 191-215.

163. E.Pardoux, Shanjian Tang, Forward-backward stochastic differential equations and quasilinear parabolic PDEs. Probab. Theory Relat. Fields 114, No.2, 123-150 (1999).

164. E.Pardoux, A.Yu. Veretennikov, Averaging of backward stochastic differential equations, with application to semilinear PDE's. Stochastics Stochastics Rep. 60, No.3-4, 255-270 (1997).

165. E.Pardoux, SG Zhang, Generalized BSDEs and nonlinear Neumann boundary value problems, Probab Theory Rel 110 (4): 535-558 Apr. 1998

166. E.Pardoux, Aurel Rascanu, Backward stochastic variational inequalities. Stochastics Stochastics Rep. 67, No.3-4, 159-167 (1999).

167. Shige Peng, *On Hamilton-Jacobi-Bellman Equation with Stochastic Coefficients,* in Proceeding of the Annual Meeting on Control Theory and It's Applications, 1989.

168. Shige Peng, *A General Stochastic Maximum Principle for Optimal Control Problems,* SIAM J. Cont. 28: 4, 966-979, 1990.

169. Shige Peng, Maximum Principle for Stochastic Optimal Control with Nonconvex Control Domain, in Analysis and Optimization of Systems, A. Bensoussan J. L. Lions eds. Lecture Notes in Control and Information Sciences, 144, (1990), 724-732.

170. Shige Peng, *Probabilistic Interpretation for Systems of Quasilinear Parabolic Partial Differential Equations,* Stochastics, 37, 61–74, 1991.

171. Shige Peng, A Generalized Hamilton-Jacobi-Bellman Equation, Lect Notes Contr Inf 159: 126-134 1991.

172. Shige Peng, *Stochastic Hamilton-Jacobi-Bellman Equations,* SIAM Control 30(2), 284-304, 1992.

173. Shige Peng, *A Generalized Dynamic Programming Principle and Hamilton-Jacobi-Bellmen equation,* Stochastics, 38, 119–134, 1992.

174. Shige Peng, *A Nonlinear Feynman–Kac Formula and Applications,* Proceedings of Symposium of System Sciences and Control theory, Chen Yong ed. 173-184, World Scientific, Singapore, 1992.

175. Shige Peng, *New Development in Stochastic Maximum Principle and Related Backward Stochastic Differential Equations,* in proceedings of 31st CDC Conference, Tucson 1992.
176. Shige Peng, Backward stochastic differential equations and applications to optimal control. Appl. Math. Optimization 27, No.2, 125-144 (1993).
177. Shige Peng, *BSDE and Exact Controllability of Stochastic Control Systems,* Progress in Natural Science, 4 3, 274–284, 1994.
178. Shige Peng, The backward stochastic differential equations and its applications. (Chinese. English summary) Adv. Math., Beijing 26, No.2, 97-112 (1997).
179. Shige Peng, *Backward Stochastic Differential Equation in Finance,* Mathematical Finance, 1997, 7, 1–71,
180. Shige Peng, *Topics in Stochastic Analysis,* (with J. Yan, S. Fang and L.M. Wu), Ch.2: BSDE and Stochastic Optimizations (Chinese vers.), Science Publication, 1997.
181. Shige Peng, Monotonic limit theorem of BSDE and nonlinear decomposition theorem of Doob-Meyer's type, Probab. Theory Rel. Fiel. 113 (4): 473-499 Apr 1999
182. Problem of Eigenvalues of Stochastic Hamiltonian Systems with Boundary Conditions, *Stochastic Processes and Their Applications,* **88,** 259–290, 2000.
183. A Stochastic Laplace Transform for Adapted Processes and Related BSDEs, in *Optimal Control and Partial Differential Equations, J.L. Menaldi et al. (Eds.) 283-292,* IOS Press, Amsterdam, 2001.
184. Shige Peng, Zhen Wu, Fully coupled forward-backward stochastic differential equations and applications to optimal control, Siam J Control Optim 37 (3): 825-843 Apr 13 1999
185. Shige Peng, Open problems on backward stochastic differential equations. Shuping Chen (ed.) et al., Control of distributed parameter and stochastic systems. Proceedings of the international conference (IFIP WG 7.2), Hangzhou, China, June 19-22, 1998. Boston, MA: Kluwer Academic Publishers. 265-273 (1999).
186. Shige Peng, Problem of eigenvalues of stochastic Hamiltonian systems with boundary conditions, Stoch Proc Appl 88 (2): 259-290 Aug 2000
187. Peng, S., Shi, Yufeng Infinite horizon forward-backward stochastic differential equations. Stochastic Processes Appl. 85, No.1, 75-92 (2000).
188. S. Peng, Yufeng Shi, A type of time-symmetric forward–backward stochastic differential equations. (English. Abridged French version) C. R., Math., Acad. Sci. Paris 336, No.9, 773-778 (2003).
189. S. Peng and Z. Wu, Fully Coupled Forward-Backward Stochastic Differential Equations and Applications to Optimal Control, *SIAM Control,* 1999.
190. Shige Peng, Yang F. Duplicating and pricing contingent claims with constrained portfolios, Prog. Nat. Sci. 8 (6): 650-659 Dec 1998
191. M.Pontier, Solutions of forward-backward stochastic differential equations. El Karoui, Nicole (ed.) et al., Backward stochastic differential equations. Harlow: Longman. Pitman Res. Notes Math. Ser. 364, 39-46 (1997).
192. F.Pradeilles, Wavefront propagation for reaction-diffusion systems and backward SDES ANN PROBAB 26 (4): 1575-1613 OCT 1998
193. S.Ramasubramanian, Reflected backward stochastic differential equations in an orthant. Proc. Indian Acad. Sci., Math. Sci. 112, No.2, 347-360 (2002).
194. C. Rainer, Backward stochastic differential equations with Azema's martingale, Stochastics Stochastics Rep. 73, No.1-2, 65-98 (2002).

195. A. Rozkosz, Backward SDEs and Cauchy problem for semilinear equations in divergence form, Probab Theory Rel. Field. 125 (3): 393-407, 2003

196. Zhiqiang Shun, The pricing problem and the existence/uniqueness of solutions to a class of backward stochastic differential equations. (Chinese. English summary) Chin. J. Appl. Probab. Stat. 14, No.4, 409-418 (1998).

197. Rong Situ, On solutions of Backward stochastic differential equations with jumps and applications, Stochastic Process and their Applications (1996)

198. Rong Situ, Yueping Wang,On solutions of backward stochastic differential equations with jumps, with unbounded stopping times as terminal and with non-Lipschitz coefficients, and probabilistic interpretation of quasi-linear elliptic type integro-differential equations. Appl. Math. Mech., Engl. Ed. 21, No.6, 659-672 (2000).

199. Rong Situ, Min Huang, On solutions of backward stochastic differential equations with jumps in Hilbert spaces. II. (Chinese. English summary) Acta Sci. Nat. Univ. Sunyatseni 40, No.4, 20-23 (2001).

200. Rong Situ, Huanyao Xu, Adapted solutions of backward stochastic evolution equations with jumps on Hilbert space. II. (Chinese. English summary) Acta Sci. Nat. Univ. Sunyatseni 40, No.2, 1-5 (2001).

201. Rong Situ, On solutions of backward stochastic differential equations with jumps and with non-Lipschitzian coefficients in Hilbert spaces and stochastic control. Stat. Probab. Lett. 60, No.3, 279-288 (2002).

202. Stoica IL A probabilistic interpretation of the divergence and BSDE's, Stoch Proc Appl 103 (1): 31-55 Jan 2003

203. M. Sirbu, G.Tessitore, Null controllability of an infinite dimensional SDE with state- and control-dependent noise Syst. Contro. Lett. 44 (5): 385-394 DEC 14 2001

204. Shangjina Tang, The maximum principle for partially observed optimal control of stochastic differential equations, Siam J Control Optim 36 (5): 1596-1617 Sep 1998

205. Shangjina Tang, Financial mean-variance problems and stochastic LQ problems: Linear stochastic Hamilton systems and backward stochastic Riccati equations. Yong, Jiongmin (ed.), Recent developments in mathematical finance. Proceedings of the international conference on mathematical finance, Shanghai, China, May 10-13, 2001. Singapore: World Scientific. 190-203 (2002).

206. Shangjina Tang, SH Hou, Optimal control of point processes with noisy observations: The maximum principle APPL MATH OPT 45 (2): 185-212 MAR-APR 2002

207. Shangjina Tang, General linear quadratic optimal stochastic control problems with random coefficients: Linear stochastic Hamilton systems and backward stochastic Riccati equations SIAM J CONTROL OPTIM 42 (1): 53-75 2003.

208. Shanjian Tang, Xunjing Li, Necessary conditions for optimal control of stochastic systems with random jumps, SIAM J.Control and optimization, Vol. 32, No. 5, 1447-1475.

209. A. Thalmaier, Martingales on Riemannian manifolds and the nonlinear heat equation Stochastic analysis and Applications, Singapore: World Scientific Press, 1996, 429-440

210. A. Thalmaier, Brownian Motion and the formation singularities in the heat flow for harmonic maps 350-366

211. Xiangjun Wang, On backward stochastic differential equations driven by a continuous semi-martingale. (Chinese. English summary) J. Math., Wuhan Univ. 19, No.1, 45-50 (1999).

212. Zhen Wu, Maximum principle for optimal control problem of fully coupled forward-backward stochastic systems. Syst. Sci. Math. Sci. 11, No.3, 249-259 (1998).

213. Zhen Wu, Adapted solution of generalized forward-backward stochastic differential equations and its dependence on parameters. (Chinese) Chin. Ann. Math., Ser. A 19, No.1, 55-62 (1998).

214. Zhen Wu, Forward-Backward Stochastic Differential Equations with Brownian Motion and Poisson process, ACTA Mathematics Application Sinica Oct. 1999 Vol.15 No.4 433–443.

215. Z. Wu The comparison theorem of FBSDE Stat. Prob. Lett. 44 (1): 1-6, 1, 1999.

216. Zhen Wu, Fully coupled FBSDE with Brownian motion and Poisson process in stopping time duration, J Aust Math Soc 74: 249-266 Part 2 Apr 2003

217. Jianming Xia, Backward stochastic differential equation with random measures. Acta Math. Appl. Sin., Engl. Ser. 16, No.3, 225-234 (2000).

218. Wensheng Xu, Stochastic maximum principle for optimal control problem of forward and backward system. J. Aust. Math. Soc., Ser. B 37, No.2, 172-185 (1995).

219. Bo Yang, Necessary conditions for optimal controls of forward-backward stochastic systems with nonsmooth cost functionals. (Chinese. English summary) J. Fudan Univ., Nat. Sci. 39, No.1, 61-67 (2000).

220. Jinchun Ye, Coupled forward-backward stochastic differential equations with random jumps. (Chinese. English summary) Chin. Ann. Math., Ser. A 23, No.6, 737-750 (2002).

221. Jiongmin Yong, Finding adapted solutions of forward-backward stochastic differential equations method of continuation, Probability Theory and Related Fields 107. 537–572 (1997)

222. Jiongmin Yong, Stochastic controls and forward-backward SDES. Chen, Shuping (ed.) et al., Control of distributed parameter and stochastic systems. Proceedings of the international conference (IFIP WG 7.2), Hangzhou, China, June 19-22, 1998. Boston, MA: Kluwer Academic Publishers. 307-314 (1999).

223. Jiongmin Yong, Xunyu Zhou, Stochastic control–Hamiltonian Systems and HJB Equations, Springer, Applications of Mathematics 43, 1999

224. Jiongmin Yong, Linear Forward-Backward stochastic differential equations, Appl. Math. Optim. 39:93–119 (1999)

225. Jiongmin Yong, European -type contingent claims in an incomplete market with constrained wealth and portfolio, Mathematical Finance, Vol. 9, No. 4 (October 1999) 387–412

226. Jiongmin Yong, Optimal portfolios in an incomplete market, Annals of Economics and Finance 1, 359–381 (2000).

227. Jiongmin Yong, Replication of American contingent claims in incomplete markets, International Journals of Theoretical and Applied Finance, Vol. 4, No. 3 (2001) 439–466

228. Jiongmin Yong (ed.), Recent developments in mathematical finance. Proceedings of the international conference on mathematical finance, Shanghai, China, May 10-13, 2001. Singapore: World Scientific. 28-38 (2002).

229. Jiongmin Yong, Forward-backward stochastic differential equation: A useful tool for mathematical finance and other related fields. Surv. Math. Ind. 10, No.3, 175-229 (2002).

230. Jiongmin Yong, A leader-follower stochastic linear quadratic differential game SIAM J CONTROL OPTIM 41 (4): 1015-1041 DEC 3 2002

231. Jiongmin Yong, Degenerate BSDEs and FESDEs with applications in mathematical finance, Insur Math Econ 32 (3): 483-483 Jul 21 2003

232. Zengting Yuan, Solution of generalized backward stochastic differential equations with jumps. (Chinese. English summary) J. Math., Wuhan Univ. 20, No.2, 217-221 (2000).

233. Nl Zaidi, N.Lanjri, D.Nualart, Backward stochastic differential equations in the plane. Potential Anal. 16, No.4, 373-386 (2002).

234. NL Zaidi, A class of two-parameter backward stochastic differential equations driven by a Brownian sheet, Stoch Anal Appl 20 (4): 883-899 Jul 2002

235. Guichang Zhang, Random walk and a discrete backward stochastic differential equation. (Chinese. English summary) Math. Appl. 15, No.2, 76-79 (2002).

236. Yinnan Zhang, Weian Zheng, Discretizing a backward stochastic differential equation. Int. J. Math. Math. Sci. 32, No.2, 103-116 (2002).

237. Liuyi Zhong, Minghao Xu, Local existence and uniqueness of adapted solutions of backward stochastic evolution equations. (Chinese. English summary) J. Math., Wuhan Univ. 16, No.4, 417-422 (1996).

238. Liuyi Zhong, Minghao Xu, Global existence and uniqueness of adapted solution of a backward stochastic evolution equation in Hilbert space. (Chinese. English summary) J. Wuhan Univ., Nat. Sci. Ed. 43, No.5, 591-597 (1997).

239. Shaofu Zhou, Zhiyuan Huang, Zigang Zhang, Development of backward stochastic differential equation and its applications.
(Chinese. English summary) Math. Appl. 15, No.2, 9-13 (2002).

Utility Maximisation
in Incomplete Markets

Walter Schachermayer[*]

Financial and Actuarial Mathematics, Vienna University of Technology,
Wiedner Hauptstrasse 8/105-1, 1040 Vienna, Austria
wschach@fam.tuwien.ac.at

Preface

In these lectures we give a short introduction to the basic concepts of Mathematical Finance, focusing on the notion of "no arbitrage", and subsequently apply these notions to the problem of optimizing dynamically a portfolio in an incomplete financial market with respect to a given utility function U.

In the first part we mainly restrict ourselves to the situation where the underlying probability space $(\Omega, \mathcal{F}, \mathbf{P})$ is finite, in order to reduce the functional-analytic difficulties to simple linear algebra. In my opinion, this allows — at least as a first step — for a clearer picture of the Mathematical Finance issues.

We then treat the problem of utility maximisation and, in particluar, its duality theory for a general semi-martingale models of financial market. Here we are rather informal and concentrate mainly on explaining the basic ideas, e.g., the notion of the asymptotic elasticity of a utility function U.

These notes are largely based on the surveys [S 03] and [S 01a] and, in particular, on the notes taken by P. Guasoni during my Cattedra Galileiana lectures at Scuola Normale Superiore in Pisa [S 04a]. We also refer to the original papers [KS 99] and [S 01] for more detailed information on the topics of the present lectures.

1 Problem Setting

We consider a model of a security market which consists of $d + 1$ assets. We denote by $S = ((S_t^i)_{1 \leq t \leq T})_{0 \leq i \leq d}$ the price process of the d stocks and

[*] Support by the Austrian Science Foundation (FWF) under the Wittgenstein-Preis program Z36 and grant P15889 and by the Austrian National Bank under grant 'Jubiläumsfondprojekt Number 9486' is gratefully acknowledged.

suppose that the price of the asset S^0, called the "bond" or "cash account", is constant, i.e., $S_t^0 \equiv 1$. The latter assumption does not restrict the generality of the model as we always may choose the bond as numéraire, i.e., we may express the values of the other assets in units of the "bond". In other words, $((S_t^i)_{0 \le t \le T})_{1 \le i \le d}$, is an $\mathbb{R}^d$-valued semi-martingale modeling the discounted price process of d risky assets.

The process S is assumed to be a semimartingale, based on and adapted to a filtered probability space $(\Omega, \mathcal{F}, (\mathcal{F}_t)_{0 \le t \le T}, \mathbf{P})$ satisfying the usual conditions of saturatedness and right continuity. As usual in mathematical finance, we consider a finite horizon T, but we remark that our results can also be extended to the case of an infinite horizon.

In section 2 we shall consider the case of finite Ω, in which case the paths of S are constant except for jumps at a finite number of times. We then can write S as $(S_t)_{t=0}^T = (S_0, S_1, \dots, S_T)$, for some $T \in \mathbb{N}$.

The assumption that the bond is constant is mainly chosen for notational convenience as it allows for a compact description of self-financing portfolios: a self-financing portfolio Π is defined as a pair (x, H), where the constant x is the initial value of the portfolio and $H = (H^i)_{1 \le i \le d}$ is a predictable S-integrable process specifying the amount of each asset held in the portfolio. The value process $X = (X_t)_{0 \le t \le T}$ of such a portfolio Π at time t is given by

$$X_t = X_0 + \int_0^t H_u dS_u, \quad 0 \le t \le T, \tag{1}$$

where $X_0 = x$ and the integral refers to stochastic integration in $\mathbb{R}^d$.

In order to rule out doubling strategies and similar schemes generating arbitrage-profits (by going deeply into the red) we follow Harrison and Pliska ([HP 81], see also [DS 94]), calling a predictable, S-integrable process *admissible*, if there is a constant $C \in \mathbb{R}_+$ such that, almost surely, we have

$$(H \cdot S)_t := \int_0^t H_u dS_u \ge -C, \quad \text{for } 0 \le t \le T. \tag{2}$$

Let us illustrate these general concepts in the case of an $\mathbb{R}^d$-valued process $S = (S_t)_{t=0}^T$ in finite, discrete time $\{0, 1, \dots, T\}$ adapted to the filtration $(\mathcal{F}_t)_{t=0}^T$. In this case each $\mathbb{R}^d$-valued process $(H_t)_{t=1}^T$, which is predictable (i.e. each H_t is $\mathcal{F}_{t-1}$-measurable), is S-integrable, and the stochastic integral reduces to a finite sum

$$(H \cdot S)_t = \int_0^t H_u dS_u \tag{3}$$

$$= \sum_{u=1}^t H_u \Delta S_u \tag{4}$$

$$= \sum_{u=1}^t H_u (S_u - S_{u-1}), \tag{5}$$

where $H_u \Delta S_u$ denotes the inner product of the vectors H_u and $\Delta S_u = S_u - S_{u-1}$ in $\mathbb{R}^d$, i.e.

$$H_u \Delta S_u = \sum_{j=1}^d H_u^j (S_u^j - S_u^{j-1}).$$ (6)

Of course, each such trading strategy H is admissible if the underlying probability space Ω is finite.

Passing again to the general setting of an $\mathbb{R}^d$-valued semi-martingale $S = (S_t)_{0 \leq t \leq T}$ we denote as in [KS 99] by $\mathcal{M}^e(S)$ (resp. $\mathcal{M}^a(S)$) the set of probability measures $\mathbf{Q}$ equivalent to $\mathbf{P}$ (resp. absolutely continuous with respect to $\mathbf{P}$) such that for each admissible integrand H, the process $H \cdot S$ is a local martingale under $\mathbf{Q}$.

We shall assume the following version of the no-arbitrage condition on S:

Assumption 1.1 *The set $\mathcal{M}^e(S)$ is not empty.*[2]

In these notes we shall mainly be interested in the case when $\mathcal{M}^e(S)$ is not reduced to a singleton, i.e., the case of an *incomplete* financial market.

After having specified the process S modeling the financial market we now define the function $U(x)$ modeling the utility of an agent's wealth x at the terminal time T.

We make the classical assumptions that $U : \mathbb{R} \to \mathbb{R} \cup \{-\infty\}$ is *increasing on $\mathbb{R}$, continuous* on $\{U > -\infty\}$, *differentiable and strictly concave* on the interior of $\{U > -\infty\}$, and that marginal utility tends to zero when wealth tends to infinity, i.e.,

$$U'(\infty) := \lim_{x \to \infty} U'(x) = 0.$$ (7)

These assumptions make good sense economically and it is clear that the requirement (7) of marginal utility decreasing to zero, as x tends to infinity, is necessary, if one is aiming for a general existence theorem for optimal investment.

---

[2] If follows from [DS 94] and [DS 98] that Assumption 1.1 is equivalent to the condition of "no free lunch with vanishing risk". This property can also be equivalently characterised in terms of the existence of a measure $\mathbf{Q} \sim \mathbf{P}$ such that the process S itself (rather than the integrals $H \cdot S$ for admissible integrands) is "something like a martingale". The precise notion in the general semi-martingale setting is that S is a sigma-martingale under $\mathbf{Q}$ (see [DS 98]); in the case when S is locally bounded (resp. bounded) the term "sigma-martingale" may be replaced by the more familiar term "local martingale" (resp. "martingale").

Readers who are not too enthusiastic about the rather subtle distinctions between martingales, local martingales and sigma-martingales may find some relief by noting that, in the case of finite Ω, or, more generally, for bounded processes S, these three notions coincide.

As regards the behavior of the (marginal) utility at the other end of the wealth scale we shall distinguish two cases.

Case 1 (negative wealth not allowed): in this setting we assume that U satifies the conditions $U(x) = -\infty$, for $x < 0$, while $U(x) > -\infty$, for $x > 0$, and the so-called *Inada condition*

$$U'(0) := \lim_{x \searrow 0} U'(x) = \infty. \tag{8}$$

Case 2 (negative wealth allowed): in this case we assume that $U(x) > -\infty$, for all $x \in \mathbb{R}$, and that

$$U'(-\infty) := \lim_{x \searrow -\infty} U'(x) = \infty. \tag{9}$$

Typical examples for case 1 are

$$U(x) = \ln(x), \quad x > 0, \tag{10}$$

or

$$U(x) = \frac{x^\alpha}{\alpha}, \quad \alpha \in (-\infty, 1) \setminus \{0\}, \quad x > 0, \tag{11}$$

whereas a typical example for case 2 is

$$U(x) = -e^{-\gamma x}, \quad \gamma > 0, \qquad x \in \mathbb{R}. \tag{12}$$

We again note that it is natural from economic considerations to require that the marginal utility tends to infinity when the wealth x tends to the infimum of its allowed values.

For later reference we summarize our assumptions on the utility function:

Assumption 1.2 (Usual Regularity Conditions) *A utility function U : $\mathbb{R} \to \mathbb{R} \cup \{-\infty\}$ satisfies the* usual regularity conditions *if it is increasing on $\mathbb{R}$, continuous on $\{U > -\infty\}$, differentiable and strictly concave on the interior of $\{U > -\infty\}$, and satisfies*

$$U'(\infty) := \lim_{x \to \infty} U'(x) = 0. \tag{13}$$

Denoting by $\mathrm{dom}(U)$ the interior of $\{U > -\infty\}$, we assume that we have one of the two following cases.

Case 1: $\mathrm{dom}(U) =]0, \infty[$ *in which case U satisfies the condition*

$$U'(0) := \lim_{x \searrow 0} U'(x) = \infty. \tag{14}$$

Case 2: $\mathrm{dom}(U) = \mathbb{R}$ *in which case U satisfies*

$$U'(-\infty) := \lim_{x \searrow -\infty} U'(x) = \infty. \tag{15}$$

We now can give a precise meaning to the problem of maximizing the expected utility of terminal wealth. Define the value function

$$u(x) := \sup_{H \in \mathcal{H}} \mathbf{E}\left[U(x + (H \cdot S)_T)\right], \quad x \in \text{dom}(U), \tag{16}$$

where H ranges through the family $\mathcal{H}$ of admissible S-integrable trading strategies. To exclude trivial cases we shall assume that the value function u is not degenerate:

Assumption 1.3

$$u(x) < \sup_{\xi} U(\xi), \quad \text{for some} \quad x \in \text{dom}(U). \tag{17}$$

Since u is clearly increasing, and $U(y) \leq U(x) + U'(x)(y-x)$ for any $y > x$, this assumption implies that

$$u(x) < \sup_{\xi} U(\xi), \quad \text{for all} \quad x \in \text{dom}(U). \tag{18}$$

Under appropriate hypotheses (e.g., when Ω is finite) Assumptions 1.1 and 1.2 already imply Assumption 1.3.

2 Models on Finite Probability Spaces

In order to reduce the technical difficulties of the theory of utility maximization to a minimum, we assume throughout this section that the probability space Ω will be finite, say, $\Omega = \{\omega_1, \omega_2, \ldots, \omega_N\}$. This assumption implies that all the differences among the spaces $L^\infty(\Omega, \mathcal{F}, \mathbf{P})$, $L^1(\Omega, \mathcal{F}, \mathbf{P})$ and $L^0(\Omega, \mathcal{F}, \mathbf{P})$ disappear, as all these spaces are simply isomorphic to $\mathbb{R}^N$. Hence all the functional analysis reduces to simple linear algebra in the setting of the present section.

Nevertheless we shall write $L^\infty(\Omega, \mathcal{F}, \mathbf{P})$, $L^1(\Omega, \mathcal{F}, \mathbf{P})$ etc. below (knowing very well that these spaces are isomorphic in the present setting) to indicate, what we shall encounter in the setting of the general theory.

Definition 2.1. *A model of a* finite financial market *is an* $\mathbb{R}^{d+1}$*-valued stochastic process* $S = (S)_{t=0}^T = (S_t^0, S_t^1, \ldots, S_t^d)_{t=0}^T$, *based on and adapted to the filtered stochastic base* $(\Omega, \mathcal{F}, (\mathcal{F})_{t=0}^T, \mathbf{P})$. *Without loss of generality we assume that* $\mathcal{F}_0$ *is trivial, that* $\mathcal{F}_T = \mathcal{F}$ *is the power set of* Ω, *and that* $\mathbf{P}[\omega_n] > 0$, *for all* $1 \leq n \leq N$. *We assume that the zero coordinate* S^0, *which we call the* cash account, *satisfies* $S_t^0 \equiv 1$, *for* $t = 0, 1, \ldots, T$. *The letter* ΔS_t *denotes the increment* $S_t - S_{t-1}$.

Definition 2.2. $\mathcal{H}$ *denotes the set of* trading strategies *for the financial market* S.

An element $H \in \mathcal{H}$ is an $\mathbb{R}^d$ - valued process $(H_t)_{t=1}^T = (H_t^1, H_t^2, \ldots, H_t^d)_{t=1}^T$ which is predictable, i.e. each H_t is $\mathcal{F}_{t-1}$ - measurable.

We then define the stochastic integral $(H \cdot S)$ as the $\mathbb{R}$-valued process $((H \cdot S)_t)_{t=0}^T$ given by

$$(H \cdot S)_t = \sum_{k=1}^t (H_k, \Delta S_k), \quad t = 0, \ldots, T, \tag{19}$$

where $(.\,,.)$ denotes the inner product in $\mathbb{R}^d$.

Definition 2.3. *We call the subspace K of $L^0(\Omega, \mathcal{F}, \mathbf{P})$ defined by*

$$K = \{(H \cdot S)_T : H \in \mathcal{H}\} \tag{20}$$

the set of contingent claims attainable at price 0.

The economic interpretation is the following: the random variables $f = (H \cdot S)_T$, for some $H \in \mathcal{H}$, are precisely those contingent claims, i.e., the payoff functions at time T depending on $\omega \in \Omega$ in an $\mathcal{F}_T$-measurable way, that an economic agent may replicate with zero initial investment, by pursuing some predictable trading strategy H.

For $a \in \mathbb{R}$, we call the *set of contingent claims attainable at price a* the affine space K_a obtained by shifting K by the constant function $a\mathbf{1}$, in other words the random variables of the form $a + (H \cdot S)_T$, for some trading strategy H. Again the economic interpretation is that these are precisely the contingent claims that an economic agent may replicate with an initial investment of a by pursuing some predictable trading strategy H.

Definition 2.4. *We call the convex cone C in $L^\infty(\Omega, \mathcal{F}, \mathbf{P})$ defined by*

$$C = \{g \in L^\infty(\Omega, \mathcal{F}, \mathbf{P}) \text{ s.t. there is } f \in K, f \geq g\}. \tag{21}$$

the set of contingent claims super-replicable at price 0.

Economically speaking, a contingent claim $g \in L^\infty(\Omega, \mathcal{F}, \mathbf{P})$ is *super-replicable at price 0*, if we can achieve it with zero net investment, subsequently pursuing some predictable trading strategy H — thus arriving at some contingent claim f — and then, possibly, "throwing away money" to arrive at g. This operation of "throwing away money" may seem awkward at this stage, but we shall see later that the set C plays an important role in the development of the theory. Observe that C is a convex cone containing the negative orthant $L_-^\infty(\Omega, \mathcal{F}, \mathbf{P})$. Again we may define C_a as the set of *contingent claims super-replicable at price a* obtained by shifting C by the constant function $a\mathbf{1}$.

Definition 2.5. *A financial market S satifies the* no-arbitrage condition *(NA) if*

$$K \cap L_+^0(\Omega, \mathcal{F}, \mathbf{P}) = \{0\} \tag{22}$$

or, equivalently,

$$C \cap L^\infty_+(\Omega, \mathcal{F}, \mathbf{P}) = \{0\} \qquad (23)$$

where 0 denotes the function identically equal to zero.

In other words we now have formalized the concept of an arbitrage possibility: it consists of the existence of a trading strategy H such that — starting from an initial investment zero — the resulting contingent claim $f = (H \cdot S)_T$ is non-negative and not identically equal to zero. If a financial market does not allow for arbitrage we say it satisfies the *no-arbitrage condition (NA)*.

Definition 2.6. *A probability measure* $\mathbf{Q}$ *on* $(\Omega, \mathcal{F})$ *is called an* equivalent martingale measure *for* S, *if* $\mathbf{Q} \sim \mathbf{P}$ *and* S *is a martingale under* $\mathbf{Q}$.

We denote by $\mathcal{M}^e(S)$ the set of equivalent martingale probability measures and by $\mathcal{M}^a(S)$ the set of all (not necessarily equivalent) martingale probability measures. The letter a stands for "absolutely continuous with respect to $\mathbf{P}$" which in the present setting (finite Ω and $\mathbf{P}$ having full support) automatically holds true, but which will be of relevance for general probability spaces $(\Omega, \mathcal{F}, \mathbf{P})$ later. We shall often identify a measure $\mathbf{Q}$ on $(\Omega, \mathcal{F})$ with its Radon-Nikodym derivative $\frac{d\mathbf{Q}}{d\mathbf{P}} \in L^1(\Omega, \mathcal{F}, \mathbf{P})$.

Lemma 2.1. *For a probability measure* $\mathbf{Q}$ *on* $(\Omega, \mathcal{F})$ *the following are equivalent:*

(i) $\mathbf{Q} \in \mathcal{M}^a(S)$,
(ii) $\mathbf{E}_{\mathbf{Q}}[f] = 0$, *for all* $f \in K$,
(iii) $\mathbf{E}_{\mathbf{Q}}[g] \leq 0$, *for all* $g \in C$.

Proof The equivalences are rather trivial, as (ii) is tantamount to the very definition of S being a martingale under $\mathbf{Q}$, and the equivalence of (ii) and (iii) is straightforward. □

After having fixed these formalities we may formulate and prove the central result of the theory of pricing and hedging by no-arbitrage, sometimes called the "fundamental theorem of asset pricing", which in its present form (i.e., finite Ω) is due to Harrison and Pliska [HP 81].

Theorem 2.1 (Fundamental Theorem of Asset Pricing). *For a financial market* S *modeled on a finite stochastic base* $(\Omega, \mathcal{F}, (\mathcal{F}_t)_{t=0}^T, \mathbf{P})$ *the following are equivalent:*

(i) S *satisfies (NA).*
(ii) $\mathcal{M}^e(S) \neq \emptyset$.

Proof (ii) $\Rightarrow$ (i): This is the obvious implication. If there is some $\mathbf{Q} \in \mathcal{M}^e(S)$ then by lemma 2.1 we have that

$$\mathbf{E}_{\mathbf{Q}}[g] \leq 0, \quad \text{for } g \in C. \tag{24}$$

On the other hand, if there were $g \in C \cap L^\infty_+, g \neq 0$, then, using the assumption that $\mathbf{Q}$ is equivalent to $\mathbf{P}$, we would have

$$\mathbf{E}_{\mathbf{Q}}[g] > 0, \tag{25}$$

a contradiction.

(i) $\Rightarrow$ (ii) This implication is the important message of the theorem which will allow us to link the no-arbitrage arguments with martingale theory. We give a functional analytic existence proof, which will be generalizable — in spirit — to more general situations.

By assumption the space K intersects L^∞_+ only at 0. We want to separate the disjoint convex sets $L^\infty_+ \setminus \{0\}$ and K by a hyperplane induced by a linear functional $\mathbf{Q} \in L^1(\Omega, \mathcal{F}, \mathbf{P})$ which is *strictly positive* on $L^\infty_+ \setminus \{0\}$. Unfortunately this is a situation, where the usual versions of the separation theorem (i.e., the Hahn-Banach Theorem) do not apply (even in finite dimensions!). Indeed, one usually assumes that one of the convex sets is compact in order to obtain a strict separation.

One way to overcome this difficulty (in finite dimension) is to consider the convex hull of the unit vectors $(\mathbf{1}_{\{\omega_n\}})_{n=1}^N$ in $L^\infty(\Omega, \mathcal{F}, \mathbf{P})$ i.e.

$$P := \left\{ \sum_{n=1}^N \mu_n \mathbf{1}_{\{\omega_n\}} : \mu_n \geq 0, \sum_{n=1}^N \mu_n = 1 \right\}. \tag{26}$$

This is a convex, compact subset of $L^1_+(\Omega, \mathcal{F}, \mathbf{P})$ and, by the *(NA)* assumption, disjoint from K. Hence we may strictly separate the sets P and K by a linear functional $\mathbf{Q} \in L^1(\Omega, \mathcal{F}, \mathbf{P})^* = L^1(\Omega, \mathcal{F}, \mathbf{P})$, i.e., find $\alpha < \beta$ such that

$$\mathbf{E}_Q[f] = \langle \mathbf{Q}, f \rangle \leq \alpha \quad \text{for} \quad f \in K, \tag{27}$$
$$\mathbf{E}_Q[h] = \langle \mathbf{Q}, h \rangle \geq \beta \quad \text{for} \quad h \in P.$$

As K is a linear space, we have $\alpha \geq 0$ and may, in fact, replace α by 0. Hence $\beta > 0$. Therefore $\langle \mathbf{Q}, \mathbf{1} \rangle > 0$, and we may normalize $\mathbf{Q}$ such that $\langle \mathbf{Q}, \mathbf{1} \rangle = 1$. As $\mathbf{Q}$ is strictly positive on each $\mathbf{1}_{\{\omega_n\}}$, we therefore have found a probability measure $\mathbf{Q}$ on $(\Omega, \mathcal{F})$ equivalent to $\mathbf{P}$ such that condition (ii) of lemma 2.1 holds true. In other words, we found an equivalent martingale measure $\mathbf{Q}$ for the process S. $\qquad\square$

Corollary 2.1. *Let S satisfy (NA) and $f \in L^\infty(\Omega, \mathcal{F}, \mathbf{P})$ be an attainable contingent claim so that*

$$f = a + (H \cdot S)_T, \tag{28}$$

for some $a \in \mathbb{R}$ and some trading strategy H.

Then the constant a and the process $(H \cdot S)$ are uniquely determined by (28) *and satisfy, for every $\mathbf{Q} \in \mathcal{M}^e(S)$,*

$$a = \mathbf{E}_{\mathbf{Q}}[f], \quad \text{and} \quad a + (H \cdot S)_t = \mathbf{E}_{\mathbf{Q}}[f|\mathcal{F}_t] \quad \text{for} \quad 0 \le t \le T. \tag{29}$$

Proof As regards the uniqueness of the constant $a \in \mathbb{R}$, suppose that there are two representations $f = a^1 + (H^1 \cdot S)_T$ and $f = a^2 + (H^2 \cdot S)_T$ with $a^1 \neq a^2$. Assuming w.l.o.g. that $a^1 > a^2$ we find an obvious arbitrage possibility: we have $a^1 - a^2 = ((H^1 - H^2) \cdot S)_T$, i.e. the trading strategy $H^1 - H^2$ produces a strictly positive result at time T, a contradiction to *(NA)*.

As regards the uniqueness or the process $H \cdot S$ we simply apply a conditional version of the previous argument: assume that $f = a + (H^1 \cdot S)_T$ and $f = a + (H^2 \cdot S)_T$ such that the processes $H^1 \cdot S$ amd $H^2 \cdot S$ are not identical. Then there is $0 < t < T$ such that $(H^1 \cdot S)_t \neq (H^2 \cdot S)_t$; w.l.g. $A := \{(H^1 \cdot S)_t > (H^2 \cdot S)_t\}$ is a non-empty event, which clearly is in $\mathcal{F}_t$. Hence, using the fact that $(H^1 \cdot S)_T = (H^2 \cdot S)_T$, the trading strategy $H := (H^2 - H^1)\chi_A \cdot \chi_{]t,T]}$ is a predictable process producing an arbitrage, as $(H \cdot S)_T = 0$ outside A, while $(H \cdot S)_T = (H^1 \cdot S)_t - (H^2 \cdot S)_t > 0$ on A, which again contradicts *(NA)*.

Finally, the equations in (29) result from the fact that, for every predictable process H and every $\mathbf{Q} \in \mathcal{M}^a(S)$, the process $H \cdot S$ is a $\mathbf{Q}$-martingale. Noting that, for a measure $\mathbf{Q} \sim \mathbf{P}$, the conditional expectation $\mathbf{E}_{\mathbf{Q}}[f|\mathcal{F}_t]$ is $\mathbf{P}$-a.s. well-defined we thus obtain (29) for each $\mathbf{Q} \in \mathcal{M}^e(S)$. □

Denote by $\text{cone}(\mathcal{M}^e(S))$ and $\text{cone}(\mathcal{M}^a(S))$ the cones generated by the convex sets $\mathcal{M}^e(S)$ and $\mathcal{M}^a(S)$ respectively. The subsequent result clarifies the polar relation between these cones and the cone C. Recall (see, e.g., [S 66]) that, for a pair (E, E') of vector spaces in separating duality via the scalar product $\langle ., . \rangle$, the polar C^0 of a set C in E is defined as

$$C^0 = \{g \in E' : \langle f, g \rangle \le 1, \text{ for all } f \in C\}. \tag{30}$$

In the case when C is closed under multiplication with positive scalars (e.g., if C is a convex cone) the polar C^0 may equivalently be defined by

$$C^0 = \{g \in E' : \langle f, g \rangle \le 0, \text{ for all } f \in C\}. \tag{31}$$

The *bipolar theorem* (see, e.g., [S 66]) states that the bipolar $C^{00} := (C^0)^0$ of a set C in E is the $\sigma(E, E')$-closed convex hull of C.

After these general considerations we pass to the concrete setting of the cone $C \subseteq L^\infty(\Omega, \mathcal{F}, \mathbf{P})$ of contingent claims super-replicable at price 0. Note that in our finite-dimensional setting this convex cone is closed as it is the algebraic sum of the closed linear space K (a linear space in $\mathbb{R}^N$ is always closed) and the closed polyhedral cone $L_-^\infty(\Omega, \mathcal{F}, \mathbf{P})$ (the verification, that the algebraic sum of a space and a polyhedral cone in $\mathbb{R}^N$ is closed, is an easy, but not completely trivial exercise). Hence we deduce from the bipolar theorem, that C equals its bipolar C^{00}.

Proposition 2.1. *Suppose that S satisfies (NA). Then the polar of C is equal to $cone(\mathcal{M}^a(S))$ and $\mathcal{M}^e(S)$ is dense in $\mathcal{M}^a(S)$. Hence the following assertions are equivalent for an element $g \in L^\infty(\Omega, \mathcal{F}, \mathbf{P})$*

(i) $g \in C$,
(ii) $\mathbf{E}_\mathbb{Q}[g] \leq 0$, for all $g \in \mathcal{M}^a(S)$,
(iii) $\mathbf{E}_\mathbb{Q}[g] \leq 0$, for all $g \in \mathcal{M}^e(S)$,

Proof   The fact that the polar C^0 and $cone(\mathcal{M}^a(S))$ coincide, follows from lemma 2.1 and the observation that $C \supseteq L^\infty_-(\Omega, \mathcal{F}, \mathbf{P})$ implies $C^0 \subseteq L^\infty_+(\Omega, \mathcal{F}, \mathbf{P})$. Hence the equivalence of (i) and (ii) follows from the bipolar theorem.

As regards the density of $\mathcal{M}^e(S)$ in $\mathcal{M}^a(S)$ we first deduce from theorem 2.1 that there is at least one $\mathbf{Q}^* \in \mathcal{M}^e(S)$. For any $\mathbf{Q} \in \mathcal{M}^a(S)$ and $0 < \mu \leq 1$ we have that $\mu\mathbf{Q}^* + (1 - \mu)\mathbf{Q} \in \mathcal{M}^e(S)$, which clearly implies the density of $\mathcal{M}^e(S)$ in $\mathcal{M}^a(S)$. The equivalence of (ii) and (iii) now is obvious. $\square$

The subsequent theorem tells us precisely what the principle of no arbitrage can tell us about the possible prices for a contingent claim f. It goes back to the work of D. Kreps [K 81] and was subsequently extended by several authors.

For given $f \in L^\infty(\Omega, \mathcal{F}, \mathbf{P})$, we call $a \in \mathbb{R}$ an *arbitrage-free price*, if in addition to the financial market S, the introduction of the contingent claim, which pays the random amount f at time $t = T$ and can be bought or sold at price a at time $t = 0$, does not create an arbitrage possibility. Mathematically speaking, this can be formalized as follows. Let $C^{f,a}$ denote the cone spanned by C and the linear space spanned by $f - a$; then a is an arbitrage-free price for f if $C^{f,a} \cap L^\infty_+(\Omega, \mathcal{F}, \mathbf{P}) = \{0\}$.

Theorem 2.2 (Pricing by No-Arbitrage). *Assume that S satisfies (NA) and let $f \in L^\infty(\Omega, \mathcal{F}, \mathbf{P})$. Define*

$$\overline{\pi}(f) = \sup \left\{ \mathbf{E}_\mathbb{Q}[f] : \mathbf{Q} \in \mathcal{M}^e(S) \right\}, \tag{32}$$

$$\underline{\pi}(f) = \inf \left\{ \mathbf{E}_\mathbb{Q}[f] : \mathbf{Q} \in \mathcal{M}^e(S) \right\}. \tag{33}$$

Either $\underline{\pi}(f) = \overline{\pi}(f)$, in which case f is attainable at price $\pi(f) := \underline{\pi}(f) = \overline{\pi}(f)$, i.e. $f = \pi(f) + (H \cdot S)_T$ for some $H \in \mathcal{H}$; therefore $\pi(f)$ is the unique arbitrage-free price for f.

Or $\underline{\pi}(f) < \overline{\pi}(f)$, in which case $\{\mathbf{E}_\mathbb{Q}[f] : \mathbf{Q} \in \mathcal{M}^e(S)\}$ equals the open interval $]\underline{\pi}(f), \overline{\pi}(f)[$, which in turn equals the set of arbitrage-free prices for the contingent claim f.

Proof First observe that the set $\{\mathbf{E}_{\mathbf{Q}}[f] : \mathbf{Q} \in \mathcal{M}^e(S)\}$ forms a bounded non-empty interval in $\mathbb{R}$, which we denote by I.

We claim that a number a is in I, iff a is an arbitrage-free price for f. Indeed, supposing that $a \in I$ we may find $\mathbf{Q} \in \mathcal{M}^e(S)$ s.t. $\mathbf{E}_{\mathbf{Q}}[f - a] = 0$ and therefore $C^{f,a} \cap L_+^\infty(\Omega, \mathcal{F}, \mathbf{P}) = \{0\}$.

Conversely suppose that $C^{f,a} \cap L_+^\infty(\Omega, \mathcal{F}, \mathbf{P}) = \{0\}$. Note that $C^{f,a}$ is a closed convex cone (it is the albegraic sum of the linear space $\mathrm{span}(K, f - a)$ and the closed, polyhedral cone $L_-^\infty(\Omega, \mathcal{F}, \mathbf{P})$). Hence by the same argument as in the proof of theorem 2.1 there exists a probability measure $\mathbf{Q} \sim \mathbf{P}$ such that $\mathbf{Q}|_{C^{f,a}} \leq 0$. This implies that $\mathbf{E}_{\mathbf{Q}}[f - a] = 0$, i.e., $a \in I$.

Now we deal with the boundary case: suppose that a equals the right boundary of I, i.e., $a = \overline{\pi}(f) \in I$, and consider the contingent claim $f - \overline{\pi}(f)$; by definition we have $\mathbf{E}_{\mathbf{Q}}[f - \overline{\pi}(f)] \leq 0$, for all $\mathbf{Q} \in \mathcal{M}^e(S)$, and therefore by proposition 2.1, that $f - \overline{\pi}(f) \in C$. We may find $g \in K$ such that $g \geq f - \overline{\pi}(f)$. If the sup in (32) is attained, i.e., if there is $\mathbf{Q}^* \in \mathcal{M}^e(S)$ such that $\mathbf{E}_{\mathbf{Q}^*}[f] = \overline{\pi}(f)$, then we have $0 = \mathbf{E}_{\mathbf{Q}^*}[g] \geq \mathbf{E}_{\mathbf{Q}^*}[f - \overline{\pi}(f)] = 0$ which in view of $\mathbf{Q}^* \sim \mathbf{P}$ implies that $f - \overline{\pi}(f) \equiv g$; in other words f is attainable at price $\overline{\pi}(f)$. This in turn implies that $\mathbf{E}_{\mathbf{Q}}[f] = \overline{\pi}(f)$, for all $\mathbf{Q} \in \mathcal{M}^e(S)$, and therefore I is reduced to the singleton $\{\overline{\pi}(f)\}$.

Hence, if $\underline{\pi}(f) < \overline{\pi}(f)$, $\overline{\pi}(f)$ connot belong to the interval I, which is therefore open on the right hand side. Passing from f to $-f$, we obtain the analogous result for the left hand side of I, which therefore equals $I =]\underline{\pi}(f), \overline{\pi}(f)[$. □

Corollary 2.2 (complete financial markets). *For a financial market S satisfying the no-arbitrage condition (NA) the following are equivalent:*

(i) $\mathcal{M}^e(S)$ consists of a single element $\mathbf{Q}$.
(ii) Each $f \in L^\infty(\Omega, \mathcal{F}, \mathbf{P})$ may be represented as

$$f = a + (H \cdot S)_T, \quad \text{for some } a \in \mathbb{R}, \text{ and } H \in \mathcal{H}. \tag{34}$$

In this case $a = \mathbf{E}_{\mathbf{Q}}[f]$, the stochastic integral $(H \cdot S)$ is unique and we have that

$$\mathbf{E}_{\mathbf{Q}}[f|\mathcal{F}_t] = \mathbf{E}_{\mathbf{Q}}[f] + (H \cdot S)_t, \quad t = 0, \ldots, T. \tag{35}$$

Proof The implication (i) $\Rightarrow$ (ii) immediately follows from the preceding theorem; for the implication (ii) $\Rightarrow$ (i), note that, (34) implies that, for elements $\mathbf{Q}_1, \mathbf{Q}_2 \in \mathcal{M}^a(S)$, we have $\mathbf{E}_{\mathbf{Q}_1}[f] = a = \mathbf{E}_{\mathbf{Q}_2}[f]$; hence it suffices to note that if $\mathcal{M}^e(S)$ contains two different elements $\mathbf{Q}_1, \mathbf{Q}_2$, we may find $f \in L^\infty(\Omega, \mathcal{F}, \mathbf{P})$ s.t. $\mathbf{E}_{\mathbf{Q}_1}[f] \neq \mathbf{E}_{\mathbf{Q}_2}[f]$. □

2.1 Utility Maximization

We are now ready to study utility maximization problems with the convex duality approach.

The complete Case (Arrow)

As a first case we analyze the situation of a *complete* financial market (Corollary 2.2 above), i.e., the set $\mathcal{M}^e(S)$ of equivalent probability measures under which S is a martingale is reduced to a singleton $\{\mathbf{Q}\}$. In this setting consider the *Arrow assets* $\mathbf{1}_{\{\omega_n\}}$, which pay 1 unit of the numéraire at time T, when ω_n turns out to be the true state of the world, and 0 otherwise. In view of our normalization of the numéraire $S_t^0 \equiv 1$, we get for the price of the Arrow assets at time $t = 0$ the relation

$$\mathbf{E}_\mathbb{Q}\left[\mathbf{1}_{\{\omega_n\}}\right] = \mathbf{Q}[\omega_n] = q_n, \tag{36}$$

and by 2.2 each Arrow asset $\mathbf{1}_{\{\omega_n\}}$ may be represented as $\mathbf{1}_{\{\omega_n\}} = \mathbf{Q}[\omega_n] + (H \cdot S)_T$, for some predictable trading strategy $H \in \mathcal{H}$.

Hence, for fixed initial endowment $x \in \mathrm{dom}(U)$, the utility maximization problem (16) above may simply be written as

$$\mathbf{E}_\mathbf{P}\left[U(X_T)\right] = \sum_{n=1}^N p_n U(\xi_n) \to \max! \tag{37}$$

$$\mathbf{E}_\mathbb{Q}[X_T] = \sum_{n=1}^N q_n \xi_n \leq x. \tag{38}$$

To verify that (37) and (38) indeed are equivalent to the original problem (16) above (in the present finite, complete case), note that by Theorem 2.2 a random variable $(X_T(\omega_n))_{n=1}^N = (\xi_n)_{n=1}^N$ can be dominated by a random variable of the form $x+(H{\cdot}S)_T = x+\sum_{t=1}^T H_t \Delta S_t$ iff $\mathbf{E}_\mathbb{Q}[X_T] = \sum_{n=1}^N q_n \xi_n \leq x$. This basic relation has a particularly evident interpretation in the present setting, as q_n is simply the price of the Arrow asset $\mathbf{1}_{\{\omega_n\}}$.

We have written ξ_n for $X_T(\omega_n)$ to stress that (37) simply is a concave maximization problem in $\mathbb{R}^N$ with one linear constraint. To solve it, we form the Lagrangian

$$L(\xi_1,\ldots,\xi_N,y) = \sum_{n=1}^N p_n U(\xi_n) - y\left(\sum_{n=1}^N q_n \xi_n - x\right) \tag{39}$$

$$= \sum_{n=1}^N p_n \left(U(\xi_n) - y\frac{q_n}{p_n}\xi_n\right) + yx. \tag{40}$$

We have used the letter $y \geq 0$ instead of the usual $\lambda \geq 0$ for the Lagrange multiplier; the reason is the dual relation between x and y which will become apparent in a moment.

Write

$$\Phi(\xi_1, \ldots, \xi_N) = \inf_{y>0} L(\xi_1, \ldots, \xi_N, y), \quad \xi_n \in \text{dom}(U), \tag{41}$$

and

$$\Psi(y) = \sup_{\xi_1, \ldots, \xi_N} L(\xi_1, \ldots, \xi_N, y), \quad y \geq 0. \tag{42}$$

Note that we have

$$\sup_{\xi_1, \ldots, \xi_N} \Phi(\xi_1, \ldots, \xi_N) = \sup_{\substack{\xi_1, \ldots, \xi_N \\ \sum_{n=1}^N q_n \xi_n \leq x}} \sum_{n=1}^N p_n U(\xi_n) = u(x). \tag{43}$$

Indeed, if $(\xi_1, \ldots, \xi_N)$ is in the admissible region $\sum_{n=1}^N q_n \xi_n \leq x$ then $\Phi(\xi_1, \ldots, \xi_N) = L(\xi_1, \ldots, \xi_N, 0) = \sum_{n=1}^N p_n U(\xi_n)$. On the other hand, if $(\xi_1, \ldots, \xi_N)$ satisfies $\sum_{n=1}^N q_n \xi_n > x$, then by letting $y \to \infty$ in (41) we note that $\Phi(\xi_1, \ldots, \xi_N) = -\infty$.

As regards the function $\Psi(y)$ we make the following pleasant observation which is the basic reason for the efficiency of the duality approach: using the form (40) of the Lagrangian and fixing $y > 0$, the optimization problem appearing in (42) splits into N independent optimization problems over $\mathbb{R}$

$$U(\xi_n) - y \tfrac{q_n}{p_n} \xi_n \to \max!, \quad \xi_n \in \mathbb{R}. \tag{44}$$

In fact, these one-dimensional optimization problems are of a very convenient form: recall (see, e.g., [R 70], [ET 76] or [KLSX 91]) that, for a concave function $U : \mathbb{R} \to \mathbb{R} \cup \{-\infty\}$, the *conjugate function* V (which is just the Legendre-transform of $x \mapsto -U(-x)$) is defined by

$$V(\eta) = \sup_{\xi \in \mathbb{R}} [U(\xi) - \eta\xi], \quad \eta > 0. \tag{45}$$

Definition 2.7. *We say that the function* $V : \mathbb{R}_+ \to \mathbb{R}$*, conjugate to the function* U*, satisfies the* usual regularity assumptions*, if* V *is finitely valued, differentiable, strictly convex on* $]0, \imath[$*, and satisfies*

$$V'(0) := \lim_{y \searrow 0} V'(y) = -\imath. \tag{46}$$

As regards the behavior of V *at infinity, we have to distinguish between case 1 and case 2 in Assumption 1.2 above:*

$$\text{case 1:} \quad \lim_{y \to \imath} V(y) = \lim_{x \to 0} U(x) \quad \text{and} \quad \lim_{y \to \imath} V'(y) = 0 \tag{47}$$

$$\text{case 2:} \quad \lim_{y \to \imath} V(y) = \imath \quad \text{and} \quad \lim_{y \to \imath} V'(y) = \imath \tag{48}$$

We have the following wellknown fact (see [R 70] or [ET 76]).

Proposition 2.2. *If U satisfies Assumption 1.2, then its conjugate function V satisfies the the inversion formula*

$$U(\xi) = \inf_{\eta} \left[V(\eta) + \eta\xi \right], \quad \xi \in \text{dom}(U) \tag{49}$$

and satisfies the regularity assumptions in Definition 2.7. In addition, $-V'(y)$ is the inverse function of $U'(x)$. Conversely, if V satisfies the regulatory assumptions of Definition 2.7, then U defined by (49) satisfies Assumption 1.2. Following [KLS 87] we denote $-V' = I$ (for "inverse" function).

Proof It follows from Assumption 1.2 that V is finitely valued on $]0, \infty[$. Note that we have that

$$U(x) \leq a + yx \quad \forall x \in \text{dom}(U) \qquad \Longleftrightarrow \qquad V(y) \leq a \tag{50}$$

which implies the inversion formula above. In turn, this formula shows that V is the supremum of affine functions, and therefore convex. Since U is strictly concave and differentiable, the maximizer $\widehat{\xi} = \xi(\mu)$ in (45) solves the first-order condition $U'(\xi(\eta)) = \eta$. Also, we have that U' is a continuous bijection between $\{U > -\infty\}$ and $_+$. This observation and the inversion formula show that V is both strictly convex, differentiable, and that $-V'$ is the inverse of U'. □

Remark 2.1. Of course, U' has a good economic interpretation as the *marginal utility* of an economic agent modeled by the utility function U.

Here are some concrete examples of pairs of conjugate functions:

$$U(x) = \ln(x), \quad x > 0, \quad V(y) = -\ln(y) - 1,$$
$$U(x) = -\frac{e^{-\gamma x}}{\gamma}, \quad x \in \mathbb{R}, \quad V(y) = \frac{y}{\gamma}(\ln(y) - 1), \quad \gamma > 0$$
$$U(x) = \frac{x^\alpha}{\alpha}, \quad x > 0, \quad V(y) = \frac{1-\alpha}{\alpha} y^{\frac{\alpha}{\alpha-1}}, \quad \alpha \in (-\infty, 1) \setminus \{0\}.$$

We now apply these general facts about the Legendre transformation to calculate $\Psi(y)$. Using definition (45) of the conjugate function V and (40), formula (42) becomes

$$\Psi(y) = \sum_{n=1}^{N} p_n V\left(y\frac{q_n}{p_n} \right) + yx \tag{51}$$

$$= \mathbf{E_P} \left[V\left(y\frac{d\mathbf{Q}}{d\mathbf{P}} \right) \right] + yx. \tag{52}$$

Denoting by $v(y)$ the dual value function

$$v(y) := \mathbf{E}_{\mathbf{P}}\left[V\left(y\tfrac{d\mathbf{Q}}{d\mathbf{P}}\right)\right] = \sum_{n=1}^{N} p_n V\left(y\tfrac{q_n}{p_n}\right), \quad y > 0, \tag{53}$$

the function v has the same qualitative properties as the function V listed in Definition 2.7, since it is a convex combination of V calculated on linearly scaled arguments.

Hence by (46), (47), and (48) we find, for fixed $x \in \mathrm{dom}(U)$, a unique $\widehat{y} = \widehat{y}(x) > 0$ such that $v'(\widehat{y}(x)) = -x$, which therefore is the unique minimizer to the dual problem

$$\Psi(y) = \mathbf{E}_{\mathbf{P}}\left[V\left(y\tfrac{d\mathbf{Q}}{d\mathbf{P}}\right)\right] + yx = \min! \tag{54}$$

Fixing the critical value $\widehat{y}(x)$, the concave function

$$(\xi_1, \ldots, \xi_N) \mapsto L(\xi_1, \ldots, \xi_N, \widehat{y}(x)) \tag{55}$$

defined in (40) assumes its unique maximum at the point $(\widehat{\xi}_1, \ldots, \widehat{\xi}_N)$ satisfying

$$U'(\widehat{\xi}_n) = \widehat{y}(x)\tfrac{q_n}{p_n} \quad \text{or, equivalently,} \quad \widehat{\xi}_n = I\left(\widehat{y}(x)\tfrac{q_n}{p_n}\right), \tag{56}$$

so that we have

$$\inf_{y>0} \Psi(y) = \inf_{y>0} (v(y) + xy) \tag{57}$$

$$= v(\widehat{y}(x)) + x\widehat{y}(x) \tag{58}$$

$$= L(\widehat{\xi}_1, \ldots, \widehat{\xi}_N, \widehat{y}(x)). \tag{59}$$

Note that $\widehat{\xi}_n$ are in $\mathrm{dom}(U)$, for $1 \leq n \leq N$, so that L is continuously differentiable at $(\widehat{\xi}_1, \ldots, \widehat{\xi}_N, \widehat{y}(x))$, which implies that the gradient of L vanishes at $(\widehat{\xi}_1, \ldots, \widehat{\xi}_N, \widehat{y}(x))$ and, in particular, that $\frac{\partial}{\partial y}L(\xi_1, \ldots, \xi_N, y)|_{(\widehat{\xi}_1, \ldots, \widehat{\xi}_N, \widehat{y}(x))} = 0$. Hence we infer from (39) and the fact that $\widehat{y}(x) > 0$ that the constraint (38) is binding, i.e.,

$$\sum_{n=1}^{N} q_n \widehat{\xi}_n = x, \tag{60}$$

and that

$$\sum_{n=1}^{N} p_n U(\widehat{\xi}_n) = L(\widehat{\xi}_1, \ldots, \widehat{\xi}_N, \widehat{y}(x)). \tag{61}$$

In particular, we obtain that

$$u(x) = \sum_{n=1}^{N} p_n U(\widehat{\xi}_n). \tag{62}$$

Indeed, the inequality $u(x) \geq \sum_{n=1}^{N} p_n U(\widehat{\xi}_n)$ follows from (60) and (43), while the reverse inequality follows from (61) and the fact that for all $\xi_1, \ldots, \xi_N$ verifying the constraint (38)

$$\sum_{n=1}^{N} p_n U(\xi_n) \leq L(\xi_1, \ldots, \xi_N, \widehat{y}(x)) \leq L(\widehat{\xi}_1, \ldots, \widehat{\xi}_N, \widehat{y}(x)). \qquad (63)$$

We shall write $\widehat{X}_T(x) \in C(x)$ for the optimizer $\widehat{X}_T(x)(\omega_n) = \widehat{\xi}_n$, $n = 1, \ldots, N$.

Combining (57), (61) and (62) we note that the value functions u and v are conjugate:

$$\inf_{y>0} (v(y) + xy) = v(\widehat{y}(x)) + x\widehat{y}(x) = u(x), \quad x \in \mathrm{dom}(U), \qquad (64)$$

which, by Proposition 32 the remarks after equation (53), implies that u inherits the properties of U listed in Assumption 1.2. The relation $v'(\widehat{y}(x)) = -x$ which was used to define $\widehat{y}(x)$, therefore translates into

$$u'(x) = \widehat{y}(x), \quad \text{for } x \in \mathrm{dom}(U). \qquad (65)$$

Let us summarize what we have proved:

Theorem 2.3 (finite Ω, complete market). *Let the financial market $S = (S_t)_{t=0}^{T}$ be defined over the finite filtered probability space $(\Omega, \mathcal{F}, (\mathcal{F})_{t=0}^{T}, \mathbf{P})$ and satisfy $\mathcal{M}^e(S) = \{\mathbf{Q}\}$, and let the utility function U satisfy Assumption 1.2.*

Denote by $u(x)$ and $v(y)$ the value functions

$$u(x) = \sup_{X_T \in C(x)} \mathbf{E}[U(X_T)], \quad x \in \mathrm{dom}(U), \qquad (66)$$

$$v(y) = \mathbf{E}\left[V\left(y\frac{d\mathbf{Q}}{d\mathbf{P}}\right)\right], \quad y > 0. \qquad (67)$$

We then have:

(i) *The value functions $u(x)$ and $v(y)$ are conjugate and u inherits the qualitative properties of U listed in Assumption 1.2.*

(ii) *The optimizer $\widehat{X}_T(x)$ in (66) exists, is unique and satisfies*

$$\widehat{X}_T(x) = I(y\tfrac{d\mathbf{Q}}{d\mathbf{P}}), \quad \text{or, equivalently,} \quad y\tfrac{d\mathbf{Q}}{d\mathbf{P}} = U'(\widehat{X}_T(x)), \qquad (68)$$

where $x \in \mathrm{dom}(U)$ and $y > 0$ are related via $u'(x) = y$ or, equivalently, $x = -v'(y)$.

(iii) *The following formulae for u' and v' hold true:*

$$u'(x) = \mathbf{E}_\mathbf{P}[U'(\widehat{X}_T(x))], \quad v'(y) = \mathbf{E}_\mathbb{Q}\left[V'\left(y\tfrac{d\mathbf{Q}}{d\mathbf{P}}\right)\right] \qquad (69)$$

$$xu'(x) = \mathbf{E}_\mathbf{P}\left[\widehat{X}_T(x)U'(\widehat{X}_T(x))\right], \quad yv'(y) = \mathbf{E}_\mathbf{P}\left[y\tfrac{d\mathbf{Q}}{d\mathbf{P}}V'\left(y\tfrac{d\mathbf{Q}}{d\mathbf{P}}\right)\right]. (70)$$

Proof Items (i) and (ii) have been shown in the preceding discussion, hence we only have to show (iii). The formulae for $v'(y)$ in (69) and (70) immediately follow by differentiating the relation

$$v(y) = \mathbf{E_P}\left[V\left(y\frac{d\mathbf{Q}}{d\mathbf{P}}\right)\right] = \sum_{n=1}^{N} p_n V\left(y\frac{q_n}{p_n}\right). \tag{71}$$

Of course, the formula for v' in (70) is an obvious reformulation of the one in (69). But we write both of them to stress their symmetry with the formulae for $u'(x)$.

The formula for u' in (69) translates via the relations exhibited in (ii) into the identity

$$y = \mathbf{E_P}\left[y\frac{d\mathbf{Q}}{d\mathbf{P}}\right], \tag{72}$$

while the formula for $u'(x)$ in (70) translates into

$$v'(y)y = \mathbf{E_P}\left[V'\left(y\frac{d\mathbf{Q}}{d\mathbf{P}}\right)y\frac{d\mathbf{Q}}{d\mathbf{P}}\right], \tag{73}$$

which we just have seen to hold true.   □

Remark 2.2. Firstly, let us recall the economic interpretation of (68)

$$U'\left(\widehat{X}_T(x)(\omega_n)\right) = y\frac{q_n}{p_n}, \quad n = 1,\ldots,N. \tag{74}$$

This equality means that, in every possible state of the world ω_n, the *marginal utility* $U'(\widehat{X}_T(x)(\omega_n))$ of the wealth of an optimally investing agent at time T is *proportional to the ratio of the price q_n of the corresponding Arrow security* $\mathbf{1}_{\{\omega_n\}}$ *and the probability of its success $p_n = \mathbf{P}[\omega_n]$.* This basic relation was analyzed in the fundamental work of K. Arrow and allows for a convincing economic interpretation: considering for a moment the situation where this proportionality relation fails to hold true, one immediately deduces from a marginal variation argument that the investment of the agent cannot be optimal. Indeed, by investing a little more in the more favorable asset and a little less in the less favorable the economic agent can strictly increase expected utility under the same budget constraint. Hence for the optimal investment the proportionality must hold true. The above result also identifies the proportionality factor as $y = u'(x)$, where x is the initial endowment of the investor. This also allows for an economic interpretation.

Theorem 2.3 indicates an easy way to solve the utility maximization at hand: calculate $v(y)$ by (67), which reduces to a simple one-dimensional computation; once we know $v(y)$, the theorem provides easy formulae to calculate all the other quantities of interest, e.g., $\widehat{X}_T(x)$, $u(x)$, $u'(x)$ etc.

Another message of the above theorem is that the value function $x \mapsto u(x)$ may be viewed as a utility function as well, sharing all the qualitative features of the original utility function U. This makes sense economically, as the "indirect utility" function $u(x)$ denotes the expected utility at time T of an agent with initial endowment x, after having optimally invested in the financial market S.

Let us also give an economic interpretation of the formulae for $u'(x)$ in item (iii) along these lines: suppose the initial endowment x is varied to $x+h$, for some small real number h. The economic agent may use the additional endowment h to finance, in addition to the optimal pay-off function $\widehat{X}_T(x)$, h units of the cash account, thus ending up with the pay-off function $\widehat{X}_T(x)+h$ at time T. Comparing this investment strategy to the optimal one corresponding to the initial endowment $x+h$, which is $\widehat{X}_T(x+h)$, we obtain

$$\lim_{h\to0}\frac{u(x+h)-u(x)}{h} = \lim_{h\to0}\frac{\mathbf{E}[U(\widehat{X}_T(x+h))-U(\widehat{X}_T(x))]}{h} \tag{75}$$

$$\geq \lim_{h\to0}\frac{\mathbf{E}[U(\widehat{X}_T(x)+h)-U(\widehat{X}_T(x))]}{h} \tag{76}$$

$$= \mathbf{E}[U'(\widehat{X}_T(x))]. \tag{77}$$

Using the fact that u is differentiable, and that h may be positive as well as negative, we must have equality in (76) and therefore have found another proof of formula (69) for $u'(x)$; the economic interpretation of this proof is that the economic agent, who is optimally investing, is indifferent of first order towards a (small) additional investment into the cash account.

Playing the same game as above, but using the additional endowment $h \in \mathbb{R}$ to finance an additional investment into the optimal portfolio $\widehat{X}_T(x)$ (assuming, for simplicity, $x \neq 0$), we arrive at the pay-off function $\frac{x+h}{x}\widehat{X}_T(x)$. Comparing this investment with $\widehat{X}_T(x+h)$, an analogous calculation as in (75) leads to the formula for $u'(x)$ displayed in (70). The interpretation now is, that the optimally investing economic agent is indifferent of first order towards a marginal variation of the investment into the optimal portfolio.

It now becomes clear that formulae (69) and (70) for $u'(x)$ are just special cases of a more general principle: for each $f \in L^\infty(\Omega,\mathcal{F},\mathbf{P})$ we have

$$\mathbf{E}_\mathbb{Q}[f]u'(x) = \lim_{h\to0}\frac{\mathbf{E}_\mathbf{P}[U(\widehat{X}_T(x)+hf)-U(\widehat{X}_T(x))]}{h}. \tag{78}$$

The proof of this formula again is along the lines of (75) and the interpretation is the following: by investing an additional endowment $h\mathbf{E}_\mathbb{Q}[f]$ to finance the contingent claim hf, the increase in expected utility is of first order equal to $h\mathbf{E}_\mathbb{Q}[f]u'(x)$; hence again the economic agent is of first order indifferent towards an additional investment into the contingent claim f.

The Incomplete Case

We now drop the assumption that the set $\mathcal{M}^e(S)$ of equivalent martingale measures is reduced to a singleton (but we still remain in the framework of a finite probability space Ω) and replace it by Assumption 1.1 requiring that $\mathcal{M}^e(S) \neq \emptyset$.

In this setting it follows from Theorem 2.2 that a random variable $X_T(\omega_n) = \xi_n$ may be dominated by a random variable of the form $x + (H \cdot S)_T$ iff $\mathbf{E}_{\mathbb{Q}}[X_T] = \sum_{n=1}^{N} q_n \xi_n \leq x$, for each $\mathbf{Q} = (q_1 \ldots, q_N) \in \mathcal{M}^a(S)$ (or equivalently, for every $\mathbf{Q} \in \mathcal{M}^e(S)$).

In order to reduce the infinitely many constraints, where $\mathbf{Q}$ runs through $\mathcal{M}^a(S)$, to a finite number, make the easy observation that $\mathcal{M}^a(S)$ is a bounded, closed, convex polytope in $\mathbb{R}^N$ and therefore the convex hull of its finitely many extreme points $\{\mathbf{Q}^1, \ldots, \mathbf{Q}^M\}$. Indeed, $\mathcal{M}^a(S)$ is given by finitely many linear constraints. For $1 \leq m \leq M$, we identify $\mathbf{Q}^m$ with the probabilites $(q_1^m, \ldots, q_N^m)$.

Fixing the initial endowment $x \in \mathrm{dom}(U)$, we therefore may write the utility maximization problem (16) similarly as in (37) as a concave optimization problem over $\mathbb{R}^N$ with finitely many linear constraints:

$$\mathbf{E_P}[U(X_T)] = \sum_{n=1}^{N} p_n U(\xi_n) \to \max! \tag{79}$$

$$\mathbf{E}_{\mathbf{Q}^m}[X_T] = \sum_{n=1}^{N} q_n^m \xi_n \leq x, \quad \text{for} \ \ m = 1, \ldots, M. \tag{80}$$

Writing again

$$C(x) = \left\{ X_T \in L^0(\Omega, \mathcal{F}, \mathbf{P}) : \mathbf{E}_{\mathbb{Q}}[X_T] \leq x, \ \ \text{for all} \ \mathbf{Q} \in \mathcal{M}^a(S) \right\} \tag{81}$$

we define the value function, for $x \in \mathrm{dom}(U)$,

$$u(x) = \sup_{H \in \mathcal{H}} \mathbf{E}[U(x + (H \cdot S)_T)] = \sup_{X_T \in C(x)} \mathbf{E}[U(X_T)]. \tag{82}$$

The Lagrangian now is given by

$$L(\xi_1, \ldots, \xi_N, \eta_1, \ldots, \eta_M) \tag{83}$$

$$= \sum_{n=1}^{N} p_n U(\xi_n) - \sum_{m=1}^{M} \eta_m \left(\sum_{n=1}^{N} q_n^m \xi_n - x \right) \tag{84}$$

$$= \sum_{n=1}^{N} p_n \left(U(\xi_n) - \sum_{m=1}^{M} \frac{\eta_m q_n^m}{p_n} \xi_n \right) + \sum_{m=1}^{M} \eta_m x, \tag{85}$$

$$\text{where} \ \ (\xi_1, \ldots, \xi_N) \in \mathrm{dom}(U)^N, \ \ (\eta_1, \ldots, \eta_M) \in \mathbb{R}_+^M. \tag{86}$$

Writing $y = \eta_1 + \ldots + \eta_M$, $\mu_m = \frac{\eta_m}{y}$, $\mu = (\mu_1, \ldots, \mu_M)$ and

$$\mathbf{Q}^\mu = \sum_{m=1}^{M} \mu_m \mathbf{Q}^m, \tag{87}$$

note that, when $(\eta_1, \ldots, \eta_M)$ runs trough $\mathbb{R}_+^M$, the pairs $(y, \mathbf{Q}^\mu)$ run through $\mathbb{R}_+ \times \mathcal{M}^a(S)$. Hence we may write the Lagrangian as

$$L(\xi_1, \ldots, \xi_N, y, \mathbf{Q}) == \quad \mathbf{E_P}[U(X_T)] - y\left(\mathbf{E_Q}[X_T - x]\right)$$

$$= \sum_{n=1}^{N} p_n \left(U(\xi_n) - \frac{yq_n}{p_n} \xi_n \right) + yx, \qquad (88)$$

where $\xi_n \in \mathrm{dom}(U)$, $y > 0$, $\mathbf{Q} = (q_1, \ldots, q_N) \in \mathcal{M}^a(S)$.

This expression is entirely analogous to (40), the only difference now being that $\mathbf{Q}$ runs through the set $\mathcal{M}^a(S)$ instead of being a fixed probability measure. Defining again

$$\Phi(\xi_1, \ldots, \xi_n) = \inf_{y>0, \mathbf{Q} \in \mathcal{M}^a(S)} L(\xi_1, \ldots, \xi_N, y, \mathbf{Q}), \qquad (89)$$

and

$$\Psi(y, \mathbf{Q}) = \sup_{\xi_1, \ldots, \xi_N} L(\xi_1, \ldots, \xi_N, y, \mathbf{Q}), \qquad (90)$$

we obtain, just as in the complete case,

$$\sup_{\xi_1, \ldots, \xi_N} \Phi(\xi_1, \ldots, \xi_N) = u(x), \quad x \in \mathrm{dom}(U), \qquad (91)$$

and

$$\Psi(y, \mathbf{Q}) = \sum_{n=1}^{N} p_n V\left(\frac{yq_n}{p_n}\right) + yx, \quad y > 0, \quad \mathbf{Q} \in \mathcal{M}^a(S), \qquad (92)$$

where $(q_1, \ldots, q_N)$ denotes the probabilities of $\mathbf{Q} \in \mathcal{M}^a(S)$. The minimization of Ψ will be done in two steps: first we fix $y > 0$ and minimize over $\mathcal{M}^a(S)$, i.e.,

$$\Psi(y) := \inf_{\mathbf{Q} \in \mathcal{M}^a(S)} \Psi(y, \mathbf{Q}), \quad y > 0. \qquad (93)$$

For fixed $y > 0$, the continuous function $\mathbf{Q} \to \Psi(y, \mathbf{Q})$ attains its minimum on the compact set $\mathcal{M}^a(S)$, and the minimizer $\widehat{\mathbf{Q}}(y)$ is unique by the strict convexity of V. Writing $\widehat{\mathbf{Q}}(y) = (\widehat{q}_1(y), \ldots, \widehat{q}_N(y))$ for the minimizer, it follows from $V'(0) = -\infty$ that $\widehat{q}_n(y) > 0$, for each $n = 1, \ldots, N$; Indeed, suppose that $\widehat{q}_n(y) = 0$, for some $1 \leq n \leq N$ and fix any equivalent martingale measure $\mathbf{Q} \in \mathcal{M}^e(S)$. Letting $\mathbf{Q}^\epsilon = \epsilon \mathbf{Q} + (1 - \epsilon)\widehat{\mathbf{Q}}$ we have that $\mathbf{Q}^\epsilon \in \mathcal{M}^e(S)$, for $0 < \epsilon < 1$, and $\Psi(y, \mathbf{Q}^\epsilon) < \Psi(y, \widehat{\mathbf{Q}})$ for $\epsilon > 0$ sufficiently small, a contradiction. In other words, $\widehat{\mathbf{Q}}(y)$ is an equivalent martingale measure for S.

Defining the dual value function $v(y)$ by

$$v(y) = \inf_{\mathbf{Q} \in \mathcal{M}^a(S)} \sum_{n=1}^{N} p_n V\left(y \frac{q_n}{p_n}\right) \qquad (94)$$

$$= \sum_{n=1}^{N} p_n V\left(y \frac{\widehat{q}_n(y)}{p_n}\right) \qquad (95)$$

we find ourselves in an analogous situation as in the complete case above: defining again $\widehat{y}(x)$ by $v'(\widehat{y}(x)) = -x$ and

$$\widehat{\xi}_n = I\left(\widehat{y}(x)\frac{\widehat{q}_n(y)}{p_n}\right), \tag{96}$$

similar arguments as above apply to show that $(\widehat{\xi}_1, \ldots, \widehat{\xi}_N, \widehat{y}(x), \widehat{\mathbf{Q}}(y))$ is the unique saddle-point of the Lagrangian (88) and that the value functions u and v are conjugate.

Let us summarize what we have found in the incomplete case:

Theorem 2.4 (finite Ω, incomplete market). *Let the financial market $S = (S_t)_{t=0}^T$ defined over the finite filtered probability space $(\Omega, \mathcal{F}, (\mathcal{F})_{t=0}^T, \mathbf{P})$ and let $\mathcal{M}^e(S) \neq \emptyset$, and the utility function U satisfies Assumptions 1.2.*

Denote by $u(x)$ and $v(y)$ the value functions

$$u(x) = \sup_{X_T \in C(x)} \mathbf{E}[U(X_T)], \qquad x \in \mathrm{dom}(U), \tag{97}$$

$$v(y) = \inf_{\mathbf{Q} \in \mathcal{M}^a(S)} \mathbf{E}\left[V\left(y\frac{d\mathbf{Q}}{d\mathbf{P}}\right)\right], \qquad y > 0. \tag{98}$$

We then have:

(i) The value functions $u(x)$ and $v(y)$ are conjugate and u shares the qualitative properties of U listed in Assumption 1.2.

(ii) The optimizers $\widehat{X}_T(x)$ and $\widehat{\mathbf{Q}}(y)$ in (97) and (98) exist, are unique, $\widehat{\mathbf{Q}}(y) \in \mathcal{M}^e(S)$, and satisfy

$$\widehat{X}_T(x) = I\left(y\frac{d\widehat{\mathbf{Q}}(y)}{d\mathbf{P}}\right), \qquad y\frac{d\widehat{\mathbf{Q}}(y)}{d\mathbf{P}} = U'(\widehat{X}_T(x)), \tag{99}$$

where $x \in \mathrm{dom}(U)$ and $y > 0$ are related via $u'(x) = y$ or, equivalently, $x = -v'(y)$.

(iii) The following formulae for u' and v' hold true:

$$u'(x) = \mathbf{E}_\mathbf{P}[U'(\widehat{X}_T(x))], \qquad v'(y) = \mathbf{E}_{\widehat{\mathbf{Q}}}\left[V'\left(y\frac{d\widehat{\mathbf{Q}}(y)}{d\mathbf{P}}\right)\right] \tag{100}$$

$$xu'(x) = \mathbf{E}_\mathbf{P}[\widehat{X}_T(x)U'(\widehat{X}_T(x))], \quad yv'(y) = \mathbf{E}_\mathbf{P}\left[y\frac{d\widehat{\mathbf{Q}}(y)}{d\mathbf{P}}V'\left(y\frac{d\widehat{\mathbf{Q}}(y)}{d\mathbf{P}}\right)\right] \tag{101}$$

Remark 2.3. Let us again interpret the formulae (100), (101) for $u'(x)$ similarly as in Remark 2.2 above. In fact, the interpretations of these formulae as well as their derivations remain in the incomplete case exactly the same.

But a new and interesting phenomenon arises when we pass to the variation of the optimal pay-off function $\widehat{X}_T(x)$ by a small unit of an arbitrary pay-off function $f \in L^\infty(\Omega, \mathcal{F}, \mathbf{P})$. Similarly as in (78) we have the formula

$$\mathbf{E}_{\widehat{\mathbf{Q}}(y)}[f]u'(x) = \lim_{h \to 0} \frac{\mathbf{E}_\mathbf{P}[U(\widehat{X}_T(x) + hf) - U(\widehat{X}_T(x))]}{h}, \tag{102}$$

the only difference being that $\mathbf{Q}$ has been replaced by $\widehat{\mathbf{Q}}(y)$ (recall that x and y are related via $u'(x) = y$).

The remarkable feature of this formula is that it does not only pertain to variations of the form $f = x + (H \cdot S)_T$, i.e, contingent claims attainable at price x, but to arbitrary contingent claims f, for which — in general — we cannot derive the price from no arbitrage considerations.

The economic interpretation of formula (102) is the following: the pricing rule $f \mapsto \mathbf{E}_{\widehat{\mathbf{Q}}(y)}[f]$ yields precisely those prices, at which an economic agent with initial endowment x, utility function U and investing optimally, is indifferent of first order towards adding a (small) unit of the contingent claim f to her portfolio $\widehat{X}_T(x)$.

In fact, one may turn the view around, and this was done by M. Davis [D 97] (compare also the work of L. Foldes [F 90]): one may *define* $\widehat{\mathbf{Q}}(y)$ by (102), verify that this indeed is an equivalent martingale measure for S, and interpret this pricing rule as "pricing by marginal utility", which is, of course, a classical and basic paradigm in economics.

Let us give a proof for (102) (under the hypotheses of Theorem 2.4). One possible strategy of proof, which also has the advantage of a nice economic interpretation, is the idea of introducing "fictitious securities" as developed in [KLSX 91]: fix $x \in \mathrm{dom}(U)$ and $y = u'(x)$ and let $(f^1, \ldots, f^k)$ be finitely many elements of $L^\infty(\Omega, \mathcal{F}, \mathbf{P})$ such that the space $K = \{(H \cdot S)_T : H \in \mathcal{H}\}$, the constant function $\mathbf{1}$, and $(f^1, \ldots, f^k)$ linearly span $L^\infty(\Omega, \mathcal{F}, \mathbf{P})$. Define the k processes

$$S_t^{d+j} = \mathbf{E}_{\widehat{\mathbf{Q}}(y)}[f^j | \mathcal{F}_t], \quad j = 1, \ldots, k, \quad t = 0, \ldots, T. \tag{103}$$

Now extend the $\mathbb{R}^{d+1}$-valued process $S = (S^0, S^1, \ldots, S^d)$ to the $\mathbb{R}^{d+k+1}$-valued process $\overline{S} = (S^0, S^1, \ldots, S^d, S^{d+1}, \ldots, S^{d+k})$ by adding these new co-ordinates. By (103) we still have that $\overline{S}$ is a martingale under $\widehat{\mathbf{Q}}(y)$, which now is the unique probability under which $\overline{S}$ is a martingale, by our choice of $(f^1, \ldots, f^k)$ and Corollary 2.2.

Hence we find ourselves in the situation of Theorem 2.3. By comparing (68) and (99) we observe that the optimal pay-off function $\widehat{X}_T(x)$ has not changed. Economically speaking this means that in the "completed" market $\overline{S}$ the optimal investment may still be achieved by trading only in the first $d + 1$ assets and without touching the "fictitious" securities $S^{d+1}, \ldots, S^{d+k}$.

In particular, we now may apply formula (78) to $\mathbf{Q} = \widehat{\mathbf{Q}}(y)$ to obtain (102).

Finally we remark that the pricing rule induced by $\widehat{\mathbf{Q}}(y)$ is precisely such that the interpretation of the optimal investment $\widehat{X}_T(x)$ defined in (99) (given in Remark 2.2 in terms of marginal utility and the ratio of Arrow prices $\widehat{q}_n(y)$ and probabilities p_n) carries over to the present incomplete setting. The above completion of the market by introducing "fictitious securities" allows for an economic interpretation of this fact.

3 The General Case

In the previous section we have analyzed the duality theory of the utility maximization problem in detail and with full proofs, for the case when the underlying probability space is finite.

We now pass to the question under which conditions the crucial features of the above Theorem 2.4 carry over to the general setting. In particular one is naturally led to ask: under which conditions

- are the optimizers $\widehat{X}_T(x)$ and $\widehat{\mathbf{Q}}(y)$ of the value functions $u(x)$ and $v(y)$ attained?
- does the basic duality formula

$$U'\left(\widehat{X}_T(x)\right) = \widehat{y}(x)\frac{d\widehat{\mathbf{Q}}(\widehat{y}(x))}{d\mathbf{P}} \tag{104}$$

or, equivalently

$$\widehat{X}_T(x) = I\left(\widehat{y}(x)\frac{d\widehat{\mathbf{Q}}(\widehat{y}(x))}{d\mathbf{P}}\right) \tag{105}$$

hold true?
- are the value functions $u(x)$ and $v(y)$ conjugate?
- does the value function $u(x)$ still inherit the qualitative properties of U listed in Assumption 1.2?
- do the formulae for $u'(x)$ and $v'(y)$ still hold true?

We shall see that we get affirmative answers to these questions under two provisos: firstly, one has to make an appropriate choice of the sets in which X_T and $\mathbf{Q}$ are allowed to vary. This choice will be different for case 1, where $\mathrm{dom}(U) = \mathbb{R}_+$, and case 2, where $\mathrm{dom}(U) = \mathbb{R}$. Secondly, the utility function U has to satisfy — in addition to Assumption 1.2 — a mild regularity condition, namely the property of "reasonable asymptotic elasticity".

3.1 The Reasonable Asymptotic Elasticity Condition

The essential message of the theorems below is that, assuming that U has "reasonable asymptotic elasticity", the duality theory works just as well as in the case of finite Ω. On the other hand, we shall see that we do not have to impose any regularity conditions on the underlying stochastic process S, except for its arbitrage-freeness in the sense made precise by Assumption 1.1. We shall also see that the assumption of reasonable asymptotic elasticity on the utility function U cannot be relaxed, even if we impose very strong assumptions on the process S (e.g., having continuous paths and defining a complete financial market), as we shall see below.

Before passing to the positive results we first analyze the notion of "reasonable asymptotic elasticity" and sketch the announced counterexample.

Definition 3.1. *A utility function U satisfying Assumption 1.2 is said to have "reasonable asymptotic elasticity" if*

$$AE_{+\infty}(U) = \limsup_{x \to \infty} \frac{xU'(x)}{U(x)} < 1, \tag{106}$$

and, in case 2 of Assumption 1.2, we also have

$$AE_{-\infty}(U) = \liminf_{x \to -\infty} \frac{xU'(x)}{U(x)} > 1. \tag{107}$$

We recall the following lemma from [KS99, Lemma 6.1], from which it follows that, for any concave function U such that the right hand side makes sense, we always have that $AE_{+\infty}(U) \le 1$. Note that, the asymptotic elasticity assumption requires that the strict inequality holds.

Lemma 3.1. *For a strictly concave, increasing, real-valued differentiable function U the asymptotic elasticity $AE(U)$ is well-defined and, depending on $U(\infty) = \lim_{x \to \infty} U(x)$, takes its values in the following sets:*

> (i) For $U(\infty) = \infty$ we have $AE(U) \in [0, 1]$,
> (ii) For $0 < U(\infty) < \infty$ we have $AE(U) = 0$,
> (iii) For $-\infty < U(\infty) \le 0$ we have $AE(U) \in [-\infty, 0]$.

Proof (i) Using the monotonicity and positivity of U' we may estimate

$$0 \le xU'(x) = (x - 1)U'(x) + U'(x)$$
$$\le [U(x) - U(1)] + U'(1)$$

hence, in the case $U(\infty) = \infty$,

$$0 \le \limsup_{x \to \infty} \frac{xU'(x)}{U(x)} \le \limsup_{x \to \infty} \frac{U(x) - U(1) + U'(1)}{U(x)} = 1.$$

(ii) For each $x_0 > 0$ we have

$$\limsup_{x \to \infty} xU'(x) = \limsup_{x \to \infty}(x - x_0)U'(x)$$
$$\le \limsup_{x \to \infty}(U(x) - U(x_0)).$$

If $U(\infty) < \infty$ we may choose x_0 such that the right hand side becomes arbitrary small.

(iii) We infer from $U(\infty) \le 0$ that $U(x) < 0$, for $x \in \mathbb{R}_+$, so that $\frac{xU'(x)}{U(x)} < 0$, for all $x \in \mathbb{R}_+$. □

Example 3.1.

- For $U(x) = \ln x$, we have $AE_{+\infty}(U) = 0$.

- For $U(x) = \frac{x^\alpha}{\alpha}$, we have $AE_{+\infty}(U) = \alpha$, for $\alpha \in (-\infty, 1) \setminus \{0\}$.
- For $U(x) = \frac{x}{\ln x}$ for $x \geq x_0$, we have $AE_{+\infty}(U) = 1$.

The asymptotic elasticity compares as follows with other conditions used in the literature [KLSX 91]:

Lemma 3.2. *Let U be a utility function, and consider the following conditions:*

i) *There exists $x_0 > 0$, $\alpha < 1$, $\beta > 1$ such that $U'(\beta x) < \alpha U'(x)$ for all $x \geq x_0$.*

ii) *$AE_{+\infty}(U) < 1$*

iii) *There exist k_1, k_2 and $\gamma < 1$ such that $U(x) \leq k_1 + k_2 x^\gamma$ for all $x \geq 0$.*

Then we have that i) $\Rightarrow$ ii) $\Rightarrow$ iii). The reverse implications do not hold true in general.

Proof $(i) \Rightarrow (ii)$ Assume (i) and let $a = \alpha\beta$ and $b = \frac{1}{\alpha} > 1$ and estimate, for $x > ax_0$:

$$
\begin{aligned}
U(bx) &= \quad U(\beta x_0) + \int_{\beta x_0}^{bx} U'(t)dt \\
&= \quad U(\beta x_0) + \beta \int_{x_0}^{x/a} U'(\beta t)dt \\
&\leq \quad U(\beta x_0) + \alpha\beta \int_{x_0}^{x/a} U'(t)dt \\
&= \quad U(\beta x_0) + aU(\tfrac{x}{a}) - aU(x_0).
\end{aligned}
$$

It follows that criterion (ii) of corollary 6.1 in [KS99] is satisfied, hence $AE(U) < 1$.

$(ii) \Rightarrow (iii)$ is immediate from assertion (i) of lemma 6.3 in [KS99].

$(ii) \not\Rightarrow (i)$: For $n \in \mathbb{N}$, let $x_n = 2^{2^n}$ and define the function $U(x)$ by letting $U(x_n) = 1 - \frac{1}{n}$ and to be linear on the intervals $[x_{n-1}, x_n]$; (for $0 < x \leq x_1$ continue $U(x)$ in an arbitrary way, so that U satisfies (2.4)).

Clearly $U(x)$ fails (i) as for any $\beta > 1$ there are arbitrary large $x \in \mathbb{R}$ with $U'(\beta x) = U'(x)$. On the other hand, we have $U(\infty) = 1$ so that $AE(U) = 0$ by Lemma 3.1. Finally, note that in this counterexample the limit $\lim_{x \to \infty} \frac{xU'(x)}{U(x)}$ exists and equals zero.

The attentive reader might object that $U(x)$ is neither strictly concave nor differentiable. But it is obvious that one can slightly change the function to "smooth out" the kinks and to "strictly concavify" the straight lines so that the above conclusion still holds true.

$(iii) \not\Rightarrow (ii)$: Let again $x_n = 2^{2^n}$ and consider the utility function $\tilde{U}(x) = x^{1/2}$. Define $U(x)$ by letting $U(x_n) = \tilde{U}(x_n)$, for $n = 0, 1, 2 \ldots$ and to be linear on the intervals $[x_n, x_{n+1}]$; (for $0 < x \leq x_1$ again continue $U(x)$ in an arbitrary way, so that U satisfies (2.4)).

Clearly $U(x)$ satisfies condition (iii) as U is dominated by $\tilde{U}(x) = x^{1/2}$.

To show that $AE(U) = 1$ let $x \in]x_{n-1}, x_n[$ and calculate the marginal utility U' at x:

$$U'(x) = \frac{U(x_n) - U(x_{n-1})}{x_n - x_{n-1}} = \frac{2^{2^{n-1}} - 2^{2^{n-2}}}{2^{2^n} - 2^{2^{n-1}}}$$

$$= \frac{2^{2^{n-1}}(1 - 2^{-2^{n-2}})}{2^{2^n}(1 - 2^{-2^{n-1}})} = 2^{-2^{n-1}}(1 + o(1)).$$

On the other hand we calculate the average utility at $x = x_n$:

$$\frac{U(x_n)}{x_n} = \frac{2^{2^{n-1}}}{2^{2^n}} = 2^{-2^{n-1}}.$$

Hence

$$AE_{+\infty}(U) = \limsup_{x \to \infty} \frac{xU'(x)}{U(x)} = 1.$$

As regards the lack of smoothness and strict concavity of U a similar remark applies as in $(ii) \not\Rightarrow (i)$ above.   □

Let us discuss the economic meaning of the notion of reasonable asymptotic elasticity: as H.-U. Gerber ponted out to us, the quantity $\frac{xU'(x)}{U(x)}$ is the elasticity of the function U at x. We are interested in its asymptotic behaviour. It easily follows from Assumption 1.2 that the limits in (106) and (107) are less (resp. bigger) than or equal to one (compare Lemma 3.1). What does it mean that $\frac{xU'(x)}{U(x)}$ tends to one, for $x \to \infty$? It means that the ratio between the *marginal utility* $U'(x)$ and the *average utility* $\frac{U(x)}{x}$ tends to one. A typical example is a function $U(x)$ which equals $\frac{x}{\ln(x)}$, for x large enough; note however, that in this example Assumption 1.2 is not violated insofar as the marginal utility still decreases to zero for $x \to \infty$, i.e., $\lim_{x \to \infty} U'(x) = 0$.

If the marginal utility $U'(x)$ is approximately equal to the average utility $\frac{U(x)}{x}$ for large x, this means that for an economic agent, modeled by the utility function U, the increase in utility by varying wealth from x to $x + 1$, when x is large, is approximately equal to the average of the increase of utility by changing wealth from n to $n + 1$, where n runs through $1, 2, \ldots, x - 1$ (we assume in this argument that x is a large natural number and, w.l.o.g., that $U(1) \approx 0$). We feel that the economic intuition behind decreasing marginal utility suggests that, for large x, the marginal utility $U'(x)$ should be substantially smaller than the average utility $\frac{U(x)}{x}$. Therefore we have denoted a utility function, where the ratio of $U'(x)$ and $\frac{U(x)}{x}$ becomes arbitrarily close to one if x tends either to $+\infty$ or $-\infty$, as being "unreasonable".

P. Guasoni observed, that there is a close connection between the asymptotic behaviour of the *elasticity* of U, and the asymptotic behaviour of the *relative risk aversion* associated to U. Recall (see, e.g., [HL88]) that the relative risk aversion of an agent with endowment x, whose preferences are described by the utility function U, equals

$$RRA(U)(x) = -\frac{xU''(x)}{U'(x)}. \tag{108}$$

A formal application of de l'Hôpital's rule yields

$$\lim_{x\to\infty}\frac{xU'(x)}{U(x)} = \lim_{x\to\infty}\frac{U'(x) + xU''(x)}{U'(x)} = 1 - \lim_{x\to\infty}\left(-\frac{xU''(x)}{U'(x)}\right) \tag{109}$$

which insinuates that the asymptotic elasticity of U is less than one iff the *"asymptotic relative risk aversion"* is strictly positive.

Turning the above formal argument into a precise statement, one easily proves the following result: if $\lim_{x\to\infty}(-\frac{xU''(x)}{U'(x)})$ exists, then $\lim_{x\to\infty}\frac{xU'(x)}{U(x)}$ exists too, and the former is strictly positive iff the latter is less than one (for details see [S 04a]). Hence *"essentially"* these two concepts coincide.

On the other hand, in general (i.e. without assuming that the above limit exists), there is no way to characterize the condition $\limsup_{x\to\infty}\frac{xU'(x)}{U(x)} < 1$ in terms of the asymptotic behaviour of $-\frac{xU''(x)}{U'(x)}$, as $x\to\infty$.

3.2 Existence Theorems

Let us now move to the positive results in the spirit of Theorem 2.3 and Theorem 2.4 above. We first consider the case where U satisfies case 1 of Assumption 1.2, which was studied in [KS 99].

Case 1: $\mathrm{dom}(U) = \mathbb{R}_+$.

The heart of the argument in the proof of Theorem 2.4 (which we now want to extend to the general case) is to find a saddlepoint for the Lagrangian. In more general situations we have to apply the minimax theorem, which is crucial in the theory of Lagrange multipliers. We want to extend the applicability of the minimax theorem to the present situation. The infinite-dimensional versions of the minimax theorem available in the literature (see, e.g, [ET 76] or [St 85]) are along the following lines: Let $\langle E, F\rangle$ be a pair of locally convex vector spaces in separating duality, $C \subseteq E$, $D \subseteq F$ a pair of convex subsets, and $L(x, y)$ a function defined on $C \times D$, concave in the first and convex in the second variable, having some (semi-)continuity property compatible with the topologies of E and F (which in turn should be compatible with the duality between E and F). If (at least) one of the sets C and D is compact and the other is complete, then one may assert the existence of a saddle point $(\widehat{\xi}, \widehat{\eta}) \in C \times D$ such that

$$L(\widehat{\xi}, \widehat{\eta}) = \sup_{\xi\in C}\inf_{\eta\in D} L(\xi, \eta) = \inf_{\eta\in D}\sup_{\xi\in C} L(\xi, \eta). \tag{110}$$

We try to apply this theorem to the analogue of the Lagrangian encountered in the proof of Theorem 2.4 above. Fixing $x > 0$ and $y > 0$ let us formally write the Lagrangian (88) in the infinite-dimensional setting,

$$L^{x,y}(X_T, \mathbf{Q}) = \mathbf{E_P}[U(X_T)] - y(\mathbf{E_Q}[X_T - x]) \qquad (111)$$

$$= \mathbf{E_P}\left[U(X_T) - y\frac{d\mathbf{Q}}{d\mathbf{P}}X_T \right] + yx, \qquad (112)$$

where X_T runs through "all" non-negative $\mathcal{F}_T$-measurable functions and $\mathbf{Q}$ through the set $\mathcal{M}^a(S)$ of absolutely continuous local martingale measures.

To restrict the set of "all" nonnegative functions to a more amenable one, note that $\inf_{y>0, \mathbf{Q}\in\mathcal{M}^a(S)} L^{x,y}(X_T, \mathbf{Q}) > -\infty$ iff

$$\mathbf{E_Q}[X_T] \leq x, \quad \text{for all } \mathbf{Q} \in \mathcal{M}^a(S). \qquad (113)$$

Using the basic result on the super-replicability of the contingent claim X_T (see [KQ 95], [J 92], [AS 94], [DS 94], and [DS 98]), we have — as encountered in Theorem 2.2 for the finite dimensional case — that a non-negative $\mathcal{F}_T$-measurable random variable X_T satisfies (113) iff there is an admissible trading strategy H such that

$$X_T \leq x + (H \cdot S)_T. \qquad (114)$$

Hence let

$$C(x) = \big\{ X_T \in L^0_+(\Omega, \mathcal{F}_T, \mathbf{P}) :$$
$$X_T \leq x + (H \cdot S)_T, \text{ for some admissible } H \big\} \qquad (115)$$
$$= \big\{ X_T \in L^0_+(\Omega, \mathcal{F}_T, \mathbf{P}) :$$
$$\mathbf{E_Q}[X_T] \leq x, \text{ for all } \mathbf{Q} \in \mathcal{M}^a(S) \big\} \qquad (116)$$

and simply write C for $C(1)$ (observe that $C(x) = xC$).

We thus have found a natural set $C(x)$ in which X_T should vary when we are mini-maxing the Lagrangian $L^{x,y}$. Dually, the set $\mathcal{M}^a(S)$ seems to be the natural domain where the measure $\mathbf{Q}$ is allowed to vary (in fact, we shall see later, that this set still has to be slightly enlarged). But what are the locally convex vector spaces E and F in separating duality into which C and $\mathcal{M}^a(S)$ are naturally embedded? As regards $\mathcal{M}^a(S)$ the natural choice seems to be $L^1(\mathbf{P})$ (by identifying a measure $\mathbf{Q} \in \mathcal{M}^a(S)$ with its Radon-Nikodym derivative $\frac{d\mathbf{Q}}{d\mathbf{P}}$); note that $\mathcal{M}^a(S)$ is a *closed* subset of $L^1(\mathbf{P})$, which is good news. On the other hand, there is no reason for C to be contained in $L^\infty(\mathbf{P})$, or even in $L^p(\mathbf{P})$, for any $p > 0$; the natural space in which C is embedded is just $L^0(\Omega, \mathcal{F}_T, \mathbf{P})$, the space of all real-valued $\mathcal{F}_T$-measurable functions endowed with the topology of convergence in probability.

The situation now seems hopeless (if we don't want to impose artificial $\mathbf{P}$-integrability assumptions on X_T and/or $\frac{d\mathbf{Q}}{d\mathbf{P}}$), as $L^0(\mathbf{P})$ and $L^1(\mathbf{P})$ are not in any reasonable duality; in fact, $L^0(\mathbf{P})$ is not even a locally convex space, hence there seems to be no hope for a good duality theory, which could serve as a basis for the application of the minimax theorem. But the good news is that the sets C and $\mathcal{M}^a(S)$ are in the *positive orthant* of $L^0(\mathbf{P})$ and $L^1(\mathbf{P})$

respectively; the crucial observation is, that for $f \in L^0_+(\mathbf{P})$ and $g \in L^1_+(\mathbf{P})$, it is possible to well-define

$$\langle f, g \rangle := \mathbf{E_P}[fg] \in [0, \infty]. \tag{117}$$

The spirit here is similar as in the very foundation of Lebesgue integration theory: For positive measurable functions the integral is always well-defined, but possibly $+\infty$. This does not cause any logical inconsistency.

Similarly the bracket $\langle .\,,.\rangle$ defined in (117) shares many of the usual properties of a scalar product. The difference is that $\langle f, g \rangle$ now may assume the value $+\infty$ and that the map $(f, g) \mapsto \langle f, g \rangle$ is not continuous on $L^0_+(\mathbf{P}) \times L^1_+(\mathbf{P})$, but only lower semi-continuous (this immediately follows from Fatou's lemma).

At this stage it becomes clear that the role of $L^1_+(\mathbf{P})$ is somewhat artificial, and it is more natural to define (117) in the general setting where f and g are both allowed to vary in $L^0_+(\mathbf{P})$. The pleasant feature of the space $L^0(\mathbf{P})$ in the context of Mathematical Finance is, that it is invariant under the passage to an equivalent measure $\mathbf{Q}$, a property only shared by $L^\infty(\mathbf{P})$, but by no other $L^p(\mathbf{P})$, for $0 < p < \infty$.

We now can turn to the polar relation between the sets C and $\mathcal{M}^a(S)$. By (114) we have, for an element $X_T \in L^0_+(\Omega, \mathcal{F}, \mathbf{P})$,

$$X_T \in C \iff \mathbf{E_Q}[X_T] = \mathbf{E_P}[X_T \tfrac{d\mathbf{Q}}{d\mathbf{P}}] \leq 1, \quad \text{for } \mathbf{Q} \in \mathcal{M}^a(S). \tag{118}$$

Denote by D the closed, convex, solid hull of $\mathcal{M}^a(S)$ in $L^0_+(\mathbf{P})$. It is easy to show (using, e.g., Lemma 3.3 below), that D equals

$$D = \{Y_T \in L^0_+(\Omega, \mathcal{F}_T, \mathbf{P}) : \text{ there is}$$
$$(\mathbf{Q}_n)^\infty_{n=1} \in \mathcal{M}^a(S) \text{ s.t. } Y_T \leq \lim_{n \to \infty} \tfrac{d\mathbf{Q}_n}{d\mathbf{P}}\}, \tag{119}$$

where the $\lim_{n \to \infty} \frac{d\mathbf{Q}_n}{d\mathbf{P}}$ is understood in the sense of almost sure convergence. We have used the letter Y_T for the elements of D to stress the dual relation to the elements X_T in C. In further analogy we write, for $y > 0$, $D(y)$ for yD, so that $D = D(1)$. By (119) and Fatou's lemma we again find that, for $X_T \in L^0_+(\Omega, \mathcal{F}, \mathbf{P})$

$$X_T \in C \iff \mathbf{E_P}[X_T Y_T] \leq 1, \quad \text{for } Y_T \in D. \tag{120}$$

Why did we pass to this enlargement D of the set $\mathcal{M}^a(S)$? The reason is that we now obtain a more symmetric relation between C and D: for $Y_T \in L^0_+(\Omega, \mathcal{F}, \mathbf{P})$ we have

$$Y_T \in D \iff \mathbf{E_P}[X_T Y_T] \leq 1, \quad \text{for } X_T \in C. \tag{121}$$

The proof of (121) relies on an adaption of the "bipolar theorem" from the theory of locally convex spaces (see, e.g., [S 66]) to the present duality $\langle L^0_+(\mathbf{P}), L^0_+(\mathbf{P})\rangle$, which was worked out in [BS 99].

Why is it important to define the enlargement D of $\mathcal{M}^a(S)$ in such a way that (121) holds true? After all, $\mathcal{M}^a(S)$ is a nice, convex, closed (w.r.t. the norm of $L^1(\mathbf{P})$) set and one may prove that, for $g \in L^1(\mathbf{P})$ *such that* $\mathbf{E}_{\mathbf{P}}[g] = 1$,

$$g \in \mathcal{M}^a(S) \Longleftrightarrow \mathbf{E}_{\mathbf{P}}[X_T g] \le 1, \quad \text{for } X_T \in C. \tag{122}$$

The reason is that, in general, the saddle point $(\widehat{X}_T, \widehat{\mathbf{Q}})$ of the Lagrangian will *not* be such that $\widehat{\mathbf{Q}}$ is a probability measure; it will only satisfy $\mathbf{E}\left[\frac{d\widehat{\mathbf{Q}}}{d\mathbf{P}}\right] \le 1$, the inequality possibly being strict. But it will turn out that $\widehat{\mathbf{Q}}$, which we identify with $\frac{d\widehat{\mathbf{Q}}}{d\mathbf{P}}$, is always in D. In fact, the passage from $\mathcal{M}^a(S)$ to D is the *crucial feature* in order to make the duality work in the present setting: even for nice utility functions U, such as the logarithm, and for nice processes, such as a continuous process $(S_t)_{0 \le t \le T}$ based on the filtration of two Brownian motions, the above described phenomenon can occur: the saddle point of the Lagrangian leads out of $\mathcal{M}^a(S)$.

The set D can be characterized in several equivalent manners. We have defined D above in the abstract way as the convex, closed, solid hull of $\mathcal{M}^a(S)$ and mentioned the description (119). Equivalently, one may define D as the set of random variables $Y_T \in L^0_+(\Omega, \mathcal{F}, \mathbf{P})$ such that there is a process $(Y_t)_{0 \le t \le T}$ starting at $Y_0 = 1$ with $(Y_t X_t)_{0 \le t \le T}$ a $\mathbf{P}$-supermartingale, for every non-negative process $(X_t)_{0 \le t \le T} = (x + (H \cdot S)_t)_{0 \le t \le T}$, where $x > 0$ and H is predictable and S-integrable. This definition was used in [KS 99]. Another equivalent characterization was used in [CSW 01]: Consider the convex, solid hull of $\mathcal{M}^a(S)$, and embed this subset of $L^1(\mathbf{P})$ into the bidual $L^1(\mathbf{P})^{**} = L^\infty(\mathbf{P})^*$; denote by $\overline{\mathcal{M}^a(S)}$ the weak-star closure of the convex solid hull of $\mathcal{M}^a(S)$ in $L^\infty(\mathbf{P})^*$. Each element of $\overline{\mathcal{M}^a(S)}$ may be decomposed into its regular part $\mu^r \in L^1(\mathbf{P})$ and its purely singular part $\mu^s \in L^\infty(\mathbf{P})^*$. It turns out that D equals the set $\{\mu^r \in L^1(\mathbf{P}) : \mu \in \overline{\mathcal{M}^a(S)}\}$, i.e. consists of the regular parts of the elements of $\overline{\mathcal{M}^a(S)}$. This description has the advantage that we may associate to the elements $\mu^r \in D$ a singular part μ^s, and it is this extra information which is crucial when extending the present results to the case of random endowment as in [CSW 01]. Compare also [HK 02], where the case of random endowment is analyzed in full generality without using the space $L^\infty(\mathbf{P})^*$.

Why are the sets C and D hopeful candidates for the minimax theorem to work out properly for a function L defined on $C \times D$? Both are closed, convex and bounded subsets of $L^0_+(\mathbf{P})$. But recall that we still need some compactness property to be able to localize the mini-maximizers (resp. maxi-minimizers) on C (resp. D). In general, neither C nor D is compact (w.r.t. the topology of convergence in measure), i.e., for a sequence $(f_n)_{n=1}^\infty$ in C (resp. $(g_n)_{n=1}^\infty$ in D) we cannot pass to a subsequence converging in measure. But C and D have a property which is close to compactness and in many applications turns out to serve just as well.

Lemma 3.3. *Let A be a closed, convex, bounded subset of $L_+^0(\Omega, \mathcal{F}, \mathbf{P})$. Then for each sequence $(h_n)_{n=1}^\infty \in A$ there exists a sequence of convex combinations $k_n \in \mathrm{conv}(h_n, h_{n+1}, \ldots)$ which converges almost surely to a function $k \in A$.*

This easy lemma (see, e.g., [DS 94, Lemma A.1.1], for a proof) is in the spirit of the celebrated theorem of Komlos [Kom 67], stating that for a bounded sequence $(h_n)_{n=1}^\infty$ in $L^1(\mathbf{P})$ there is a subsequence converging in Cesaro-mean almost surely. The methodology of finding pointwise limits by using convex combinations has turned out to be extremely useful as a surrogate for compactness. For an extensive discussion of more refined versions of the above lemma and their applications to Mathematical Finance we refer to [DS 99].

The application of the above lemma is the following: by passing to convex combinations of optimizing sequences $(f_n)_{n=1}^\infty$ in C (resp. $(g_n)_{n=1}^\infty$ in D), we can always find limits $f \in C$ (resp. $g \in D$) w.r.t. almost sure convergence. Note that the passage to convex combinations does not cost more than passing to a subsequence in the application to convex optimization.

We have now given sufficient motivation to state the central result of [KS 99], which is the generalization of Theorem 2.4 to the semi-martingale setting under Assumption 1.2, case 1, and having reasonable asymptotic elasticity.

Theorem 3.1 ([KS 99], Theorem 2.2).
Let the semi-martingale $S = (S_t)_{0 \le t \le T}$ and the utility function U satisfy Assumptions 1.1, 1.2 case 1 and 1.3; suppose in addition that U has reasonable asymptotic elasticity. Define

$$u(x) = \sup_{X_T \in C(x)} \mathbf{E}[U(X_T)], \quad v(y) = \inf_{Y_T \in D(y)} \mathbf{E}[V(Y_T)]. \qquad (123)$$

Then we have:

(i) The value functions $u(x)$ and $v(y)$ are conjugate; they are continuously differentiable, strictly concave (resp. convex) on $]0, \infty[$ and satisfy

$$u'(0) = -v'(0) = \infty, \quad u'(\infty) = v'(\infty) = 0. \qquad (124)$$

(ii) The optimizers $\widehat{X}_T(x)$ and $\widehat{Y}_T(y)$ in (123) exist, are unique and satisfy

$$\widehat{X}_T(x) = I(\widehat{Y}_T(y)), \quad \widehat{Y}_T(y) = U'(\widehat{X}_T(x)), \qquad (125)$$

where $x > 0$, $y > 0$ are related via $u'(x) = y$ or equivalently $x = -v'(y)$.
(iii) We have the following relations between u', v' and $\widehat{X}_T, \widehat{Y}_T$ respectively:

$$u'(x) = \mathbf{E}\left[\frac{\widehat{X}_T(x) U'(\widehat{X}_T(x))}{x}\right], \; x > 0, \quad v'(y) = \mathbf{E}\left[\frac{\widehat{Y}_T(y) V'(\widehat{Y}_T(y))}{y}\right], \; y > 0. \qquad (126)$$

For the full proof of the theorem we refer to [KS 99].

How severe is the fact that the dual optimizer $\widehat{Y}_T(1)$ may fail to be the density of a probability measure (or that $\mathbf{E}[\widehat{Y}_T(y)] < y$, for $y > 0$, which amounts to the same thing)? In fact, in many respects it does not bother us at all: we still have the basic duality relation between the primal and the dual optimizer displayed in Theorem 3.1 (ii). Even more is true: using the terminology from [KS 99] the product $(\widehat{X}_t(x)\widehat{Y}_t(y))_{0 \le t \le T}$, where x and y satisfy $u'(x) = y$, is a uniformly integrable martingale. This fact can be interpreted in the following way: by taking the optimal portfolio $(\widehat{X}_t(x))_{0 \le t \le T}$ as numéraire instead of the original cash account, the pricing rule obtained from the dual optimizer $\widehat{Y}_T(y)$ then is induced by an equivalent martingale measure. We refer to ([KS 99], p. 912) for a thorough discussion of this argument.

Finally we want to draw the attention of the reader to the fact that — comparing item (iii) of Theorem 3.1 to the corresponding item of Theorem 2.4 — we only asserted one pair of formulas for $u'(x)$ and $v'(y)$. The reason is that, in general, the formulae (100) do not hold true any more, the reason again being precisely that for the dual optimizer $\widehat{Y}_T(y)$ we may have $\mathbf{E}[\widehat{Y}_T(y)] < y$. Indeed, the validity of $u'(x) = \mathbf{E}[U'(\widehat{X}_T(x))]$ is tantamount to the validity of $y = \mathbf{E}[\widehat{Y}_T(y)]$.

Case 2: $\mathrm{dom}(U) = \mathbb{R}$

We now pass to the case of a utility function U satisfying Assumption 1.2 case 2 which is defined and finitely valued on all of $\mathbb{R}$. The reader should have in mind the exponential utility $U(x) = -e^{-\gamma x}$, for $\gamma > 0$, as the typical example.

We want to obtain a result analogous to Theorem 3.1 also in this setting. Roughly speaking, we get the same theorem, but the sets C and D considered above have to be chosen in a somewhat different way, as the optimal portfolio $\widehat{X}_T$ now may assume negative values too.

Firstly, we have to assume throughout the rest of this section that the semimartingale S is *locally bounded*. The case of non locally bounded processes is not yet understood and waiting for future research.

Next we turn to the question; what is the proper definition of the set $C(x)$ of terminal values X_T dominated by a random variable $x + (H \cdot S)_T$, where H is an "allowed" trading strategy? On the one hand we cannot be too liberal in the choice of "allowed" trading strategies as we have to exclude doubling strategies and similar schemes. We therefore maintain the definition of the value function $u(x)$ unchanged

$$u(x) = \sup_{H \in \mathcal{H}} \mathbf{E}\left[U\left(x + (H \cdot S)_T\right)\right], \quad x \in \mathbb{R}, \tag{127}$$

where we still confine H to run through the set $\mathcal{H}$ of admissible trading strategies, i.e., such that the process $((H \cdot S)_t)_{0 \le t \le T}$ is uniformly bounded from below. This notion makes good sense economically as it describes the strategies possible for an agent having a finite credit line.

On the other hand, in general, we have no chance to find the minimizer $\widehat{H}$ in (127) within the set of admissible strategies: already in the classical cases studied by Merton ([M 69] and [M 71] where, in particular, the case of exponential utility is solved for the Black-Scholes model) the optimal solution $x + (\widehat{H} \cdot S)_T$ to (127) is *not* uniformly bounded from below; this random variable typically assumes low values with very small probability, but its essential infimum typically is minus infinity.

In [S 01] the following approach was used to cope with this difficulty: fix the utility function $U : \mathbb{R} \to \mathbb{R}$ and first define the set $C_U^b(x)$ to consist of all random variables G_T dominated by $x + (H \cdot S)_T$, for some *admissible* trading strategy H and such that $\mathbf{E}[U(G_T)]$ makes sense:

$$C_U^b(x) = \{ G_T \in L^0(\Omega, \mathcal{F}_T, \mathbf{P}) : \text{ there is } H \text{ admissible s.t.} \tag{128}$$

$$G_T \leq x + (H \cdot S)_T \text{ and } \mathbf{E}[|U(G_T)|] < \infty \}. \tag{129}$$

Next we define $C_U(x)$ as the set of $\mathbb{R} \cup \{+\infty\}$-valued random variables X_T such that $U(X_T)$ can be approximated by $U(G_T)$ in the norm of $L^1(\mathbf{P})$, when G_T runs through $C_U^b(x)$:

$$C_U(x) = \{ X_T \in L^0(\Omega, \mathcal{F}_T, \mathbf{P}; \mathbb{R} \cup \{+\infty\}) : U(X_T) \text{ is in} \tag{130}$$

$$L^1(\mathbf{P})\text{-closure of } \{ U(G_T) : G_T \in C_U^b(x) \} \}. \tag{131}$$

The optimization problem (127) now reads

$$u(x) = \sup_{X_T \in C_U(x)} \mathbf{E}[U(X_T)], \quad x \in \mathbb{R}. \tag{132}$$

The set $C_U(x)$ was chosen in such a way that the value functions $u(x)$ defined in (127) and (132) coincide; but now we have much better chances to find the maximizer to (132) in the set $C_U(x)$.

Two features of the definition of $C_U(x)$ merit some comment: firstly, we have allowed $X_T \in C_U(x)$ to attain the value $+\infty$; indeed, in the case when $U(\infty) < \infty$ (e.g., the case of exponential utility), this is natural, as the set $\{ U(X_T) : X_T \in C_U(x) \}$ should equal the $L^1(\mathbf{P})$-closure of the set $\{ U(G_T) : G_T \in C_U^b(x) \}$. But we shall see that — under appropriate assumptions — the optimizer $\widehat{X}_T$, which we are going to find in $C_U(x)$, will almost surely be finite.

Secondly, the elements X_T of $C_U(x)$ are only *random variables* and, at this stage, they are not related to a *process* of the form $x + (H \cdot S)$. Of course, we finally want to find for each $X_T \in C_U(x)$, or at least for the optimizer $\widehat{X}_T$, a predictable, S-integrable process H having "allowable" properties (in order to exclude doubling strategies) and such that $X_T \leq x + (H \cdot S)_T$. We shall prove later that — under appropriate assumptions — this is possible and give a precise meaning to the word "allowable".

After having specified the proper domain $C_U(x)$ for the primal optimization problem (132), we now pass to the question of finding the proper domain for the dual optimization problem. Here we find a pleasant surprise: contrary to case 1 above, where we had to pass from the set $\mathcal{M}^a(S)$ to its closed, solid hull D, it turns out that, in the present case 2, the dual optimizer always lies in $\mathcal{M}^a(S)$. This fact was first proved by F. Bellini and M. Frittelli ([BF 02]).

We now can state the main result of [S 01]:

Theorem 3.2. *[S 01, Theorem 2.2] Let the locally bounded semi - martingale $S = (S_t)_{0 \le t \le T}$ and the utility function U satisfy Assumptions 1.1, 1.2 case 2 and 1.3; suppose in addition that U has reasonable asymptotic elasticity. Define*

$$u(x) = \sup_{X_T \in C_U(x)} \mathbf{E}[U(X_T)], \quad v(y) = \inf_{\mathbf{Q} \in \mathcal{M}^a(S)} \mathbf{E}\left[V\left(y\frac{d\mathbf{Q}}{d\mathbf{P}}\right)\right]. \quad (133)$$

Then we have:

(i) *The value functions $u(x)$ and $v(y)$ are conjugate; they are continuously differentiable, strictly concave (resp. convex) on $\mathbb{R}$ (resp. on $]0, \infty[$) and satisfy*

$$u'(-\infty) = -v'(0) = v'(\infty) = \infty, \quad u'(\infty) = 0. \quad (134)$$

(ii) *The optimizers $\widehat{X}_T(x)$ and $\widehat{\mathbf{Q}}(y)$ in (133) exist, are unique and satisfy*

$$\widehat{X}_T(x) = I\left(y\frac{d\widehat{\mathbf{Q}}(y)}{d\mathbf{P}}\right), \quad y\frac{d\widehat{\mathbf{Q}}(y)}{d\mathbf{P}} = U'(\widehat{X}_T(x)), \quad (135)$$

where $x \in \mathbb{R}$ and $y > 0$ are related via $u'(x) = y$ or equivalently $x = -v'(y)$.

(iii) *We have the following relations between u', v' and $\widehat{X}, \widehat{\mathbf{Q}}$ respectively:*

$$u'(x) = \mathbf{E}_{\mathbf{P}}[U'(\widehat{X}_T(x))], \quad v'(y) = \mathbf{E}_{\widehat{\mathbf{Q}}}\left[V'\left(y\frac{d\widehat{\mathbf{Q}}(y)}{d\mathbf{P}}\right)\right] \quad (136)$$

$$xu'(x) = \mathbf{E}_{\mathbf{P}}[\widehat{X}_T(x)U'(\widehat{X}_T(x))], \quad yv'(y) = \mathbf{E}_{\mathbf{P}}\left[y\frac{d\widehat{\mathbf{Q}}(y)}{d\mathbf{P}}V'\left(y\frac{d\widehat{\mathbf{Q}}(y)}{d\mathbf{P}}\right)\right]. \quad (137)$$

(iv) *If $\widehat{\mathbf{Q}}(y) \in \mathcal{M}^e(S)$ and $x = -v'(y)$, then $\widehat{X}_T(x)$ equals the terminal value of a process of the form $\widehat{X}_t(x) = x + (H \cdot S)_t$, where H is predictable and S-integrable, and such that $\widehat{X}$ is a uniformly integrable martingale under $\widehat{\mathbf{Q}}(y)$.*

We refer to [S 01] for a proof of this theorem and further related results. We cannot go into the technicalities here, but a few comments on the proof of the above theorem are in order: the technique is to reduce case 2 to case 1 by approximating the utility function $U : \mathbb{R} \to \mathbb{R}$ by a sequence $(U^{(n)})_{n=1}^{\infty}$ of utility functions $U^{(n)} : \mathbb{R} \to \mathbb{R} \cup \{-\infty\}$ such that $U^{(n)}$ coincides with U on $[-n, \infty[$ and equals $-\infty$ on $] - \infty, -(n+1)]$. For fixed initial endowment $x \in \mathbb{R}$, we then apply Theorem 3.1 to find for each $U^{(n)}$ the saddle-point

$(\widehat{X}_T^{(n)}(x), \widehat{Y}_T^{(n)}(\widehat{y}_n)) \in C_U^b(x) \times D(\widehat{y}_n)$; finally we show that this sequence converges to some $(\widehat{X}_T(x), \widehat{y}\widehat{Q}_T) \in C_U(x) \times \widehat{y}\mathcal{M}^a(S)$, which then is shown to be the saddle-point for the present problem. The details of this construction are rather technical and lengthy (see [S 01]).

We have assumed in item (iv) that $\widehat{Q}(y)$ is equivalent to $\mathbf{P}$ and left open the case when $\widehat{Q}(y)$ is only absolutely continuous to $\mathbf{P}$. F. Bellini and M. Frittelli have observed ([BF 02]) that, in the case $U(\infty) = \infty$ (or, equivalently, $V(0) = \infty$), it follows from (133) that $\widehat{Q}(y)$ is equivalent to $\mathbf{P}$. But there are also other important cases where we can assert that $\widehat{Q}(y)$ is equivalent to $\mathbf{P}$: for example, for the case of the exponential utility $U(x) = -e^{-\gamma x}$, in which case the dual optimization becomes the problem of finding $\widehat{Q} \in \mathcal{M}^a(S)$ minimizing the relative entropy with respect $\mathbf{P}$, it follows from the work of Csiszar [C 75] (compare also [R 84], [F 00], [GR 01]) that the dual optimizer $\widehat{Q}(y)$ is equivalent to $\mathbf{P}$, provided only that there is at least one $\mathbf{Q} \in \mathcal{M}^e(S)$ with finite relative entropy.

Under the condition $\widehat{Q}(y) \in \mathcal{M}^e(S)$, item (iv) tells us that the optimizer $\widehat{X}_T \in C_U(x)$ is almost surely finite and equals the terminal value of a process $x+(H\cdot S)$, which is a uniformly integrable martingale under $\widehat{Q}(y)$; this property qualifies H to be a "allowable", as it certainly excludes doubling strategies and related schemes. One may turn the point of view around and take this as the *definition* of the "allowable" trading strategies; this was done in [DGRSSS 02] for the case of exponential utility, where this approach is thoroughly studied and some other definitions of "allowable" trading strategies, over which the primal problem may be optimized, are also investigated. Further results on these lines were obtained in [KS 02] for the case of exponential utility, and in [S 03a] for general utility functions.

References

[AS 94]   J.P. Ansel, C. Stricker, (1994), *Couverture des actifs contingents et prix maximum*. Ann. Inst. Henri Poincaré, Vol. 30, pp. 303–315.

[A 97]   P. Artzner, (1997), *On the numeraire portfolio*. Mathematics of Derivative Securities, M. Dempster and S. Pliska, eds., Cambridge University Press, pp. 53–60.

[B 01]   D. Becherer, (2001), *The numeraire portfolio for unbounded semimartingales*. Finance and Stochastics, Vol. 5, No. 3, pp. 327–341.

[BF 02]   F. Bellini, M. Frittelli, (2002), *On the existence of minimax martingale measures*. Mathematical Finance, Vol. 12, No. 1, pp. 1–21.

[BS 99]   W. Brannath, W. Schachermayer, (1999), *A Bipolar Theorem for Subsets of $L_+^0(\Omega, \mathcal{F}, P)$*. Séminaire de Probabilités, Vol. XXXIII, pp. 349–354.

[CH 00]   P. Collin-Dufresne, J.-N. Huggonnier, (2000), *Utility-based pricing of contingent claims subject to counterparty credit risk*. Working paper GSIA & Department of Mathematics, Carnegie Mellon University.

[CH 89]   J.C. Cox, C.F. Huang, (1989), *Optimal consumption and portfolio policies when asset prices follow a diffusion process.* Journal of Economic Theory, Vol. 49, pp. 33–83.

[CH 91]   J.C. Cox, C.F. Huang, (1991), *A variational problem arising in financial economics.* Jorunal of Mathematical Economics, Vol. 20, No. 5, pp. 465–487.

[C 75]   I. Csiszar, (1975), *I-Divergence Geometry of Probability Distributions and Minimization Problems.* Annals of Probability, Vol. 3, No. 1, pp. 146–158.

[C 00]   Jakša Cvitanić, (2000) *Minimizing expected loss of hedging in incomplete and constrained markets,* SIAM Journal on Control and Optimization, Vol. 38, No. 4, pp. 1050–1066 (electronic).

[CK 96]   J. Cvitanic, I. Karatzas, (1996), *Hedging and portfolio optimization under transaction costs: A martingale approach.* Mathematical Finance, Vol. 6, No. 2, pp. 133–165.

[CSW 01]   J. Cvitanic, W. Schachermayer, H. Wang, (2001), *Utility Maximization in Incomplete Markets with Random Endowment.* Finance and Stochastics, Vol. 5, No. 2, pp. 259–272.

[CW 01]   J. Cvitanic, H. Wang, (2001), *On optimal terminal wealth under transaction costs.* Jorunal of Mathematical Economics, Vol. 35, No. 2, pp. 223–231.

[DMW 90]   R.C. Dalang, A. Morton, W. Willinger, (1990), *Equivalent martingale measures and no-arbitrage in stochastic.* Stochastics and Stochastics Reports, Vol. 29, pp. 185–201.

[D 97]   M. Davis, (1997), *Option pricing in incomplete markets.* Mathematics of Derivative Securities, eds. M.A.H. Dempster and S.R. Pliska, Cambridge University Press, pp. 216–226.

[D 00]   Mark Davis, (2000), *Option valuation and hedging with basis risk,* System theory: modeling, analysis and control (Cambridge, MA, 1999), Kluwer Internat. Ser. Engrg. Comput. Sci., vol. 518, Kluwer Acad. Publ., Boston, MA, pp. 245–254.

[DPT 01]   G. Deelstra, H. Pham, N. Touzi, (2001), *Dual formulation of the utility maximisation problem under transaction costs.* Annals of Applied Probability, Vol. 11, No. 4, pp. 1353–1383.

[DGRSSS 02] F. Delbaen, P. Grandits, T. Rheinländer, D. Samperi, M. Schweizer, C. Stricker, (2002), *Exponential hedging and entropic penalties.* Mathematical Finance, Vol. 12, No. 2, pp. 99–123.

[DS 94]   F. Delbaen, W. Schachermayer, (1994), *A General Version of the Fundamental Theorem of Asset Pricing.* Math. Annalen, Vol. 300, pp. 463–520.

[DS 95]   F. Delbaen, W. Schachermayer, (1995), *The No-Arbitrage Property under a change of numéraire.* Stochastics and Stochastic Reports, Vol. 53, pp. 213–226.

[DS 98]   F. Delbaen, W. Schachermayer, (1998), *The Fundamental Theorem of Asset Pricing for Unbounded Stochastic Processes.* Mathematische Annalen, Vol. 312, pp. 215–250.

[DS 98a]   F. Delbaen, W. Schachermayer, (1998), *A Simple Counter-example to Several Problems in the Theory of Asset Pricing, which arises in many incomplete markets.* Mathematical Finance, Vol. 8, pp. 1–12.

[DS 99]   F. Delbaen, W. Schachermayer, (1999), *A Compactness Principle for Bounded Sequences of Martingales with Applications*. Proceedings of the Seminar of Stochastic Analysis, Random Fields and Applications, Progress in Probability, Vol. 45, pp. 137–173.

[ET 76]   I. Ekeland, R. Temam, (1976), *Convex Analysis and Variational Problems*. North Holland.

[E 80]   M. Emery, (1980), *Compensation de processus à variation finie non localement intégrables*. Séminaire de Probabilités XIV, Springer Lecture Notes in Mathematics, Vol. 784, pp. 152–160.

[F 90]   L.P. Foldes, (1990), *Conditions for optimality in the infinite-horizon portfolio-cum-savings problem with semimartingale investments*. Stochastics and Stochastics Report, Vol. 29, pp. 133–171.

[FL 00]   H. Föllmer, P. Leukert, (2000), *Efficient Hedging: Cost versus Shortfall Risk*. Finance and Stochastics, Vol. 4, No. 2, pp. 117–146.

[FS 91]   H. Föllmer, M. Schweizer, (1991), *Hedging of contingent claims under incomplete information*. Applied Stochastic Analysis, Stochastic Monographs, M.H.A. Davis and R.J. Elliott, eds., Gordon and Breach, London New York, Vol. 5, pp. 389–414.

[F 00]   M. Frittelli, (2000), *The minimal entropy martingale measure and the valuation problem in incomplete markets*. Mathematical Finance, Vol. 10, No. 1, pp. 39–52.

[GK 00]   T. Goll, J. Kallsen, (2000), *Optimal portfolios for logarithmic utility*. Stochastic Processes and Their Applications, Vol. 89, pp. 31–48.

[GR 01]   T. Goll, L. Rüschendorf, (2001), *Minimax and minimal distance martingale measures and their relationship to portfolio optimization*. Finance and Stochastics, Vol. 5, No. 4, pp. 557–581.

[HP 81]   J.M. Harrison, S.R. Pliska, (1981), *Martingales and stochastic integrals in the theory of continuous trading*. Stochastic Processes and Applications, Vol. 11, pp. 215–260.

[HP 91a]   H. He, N.D. Pearson, (1991), *Consumption and Portfolio Policies with Incomplete Markets and Short-Sale Constraints: The Finite-Dimensional Case*. Mathematical Finance, Vol. 1, pp. 1–10.

[HP 91b]   H. He, N.D. Pearson, (1991), *Consumption and Portfolio Policies with Incomplete Markets and Short-Sale Constraints: The Infinite-Dimensional Case*. Journal of Economic Theory, Vol. 54, pp. 239–250.

[HN 89]   S.D. Hodges, A. Neuberger, (1989), *Optimal replication of contingent claims under transaction costs*. Review of Futures Markets, Vol. 8, pp. 222–239.

[HL88]   C.-F. Huang, R.H. Litzenberger, (1988), *Foundations for Financial Economics*. North-Holland Publishing Co. New York.

[HK 02]   J. Hugonnier, D. Kramkov, (2002), *Optimal investment with random endowments in incomplete markets*. To appear in The Annals of Applied Probability.

[J 92]   S.D. Jacka, (1992), *A martingale representation result and an application to incomplete financial markets*. Mathematical Finance, Vol. 2, pp. 239–250.

[KS 01]   Yu.M. Kabanov, Ch. Stricker, (2001), *A teachers' note on no-arbitrage criteria*. Séminaire de Probabilités XXXV, Springer Lecture Notes in Mathematics, Vol. 1755, pp. 149–152.

[KS 02]   Yu.M. Kabanov, C. Stricker, (2002), *On the optimal portfolio for the exponential utility maximization: remarks to the six-author paper.* Mathematical Finance, Vol. 12, No. 2, pp. 125–134.

[K 00]   J. Kallsen, (2000), *Optimal portfolios for exponential Lévy processes* Mathematical Methods of Operation Research, Vol. 51, No. 3, pp. 357–374.

[KLS 87]   I. Karatzas, J.P. Lehoczky, S.E. Shreve, (1987), *Optimal portfolio and consumption decisions for a "small investo" on a finite horizon.* SIAM Journal of Control and Optimization, Vol. 25, pp. 1557–1586.

[KLSX 91]   I. Karatzas, J.P. Lehoczky, S.E. Shreve, G.L. Xu, (1991), *Martingale and duality methods for utility maximization in an incomplete market.* SIAM Journal of Control and Optimization, Vol. 29, pp. 702–730.

[KJ 98]   N. El Karoui, M. Jeanblanc, (1998), *Optimization of consumptions with labor income.* Finance and Stochastics, Vol. 4, pp. 409–440.

[KQ 95]   N. El Karoui, M.-C. Quenez, (1995), *Dynamic programming and pricing of contingent claims in an incomplete market.* SIAM Journal on Control and Optimization, Vol. 33, pp. 29–66.

[KR 00]   N. El Karoui, R. Rouge, (2000), *Pricing via utility maximization and entropy.* Mathematical Finance, Vol. 10, No. 2, pp. 259–276.

[Kom 67]   J. Komlos, (1967), *A generalization of a problem of Steinhaus.* Acta Math. Sci. Hung., Vol. 18, pp. 217–229.

[KS 99]   D. Kramkov, W. Schachermayer, (1999), *The Asymptotic Elasticity of Utility Functions and Optimal Investment in Incomplete Markets.* Annals of Applied Probability, Vol. 9, No. 3, pp. 904–950.

[K 81]   D.M. Kreps, (1981), *Arbitrage and equilibrium in economies with infinitely many commodities.* Journal of Mathematical Economics, Vol. 8, pp. 15–35.

[L 00]   P. Lakner, (2000), *Portfolio Optimization with an Insurance Constraint.* Preprint of the NYU, Dept. of Statistics and Operation Research.

[L 90]   J.B. Long, (1990), *The numeraire portfolio.* Journal of Financial Economics, Vol. 26, pp. 29–69.

[M 69]   R.C. Merton, (1969), *Lifetime portfolio selection under uncertainty: the continuous-time model.* Rev. Econom. Statist., Vol. 51, pp. 247–257.

[M 71]   R.C. Merton, (1971), *Optimum consumption and portfolio rules in a continuous-time model.* Journal of Economic Theory, Vol. 3, pp. 373–413.

[M 90]   R.C. Merton, (1990), *Continuous-Time Finance.* Basil Blackwell, Oxford.

[P 86]   S.R. Pliska, (1986), *A stochastic calculus model of continuous trading: optimal portfolios.* Math. Oper. Res., Vol. 11, pp. 371–382.

[R 70]   R.T. Rockafellar, (1970), *Convex Analysis.* Princeton University Press, Princeton, New Jersey.

[R 84]   L. Rüschendorf, (1984), *On the minimum discrimination information theorem.* Statistics & Decisions Supplement Issue, Vol. 1, pp. 263–283.

[S 69]   P.A. Samuelson, (1969), *Lifetime portfolio selection by dynamic stochastic programming.* Rev. Econom. Statist., Vol. 51, pp. 239–246.

[S 01]   W. Schachermayer, (2001), *Optimal Investment in Incomplete Markets when Wealth may Become Negative*. Annals of Applied Probability, Vol. 11, No. 3, pp. 694–734.

[S 01a]   W. Schachermayer, (2001), *Optimal Investment in Incomplete Financial Markets*. Mathematical Finance: Bachelier Congress 2000 (H. Geman, D. Madan, St.R. Pliska, T. Vorst, editors), Springer, pp. 427–462.

[S 03]   W. Schachermayer, (2003), *Introduction to the Mathematics of Financial Markets*. In: S. Albeverio, W. Schachermayer, M. Talagrand: Lecture Notes in Mathematics 1816 - Lectures on Probability Theory and Statistics, Saint-Flour summer school 2000 (Pierre Bernard, editor), Springer Verlag, Heidelberg, pp. 111–177.

[S 03a]   W. Schachermayer, (2003), *A Super-Martingale Property of the Optimal Portfolio Process*. Finance and Stochastics, Vol. 7, No. 4, pp. 433–456.

[S 04]   W. Schachermayer, (2004), *The Fundamental Theorem of Asset Pricing under Proportional Transaction Costs in Finite Discrete Time*. Mathematical Finance, Vol. 14, No. 1, pp. 19–48.

[S 04a]   W. Schachermayer, (2004), *Portfolio Optimization in Incomplete Financial Markets*, to appear as Lecture Notes of "Scuola Normale Superiore di Pisa".

[S 66]   H.H. Schäfer, (1966), *Topological Vector Spaces*. Graduate Texts in Mathematics.

[St 85]   H. Strasser, (1985), *Mathematical theory of statistics: statistical experiments and asymptotic decision theory*. De Gruyter studies in mathematics, Vol. 7.

List of Participants

1 Astic Fabian
  CREST, Université Paris-Dauphine, France
  fabian.astic@ensae.fr

2 Back Kerry (lecturer)
  Washington University, USA
  back@olin.wustl.edu

3 Backhaus Jochen
  University of Leipzig/Math Institut, Germany
  jochen.backhaus@math.uni-leipzig.de

4 Baran Michail
  Polish Academy of Science, Warsaw, Poland
  mb@impan.gov.pl

5 Battauz Anna
  Bocconi University, Milano, Italy
  anna.battauz@uni-bocconi.it

6 Bauer Wolfgang
  Univerisity Zürich, Switzerland
  wobauer@isb.unizh.ch

7 Bavouzet Marie-Pierre
  INRIA, France
  marie-pierre.bavouzet@inria.fr

8 Bernyk Violetta
  Ecole Polythecnique Federale de Lowsanne, Switzerland
  violetta.hamsag-bernyk@epfl.ch

9 Bershadsky Andrew
  Moscow Institute of Physics and Tecnhology, Russia
  avb@cslab.ptci.ru

10 Beutner Eric
   RWTH Aachen, Germany
   beutner@stochastik.rwth-aachen.de

11   Biagini Francesca
      University of Bologna, Dep.of Math., Italy
      biagini@dm.unibo.it

12   Biagini Sara
      University of Perugia, Italy
      s.biagini@unipg.it

13   Bielecki Tomasz (lecturer)
      Illinois Institute of Technology
      bielecki@iit.edu

14   Borovkov Constantin
      University of Melbourne, Australia
      kostya@ms.unimeb.edu.au

15   Bouhari Arouna
      Ecole Nationale des Ponts et Chaussees, France
      arouna@cermics.enpc.fr

16   Buffett Emanuel
      Dublin City University, Ireland
      emanuel.buffet@dcu.ie

17   Burd Oleg
      Karlruhe University, Germany
      burd@gmx.net

18   Campi Luciano
      Université Paris, France and Univ."G. Di Annunzio", Chieti, Italy
      campi@ccr.jussieu.fr

19   Carroll Tom
      National University of Ireland
      t.carroll@ucc.ie

20   Catanese Elena
      Scuola normale superiore, Italy
      catanese@sns.it

21   Cerqueti Roy
      University of Rome "La Sapienza", Italy
      roy.cerqueti@uniroma1.it

22   Coculescu Delia
      University Paris IX. Dauphine, France
      delia_coculescu@yahoo.fr

23   Corcuera Jose Manuel
      University of Barcelona, Spain
      corcuera@mat.ub.es

24   Corsi Marco
      University of Padova, Italy
      mcorsi@math.unipd.it

25   Costa Vincenzo
     University of Naples "Federico II ", Italy
     vincenzo.costa@uniroma1.it

26   Cretarola Alessandra
     University of Bologna, Italy
     sagittale@yahoo.com

27   Cucicea Mihaela
     University of Padova, Dep.of Math., Italy
     cucicea@math.unipd.it

28   Dana Rose-Anne
     Universite Paris Dauphine, France
     dana@ceremade.dauphine.fr

29   De Donno Marzia
     University of Pisa, Italy
     mdedonno@dm.unipi.it

30   Deelstra Griselda
     ULB /Dept.of Mathematics, Belgium
     griselda.deelstra@ulb.ac.be

31   Di Graziano Giuseppe
     King's College, London, U.K.
     pippoh@inwind.it

32   Fabbretti Annalisa
     University of Rome "La Sapienza", Italy
     annlaisa83@hotmail.com

33   Favero Gino
     University of Padova, Italy
     favero@math.unipd.it

34   Frittelli Marco (editor)
     University of Florence, Dep.of Math., Italy
     Marco.Frittelli@dmd.unifi.it

35   Galeotti Marcello
     University of Firenze, DIMAD, Italy
     marcello.galeotti@dmd.unifi.it

36   Galiani Stefano
     King's College, London, U.K.
     stefano.galliani@kcl.ac.uk

37   Gasco Loretta
     PUCP, Lima, Peru
     lgasco@pucp.edu.pe

38   Gaygisiz Esma
     Middle East Technical University, Ankara, Turkey
     esma@metu.edu.tr

39 Gianfreda Angelica
    University of Lecce/Fac.of Economics, Italy
    a.giangreda@economia.unile.it

40 Gliklikh Andrei
    Voronezh State University, Russia
    yuri@yeg.vsu.ru

41 Grasselli Martino
    University of Verona, Italy
    martino.graselli@univr.it

42 Guerra Joao
    University of Barcelona, Spain
    jguerra@telepolis.cos

43 Halaj Grzegorz
    Warsaw School of Economics, Poland
    grzegorz.halaj@n-s.pl

44 Hallulli Vera
    University of Padova, Italy
    hallulli@math.unipd.it

45 Hammarlid Ola
    Stockholm University, Sweden
    olah@math.su.se

46 Hamza Kais
    Monash University, Australia
    kais.hamza@sci.monash.edu.au

47 Hipp Christian (lecturer)
    University of Karlsruhe, Germany
    christian.hipp@wiwi.uni-karlsruhe.de

48 Jonsson Henrik
    Mlardalen University, Sweden
    henrik.jonsson@mdh.se

49 Klöppel Susanne
    Ludwig Maximilian Universität, München, Germany
    susanna.kloeppel@web.de

50 Kordzakhia Nino
    Commonwealth Bank of Australia
    kordzani@cba.com.au

51 Koval Nataliya
    University of Freiburg, Germany
    koval@stochastik.uni-freiburg.de

52 Kovalchuk Anatoly
    St.Petersburg State University, Russia
    tolikov@front.ru

53 La Chioma Claudia
    IAC CNR, University of Rome, Italy
    c.lachioma@iac.cnr.it

54 Lasserre Guillaume
    Paris VII. and CREST, France
    lasserre@ensae.fr

55 Levit Dina Central Eu. Univ, Hungary
    dina-levit@yandex.ru

56 Liinev Jan
    Ghent University, Belgium
    jan.liinev@ugent.be

57 Longo Michele
    University of Florence, Italy
    michele.longo@dmd.unifi.it

58 Lueteta Jean-Pierre
    Universite d'Anvers, Belgium
    lueteta@yahoo.fr

59 Macrina Andrea
    King's College, London, U.K.
    andrea.macrina@kcl.ac.uk

60 Marinelli Carlo
    Columbia University, New York
    cm788@columbia.edu

61 Masetti Massimo
    Carnegie Mellon University,
    Pittsburgh and University of Bergamo, Italy
    mmasetti@andrew.cmu.edu

62 Massi Benedetti Saverio
    University of Perugia, Italy
    saverio.massi@stat.unipg.it

63 Meyer-Brands Thilo
    Dep.of Mathematics, Oslo, Norway
    meyerbr@math.uio.no

64 Moreni Nicola
    C.E.R.M.I.C.S., France
    moreni@cermics.enpc.fr

65 Mulinacci Sabrina
    Università Cattolica Milano, Italy
    sabrina.mulinacci@mi.unicatt.it

66 Novikov Alex
    UTS, Sidney, Australia
    alex.novikov@uts.edu.au

67 Obukhovski Andrei
    Voronezh State University, Russia
    avo@mathd.vsu.ru

68 Oertel Frank
    Zurich University of Applied Sciences, Switzerland
    frank.oertel@zhwin.ch

69 Palczewski Jan
    Polish Academy of Science, Warsaw, Poland
    jp@impan.gov.pl

70 Panchenko Valentyn
    UVA, Amsterdam, Netherland
    v.panchenko@uva.nl

71 Papi Marco
    Instituto Per Le Applicazioni Del Calcolo, Roma, Italy
    papi@iac.rm.cnr.it

72 Peng Shige (lecturer)
    School of Mathematics, Shandong, China
    peng@sdu.edu.cn

73 Perrotta Annalisa
    University of Rome "La Sapienza",Italy
    anna-perrotta@yahoo.com

74 Platania Alessandro
    University of Padova, Italy
    platania@stat.unipd.it

75 Polte Ulrike
    Math Insitut, Universität Leipzig, Germany
    ulrike_polte@web.de

76 Popa Cristian
    Technical University of Vienna, Austria
    popa@fam.tuwien.ac.at

77 Popovici Stefan
    University of Bonn, Germany
    popovici@wiener.iam.uni-bonn.de

78 Porchia Paolo
    University of South Switzerland
    porchiap@lu.unisi.ch

79 Pricop Mihaela
    University "Petrol Gaze" Ploiesti, Romania
    pri_mihaela@hotmail.com

80 Provenzano Davide
    University of Palermo, Italy
    davidepro@yahoo.com

81   Rafaidilis Avraam
     King's College, London, U.K.
     avraam@mth.kcl.ac.uk

82   Ravanelli Claudia
     Università della Svizzera Italiana, Switzerland
     claudia.ravanelli@lu.unisi.ch

83   Roose Frederik
     King's College, London, U.K.
     fredr@mth.kcl.ac.uk

84   Rosazza Gianin Emanuela
     University of Milano Bicocca, Italy
     emanuela.rosazza@unimib.it

85   Runggaldier Wolfgang (editor)
     University of Padova, Dep.of Math., Italy
     runggal@math.unipd.it

86   Scandolo Giacomo
     University of Milano, Italy
     scandolo@email.it

87   Scarsini Marco
     University of Torino, Italy
     marco.scarsini@unito.it

88   Schachermayer Walter (lecturer)
     Vienna University of Technology, Austria
     wschach@fam.tuwien.ac.at

89   Scolozzi Donato
     University of Lecce, Italy
     scolozzi@economia.unile.it

90   Sgarra Carlo
     Politecnico Milano, Dep.of Mathematics
     sgarra@mate.polimi.it

91   Singh Surbjeet
     University of Cambridge, U.K.
     s.singh@staslab.cam.ac.uk

92   Sottinen Tommi
     University of Helsinki, Dep.of Math., Finland
     tommi.sottinene@helsinki.fi

93   Stabile Gabriele
     University of Rome "La Sapienza", Italy
     gabriele.stabile@uniroma1.it

94   Stemberg Fredrik
     Mälardalen University, Sweden
     fredrik.stenberg@mdh.se

95 Tebaldi Claudio
    University of Verona, Italy
    claudio.tebaldi@univr.it

96 Timofeeva Galina
    Ural State University of the Railway Trasport, Russia
    gtimofeva@mail.ru

97 Tolotti Marco
    Scuola normale superiore, Italy
    m.tolotti@sns.it

98 Vanmaele Michele
    Ghent University, Belgium
    michele.vanmaele@UGent.be

99 Vargiolu Tiziano
    University of Padova, Italy
    vargiolu@math.unipd.it

100 Voloshyna Olena
    Central Eu. Univ, Hungary
    c01voo01@ceu.hu

101 Volpe Valeria
    Università Svizzera Italiana, Lugano, Switzerland
    valeria.volpe@lu.unisi.ch

102 Vuorenmaa Tommi
    University of Helsinki, Dep.of Math., Finland
    tvuorenm@cc.helsinki.fi

103 Williams John
    Oxford University, U.K.
    john.willaims@comlab.ox.ac.uk

104 Xu Mingyu
    Shandong University, China and Université du Maine, France
    mingyu.xu@univ-lemans.fr

LIST OF C.I.M.E. SEMINARS

1954	1. Analisi funzionale	C.I.M.E
	2. Quadratura delle superficie e questioni connesse	"
	3. Equazioni differenziali non lineari	"
1955	4. Teorema di Riemann-Roch e questioni connesse	"
	5. Teoria dei numeri	"
	6. Topologia	"
	7. Teorie non linearizzate in elasticità,	"
	idrodinamica, aerodinamic	
	8. Geometria proiettivo-differenziale	"
1956	9. Equazioni alle derivate parziali a caratteristiche	"
	reali	
	10. Propagazione delle onde elettromagnetiche	"
	11. Teoria della funzioni di più variabili complesse e	"
	delle funzioni automorfe	
1957	12. Geometria aritmetica e algebrica (2 vol.)	"
	13. Integrali singolari e questioni connesse	"
	14. Teoria della turbolenza (2 vol.)	"
1958	15. Vedute e problemi attuali in relatività generale	"
	16. Problemi di geometria differenziale in grande	"
	17. Il principio di minimo e le sue applicazioni alle	"
	equazioni funzionali	
1959	18. Induzione e statistica	"
	19. Teoria algebrica dei meccanismi automatici (2 vol.)	"
	20. Gruppi, anelli di Lie e teoria della coomologia	"
1960	21. Sistemi dinamici e teoremi ergodici	"
	22. Forme differenziali e loro integrali	"
1961	23. Geometria del calcolo delle variazioni (2 vol.)	"
	24. Teoria delle distribuzioni	"
	25. Onde superficiali	"
1962	26. Topologia differenziale	"
	27. Autovalori e autosoluzioni	"
	28. Magnetofluidodinamica	"
1963	29. Equazioni differenziali astratte	"
	30. Funzioni e varietà complesse	"
	31. Proprietà di media e teoremi di confronto in	"
	Fisica Matematica	
1964	32. Relatività generale	"
	33. Dinamica dei gas rarefatti	"
	34. Alcune questioni di analisi numerica	"
	35. Equazioni differenziali non lineari	"
1965	36. Non-linear continuum theories	"
	37. Some aspects of ring theory	"
	38. Mathematical optimization in economics	"

1966	39. Calculus of variations	Ed. Cremonese, Firenze
	40. Economia matematica	"
	41. Classi caratteristiche e questioni connesse	"
	42. Some aspects of diffusion theory	"
1967	43. Modern questions of celestial mechanics	"
	44. Numerical analysis of partial differential equations	"
	45. Geometry of homogeneous bounded domains	"
1968	46. Controllability and observability	"
	47. Pseudo-differential operators	"
	48. Aspects of mathematical logic	"
1969	49. Potential theory	"
	50. Non-linear continuum theories in mechanics and physics and their applications	"
	51. Questions of algebraic varieties	"
1970	52. Relativistic fluid dynamics	"
	53. Theory of group representations and Fourier analysis	"
	54. Functional equations and inequalities	"
	55. Problems in non-linear analysis	"
1971	56. Stereodynamics	"
	57. Constructive aspects of functional analysis (2 vol.)	"
	58. Categories and commutative algebra	"
1972	59. Non-linear mechanics	"
	60. Finite geometric structures and their applications	"
	61. Geometric measure theory and minimal surfaces	"
1973	62. Complex analysis	"
	63. New variational techniques in mathematical physics	"
	64. Spectral analysis	"
1974	65. Stability problems	"
	66. Singularities of analytic spaces	"
	67. Eigenvalues of non linear problems	"
1975	68. Theoretical computer sciences	"
	69. Model theory and applications	"
	70. Differential operators and manifolds	"
1976	71. Statistical Mechanics	Ed. Liguori, Napoli
	72. Hyperbolicity	"
	73. Differential topology	"
1977	74. Materials with memory	"
	75. Pseudodifferential operators with applications	"
	76. Algebraic surfaces	"
1978	77. Stochastic differential equations	Ed. Liguori, Napoli & Birkhäuser
	78. Dynamical systems	"
1979	79. Recursion theory and computational complexity	"
	80. Mathematics of biology	"
1980	81. Wave propagation	"
	82. Harmonic analysis and group representations	"
	83. Matroid theory and its applications	"

1981	84. Kinetic Theories and the Boltzmann Equation	(LNM 1048)	Springer-Verlag
	85. Algebraic Threefolds	(LNM 947)	"
	86. Nonlinear Filtering and Stochastic Control	(LNM 972)	"
1982	87. Invariant Theory	(LNM 996)	"
	88. Thermodynamics and Constitutive Equations	(LN Physics 228)	"
	89. Fluid Dynamics	(LNM 1047)	"
1983	90. Complete Intersections	(LNM 1092)	"
	91. Bifurcation Theory and Applications	(LNM 1057)	"
	92. Numerical Methods in Fluid Dynamics	(LNM 1127)	"
1984	93. Harmonic Mappings and Minimal Immersions	(LNM 1161)	"
	94. Schrödinger Operators	(LNM 1159)	"
	95. Buildings and the Geometry of Diagrams	(LNM 1181)	"
1985	96. Probability and Analysis	(LNM 1206)	"
	97. Some Problems in Nonlinear Diffusion	(LNM 1224)	"
	98. Theory of Moduli	(LNM 1337)	"
1986	99. Inverse Problems	(LNM 1225)	"
	100. Mathematical Economics	(LNM 1330)	"
	101. Combinatorial Optimization	(LNM 1403)	"
1987	102. Relativistic Fluid Dynamics	(LNM 1385)	"
	103. Topics in Calculus of Variations	(LNM 1365)	"
1988	104. Logic and Computer Science	(LNM 1429)	"
	105. Global Geometry and Mathematical Physics	(LNM 1451)	"
1989	106. Methods of nonconvex analysis	(LNM 1446)	"
	107. Microlocal Analysis and Applications	(LNM 1495)	"
1990	108. Geometric Topology: Recent Developments	(LNM 1504)	"
	109. H_∞ Control Theory	(LNM 1496)	"
	110. Mathematical Modelling of Industrial Processes	(LNM 1521)	"
1991	111. Topological Methods for Ordinary Differential Equations	(LNM 1537)	"
	112. Arithmetic Algebraic Geometry	(LNM 1553)	"
	113. Transition to Chaos in Classical and Quantum Mechanics	(LNM 1589)	"
1992	114. Dirichlet Forms	(LNM 1563)	"
	115. D-Modules, Representation Theory, and Quantum Groups	(LNM 1565)	"
	116. Nonequilibrium Problems in Many-Particle Systems	(LNM 1551)	"
1993	117. Integrable Systems and Quantum Groups	(LNM 1620)	"
	118. Algebraic Cycles and Hodge Theory	(LNM 1594)	"
	119. Phase Transitions and Hysteresis	(LNM 1584)	"
1994	120. Recent Mathematical Methods in Nonlinear Wave Propagation	(LNM 1640)	"
	121. Dynamical Systems	(LNM 1609)	"
	122. Transcendental Methods in Algebraic Geometry	(LNM 1646)	"
1995	123. Probabilistic Models for Nonlinear PDE's	(LNM 1627)	"
	124. Viscosity Solutions and Applications	(LNM 1660)	"
	125. Vector Bundles on Curves. New Directions	(LNM 1649)	"

1996	126.	Integral Geometry, Radon Transforms and Complex Analysis	(LNM 1684)	Springer-Verlag
	127.	Calculus of Variations and Geometric Evolution Problems	(LNM 1713)	"
	128.	Financial Mathematics	(LNM 1656)	"
1997	129.	Mathematics Inspired by Biology	(LNM 1714)	"
	130.	Advanced Numerical Approximation of Nonlinear Hyperbolic Equations	(LNM 1697)	"
	131.	Arithmetic Theory of Elliptic Curves	(LNM 1716)	"
	132.	Quantum Cohomology	(LNM 1776)	"
1998	133.	Optimal Shape Design	(LNM 1740)	"
	134.	Dynamical Systems and Small Divisors	(LNM 1784)	"
	135.	Mathematical Problems in Semiconductor Physics	(LNM 1823)	"
	136.	Stochastic PDE's and Kolmogorov Equations in Infinite Dimension	(LNM 1715)	"
	137.	Filtration in Porous Media and Industrial Applications	(LNM 1734)	"
1999	138.	Computational Mathematics driven by Industrial Applications	(LNM 1739)	"
	139.	Iwahori-Hecke Algebras and Representation Theory	(LNM 1804)	"
	140.	Theory and Applications of Hamiltonian Dynamics	to appear	"
	141.	Global Theory of Minimal Surfaces in Flat Spaces	(LNM 1775)	"
	142.	Direct and Inverse Methods in Solving Nonlinear Evolution Equations	(LNP 632)	"
2000	143.	Dynamical Systems	(LNM 1822)	"
	144.	Diophantine Approximation	(LNM 1819)	"
	145.	Mathematical Aspects of Evolving Interfaces	(LNM 1812)	"
	146.	Mathematical Methods for Protein Structure	(LNCS 2666)	"
	147.	Noncommutative Geometry	(LNM 1831)	"
2001	148.	Topological Fluid Mechanics	to appear	"
	149.	Spatial Stochastic Processes	(LNM 1802)	"
	150.	Optimal Transportation and Applications	(LNM 1813)	"
	151.	Multiscale Problems and Methods in Numerical Simulations	(LNM 1825)	"
2002	152.	Real Methods in Complex and CR Geometry	(LNM 1848)	"
	153.	Analytic Number Theory	to appear	"
	154.	Imaging	to appear	"
2003	155.	Stochastic Methods in Finance	(LNM 1856)	"
	156.	Hyperbolic Systems of Balance Laws	to appear	"
	157.	Symplectic 4-Manifolds and Algebraic Surfaces	to appear	"
	158.	Mathematical Foundation of Turbulent Viscous Flows	to appear	"
2004	159.	Representation Theory and Complex Analysis	announced	"
	160.	Nonlinear and Optimal Control Theory	announced	"
	161.	Stochastic Geometry	announced	"

Fondazione C.I.M.E.

Centro Internazionale Matematico Estivo
International Mathematical Summer Center
http://www.cime.unifi.it
cime@math.unifi.it

2005 COURSES LIST

Enumerative Invariants
in Algebraic Geometry and String Theory

June 6–11, Cetraro

Course Directors:

Prof. Kai Behrend (University of British Columbia, Vancouver, Canada)
Prof. Barbara Mantechi (SISSA, Trieste, Italy)

Calculus of Variations
and Non-linear Partial Differential Equations

June 27–July 2, Cetraro

Course Directors:

Prof. Bernard Dacorogna (EPFL, Lousanne, Switzerland)
Prof. Paolo Marcellini (Università di Firenze, Italy)

SPDE in Hydrodynamics:
Recent Progress and Prospects

August 29–September 3, Cetraro

Course Directors:

Prof. Giuseppe Da Prato (Scuola Normale Superiore, Pisa, Italy)
Prof. Michael Rockner (Bielefeld University, Germany)

Vol. 1775: W. H. Meeks, A. Ros, H. Rosenberg, The Global Theory of Minimal Surfaces in Flat Spaces. Martina Franca 1999. Editor: G. P. Pirola. X, 117 pages. 2002.

Vol. 1776: K. Behrend, C. Gomez, V. Tarasov, G. Tian, Quantum Comohology. Cetraro 1997. Editors: P. de Bartolomeis, B. Dubrovin, C. Reina. VIII, 319 pages. 2002.

Vol. 1777: E. García-Río, D. N. Kupeli, R. Vázquez-Lorenzo, Osserman Manifolds in Semi-Riemannian Geometry. XII, 166 pages. 2002.

Vol. 1778: H. Kiechle, Theory of K-Loops. X, 186 pages. 2002.

Vol. 1779: I. Chueshov, Monotone Random Systems. VIII, 234 pages. 2002.

Vol. 1780: J. H. Bruinier, Borcherds Products on O(2,1) and Chern Classes of Heegner Divisors. VIII, 152 pages. 2002.

Vol. 1781: E. Bolthausen, E. Perkins, A. van der Vaart, Lectures on Probability Theory and Statistics. Ecole d' Eté de Probabilités de Saint-Flour XXIX-1999. Editor: P. Bernard. VIII, 466 pages. 2002.

Vol. 1782: C.-H. Chu, A. T.-M. Lau, Harmonic Functions on Groups and Fourier Algebras. VII, 100 pages. 2002.

Vol. 1783: L. Grüne, Asymptotic Behavior of Dynamical and Control Systems under Perturbation and Discretization. IX, 231 pages. 2002.

Vol. 1784: L.H. Eliasson, S. B. Kuksin, S. Marmi, J.-C. Yoccoz, Dynamical Systems and Small Divisors. Cetraro, Italy 1998. Editors: S. Marmi, J.-C. Yoccoz. VIII, 199 pages. 2002.

Vol. 1785: J. Arias de Reyna, Pointwise Convergence of Fourier Series. XVIII, 175 pages. 2002.

Vol. 1786: S. D. Cutkosky, Monomialization of Morphisms from 3-Folds to Surfaces. V, 235 pages. 2002.

Vol. 1787: S. Caenepeel, G. Militaru, S. Zhu, Frobenius and Separable Functors for Generalized Module Categories and Nonlinear Equations. XIV, 354 pages. 2002.

Vol. 1788: A. Vasil'ev, Moduli of Families of Curves for Conformal and Quasiconformal Mappings.IX, 211 pages. 2002.

Vol. 1789: Y. Sommerhäuser, Yetter-Drinfel'd Hopf algebras over groups of prime order. V, 157 pages. 2002.

Vol. 1790: X. Zhan, Matrix Inequalities. VII, 116 pages. 2002.

Vol. 1791: M. Knebusch, D. Zhang, Manis Valuations and Prüfer Extensions I: A new Chapter in Commutative Algebra. VI, 267 pages. 2002.

Vol. 1792: D. D. Ang, R. Gorenflo, V. K. Le, D. D. Trong, Moment Theory and Some Inverse Problems in Potential Theory and Heat Conduction. VIII, 183 pages. 2002.

Vol. 1793: J. Cortés Monforte, Geometric, Control and Numerical Aspects of Nonholonomic Systems. XV, 219 pages. 2002.

Vol. 1794: N. Pytheas Fogg, Substitution in Dynamics, Arithmetics and Combinatorics. Editors: V. Berthé, S. Ferenczi, C. Mauduit, A. Siegel. XVII, 402 pages. 2002.

Vol. 1795: H. Li, Filtered-Graded Transfer in Using Noncommutative Gröbner Bases. IX, 197 pages. 2002.

Vol. 1796: J.M. Melenk, hp-Finite Element Methods for Singular Perturbations. XIV, 318 pages. 2002.

Vol. 1797: B. Schmidt, Characters and Cyclotomic Fields in Finite Geometry. VIII, 100 pages. 2002.

Vol. 1798: W.M. Oliva, Geometric Mechanics. XI, 270 pages. 2002.

Vol. 1799: H. Pajot, Analytic Capacity, Rectifiability, Menger Curvature and the Cauchy Integral. XII,119 pages. 2002.

Vol. 1800: O. Gabber, L. Ramero, Almost Ring Theory. VI, 307 pages. 2003.

Vol. 1801: J. Azéma, M. Émery, M. Ledoux, M. Yor (Eds.), Séminaire de Probabilités XXXVI. VIII, 499 pages. 2003.

Vol. 1802: V. Capasso, E. Merzbach, B.G. Ivanoff, M. Dozzi, R. Dalang, T. Mountford, Topics in Spatial Stochastic Processes. Martina Franca, Italy 2001. Editor: E. Merzbach. VIII, 253 pages. 2003.

Vol. 1803: G. Dolzmann, Variational Methods for Crystalline Microstructure - Analysis and Computation. VIII, 212 pages. 2003.

Vol. 1804: I. Cherednik, Ya. Markov, R. Howe, G. Lusztig, Iwahori-Hecke Algebras and their Representation Theory. Martina Franca, Italy 1999. Editors: V. Baldoni, D. Barbasch. X, 103 pages. 2003.

Vol. 1805: F. Cao, Geometric Curve Evolution and Image Processing. X, 187 pages. 2003.

Vol. 1806: H. Broer, I. Hoveijn. G. Lunther, G. Vegter, Bifurcations in Hamiltonian Systems. Computing Singularities by Gröbner Bases. XIV, 169 pages. 2003.

Vol. 1807: V. D. Milman, G. Schechtman (Eds.), Geometric Aspects of Functional Analysis. Israel Seminar 2000-2002. VIII, 429 pages. 2003.

Vol. 1808: W. Schindler, Measures with Symmetry Properties.IX, 167 pages. 2003.

Vol. 1809: O. Steinbach, Stability Estimates for Hybrid Coupled Domain Decomposition Methods. VI, 120 pages. 2003.

Vol. 1810: J. Wengenroth, Derived Functors in Functional Analysis. VIII, 134 pages. 2003.

Vol. 1811: J. Stevens, Deformations of Singularities. VII, 157 pages. 2003.

Vol. 1812: L. Ambrosio, K. Deckelnick, G. Dziuk, M. Mimura, V. A. Solonnikov, H. M. Soner, Mathematical Aspects of Evolving Interfaces. Madeira, Funchal, Portugal 2000. Editors: P. Colli, J. F. Rodrigues. X, 237 pages. 2003.

Vol. 1813: L. Ambrosio, L. A. Caffarelli, Y. Brenier, G. Buttazzo, C. Villani, Optimal Transportation and its Applications. Martina Franca, Italy 2001. Editors: L. A. Caffarelli, S. Salsa. X, 164 pages. 2003.

Vol. 1814: P. Bank, F. Baudoin, H. Föllmer, L.C.G. Rogers, M. Soner, N. Touzi, Paris-Princeton Lectures on Mathematical Finance. X,172 pages. 2003.

Vol. 1815: A. M. Vershik (Ed.), Asymptotic Combinatorics with Applications to Mathematical Physics. St. Petersburg, Russia 2001. IX, 246 pages. 2003.

Vol. 1816: S. Albeverio, W. Schachermayer, M. Talagrand, Lectures on Probability Theory and Statistics. Ecole d'Eté de Probabilités de Saint-Flour XXX-2000. Editor: P. Bernard. VIII, 296 pages. 2003.

Vol. 1817: E. Koelink (Ed.), Orthogonal Polynomials and Special Functions. Leuven 2002. X, 249 pages. 2003.

Vol. 1818: M. Bildhauer, Convex Variational Problems with Linear, nearly Linear and/or Anisotropic Growth Conditions. X, 217 pages. 2003.

Vol. 1819: D. Masser, Yu. V. Nesterenko, H. P. Schlickewei, W. M. Schmidt, M. Waldschmidt, Diophantine Approximation. Cetraro, Italy 2000. Editors: F. Amoroso, U. Zannier. XI,353 pages. 2003.

Vol. 1820: F. Hiai, H. Kosaki, Means of Hilbert Space Operators. VIII, 148 pages. 2003.

Vol. 1821: S. Teufel, Adiabatic Perturbation Theory in Quantum Dynamics. VI, 236 pages. 2003.

Vol. 1822: S.-N. Chow, R. Conti, R. Johnson, J. Mallet-Paret, R. Nussbaum, Dynamical Systems. Cetraro, Italy 2000. Editors: J. W. Macki, P. Zecca. XIII, 353 pages. 2003.

Vol. 1823: A. M. Anile, W. Allegretto, C. Ringhofer, Mathematical Problems in Semiconductor Physics. Cetraro, Italy 1998. Editor: A. M. Anile. X, 143 pages. 2003.

Vol. 1824: J. A. Navarro González, J. B. Sancho de Salas, C^∞ - Differentiable Spaces. XIII, 188 pages. 2003.

Vol. 1825: J. H. Bramble, A. Cohen, W. Dahmen, Multiscale Problems and Methods in Numerical Simulations, Martina Franca, Italy 2001. Editor: C. Canuto. XIII, 163 pages. 2003.

Vol. 1826: K. Dohmen, Improved Bonferroni Inequalities via Abstract Tubes. Inequalities and Identities of Inclusion-Exclusion Type. VIII, 113 pages, 2003.

Vol. 1827: K. M. Pilgrim, Combinations of Complex Dynamical Systems. IX, 118 pages, 2003.

Vol. 1828: D. J. Green, Gröbner Bases and the Computation of Group Cohomology. XII, 138 pages, 2003.

Vol. 1829: E. Altman, B. Gaujal, A. Hordijk, Discrete-Event Control of Stochastic Networks: Multimodularity and Regularity. XIV, 313 pages, 2003.

Vol. 1830: M. I. Gil', Operator Functions and Localization of Spectra. XIV, 256 pages, 2003.

Vol. 1831: A. Connes, J. Cuntz, E. Guentner, N. Higson, J. E. Kaminker, Noncommutative Geometry, Martina Franca, Italy 2002. Editors: S. Doplicher, L. Longo. XIV, 349 pages. 2004.

Vol. 1832: J. Azéma, M. Émery, M. Ledoux, M. Yor (Eds.), Séminaire de Probabilités XXXVII. XIV, 448 pages. 2003.

Vol. 1833: D.-Q. Jiang, M. Qian, M.-P. Qian, Mathematical Theory of Nonequilibrium Steady States. On the Frontier of Probability and Dynamical Systems. IX, 280 pages, 2004.

Vol. 1834: Yo. Yomdin, G. Comte, Tame Geometry with Application in Smooth Analysis. VIII, 186 pages, 2004.

Vol. 1835: O.T. Izhboldin, B. Kahn, N.A. Karpenko, A. Vishik, Geometric Methods in the Algebraic Theory of Quadratic Forms. Summer School, Lens, 2000. Editor: J.-P. Tignol. XIV, 190 pages. 2004.

Vol. 1836: C. Năstăsescu, F. Van Oystaeyen, Methods of Graded Rings. XIII, 304 pages, 2004.

Vol. 1837: S. Tavaré, O. Zeitouni, Lectures on Probability Theory and Statistics. Ecole d'Eté de Probabilités de Saint-Flour XXXI-2001. Editor: J. Picard. VII, 315 pages. 2004.

Vol. 1838: A.J. Ganesh, N.W. O'Connell, D.J. Wischik, Big Queues. XII, 254 pages, 2004.

Vol. 1839: R. Gohm, Noncommutative Stationary Processes. VIII, 170 pages, 2004.

Vol. 1840: B. Tsirelson, W. Werner, Lectures on Probability Theory and Statistics. Ecole d'Eté de Probabilités de Saint-Flour XXXII-2002. Editor: J. Picard. VII, 200 pages. 2004.

Vol. 1841: W. Reichel, Uniqueness Theorems for Variational Problems by the Method of Transformation Groups, XIII, 152 pages. 2004.

Vol. 1842: T. Johnsen, A.L. Knutsen, K3 Projective Models in Scrolls. VIII, 164 pages. 2004

Vol. 1843: B. Jefferies, Spectral Properties of Noncommuting Operators, VIII, 184 pages. 2004

Vol. 1844: K.F. Siburg, The Principle of Least Action in Geometry and Dynamics, XII, 128 pages. 2004.

Vol. 1845: Min Ho Lee, Mixed Automorphic Forms, Torus Bundles, and Jacobi Forms, X, 239 pages. 2004

Vol. 1846: h. Ammari, H. Kang, Reconstruction of Small Inhomogeneities from Boundary Measurements, IX, 238 pages. 2004

Vol. 1848: M. Abate, J. E. Fornaess, X. Huang, J. P. Rosay, A. Tumanov, Real Methods in Complex and CR Geometry, Martina Franca, Italy 2002. Editors: D. Zaitsev, G. Zampieri. IX, 219 pages. 2004.

Vol. 1849: Martin L. Brown, Heegner Modules and Elliptic Curves, X, 517 pages. 2004

Vol. 1850: V. D. Milman, G. Schechtman (Eds.), Geometric Aspects of Functional Analysis. Israel Seminar 2002-2003. X, 301 pages. 2004.

Vol. 1851: O. Catoni, Statistical Learning Theory and Stochastic Optimization, VIII, 273 pages. 2004.

Vol. 1852: A.S. Kechris, B.D. Miller, Topics in Orbit Equivalence, X, 134 pages. 2004.

Vol. 1853: Ch. Favre, M. Jonsson, The Valuative Tree, XV, 234 pages. 2004.

Vol. 1854: O. Saeki, Topology of Singular Fibers of Differential Maps, X, 145 pages. 2004.

Vol. 1855: G. Da Prato, P.C. Kunstmann, I. Lasiecka, A. Lunardi, R. Schnaubelt, L. Weis, Functional Analytic Methods for Evolution Equations, Levico Terme, Italy 2001. Editors: M. Iannelli, R. Nagel, S. Piazzera. VIII, 472 pages. 2004.

Vol. 1856: K. Back T.R. Bielecki, C. Hipp, S. Peng, W. Schachermayer, Stochastic Methods in Finance. Bressanone/Brixen, Italy 2003. Editors: M. Frittelli, W. Runggaldier. XIII, 307 pages. 2004.

Recent Reprints and New Editions

Vol. 1200: V. D. Milman, G. Schechtman, Asymptotic Theory of Finite Dimensional Normed Spaces. 1986. – Corrected Second Printing. X, 156 pages. 2001.

Vol. 1471: M. Courtieu, A.A. Panchishkin, Non-Archimedean L-Functions and Arithmetical Siegel Modular Forms. – Second Edition. VII. 196 pages. 2003.

Vol. 1618: G. Pisier, Similarity Problems and Completely Bounded Maps. 1995 – Second, Expanded Edition VII, 198 pages. 2001.

Vol. 1629: J.D. Moore, Lectures on Seiberg-Witten Invariants. 1997 – Second Edition. VIII, 121 pages. 2001.

Vol. 1638: P. Vanhaecke, Integrable Systems in the realm of Algebraic Geometry. 1996 – Second Edition. X, 256 pages. 2001.

Vol. 1702: J. Ma, J. Yong, Forward-Backward Stochastic Differential Equations and Their Applications. 1999. – Corrected Second Printing. XIII, 270 pages. 2000.

Printing and Binding: Strauss GmbH, Mörlenbach

4. For evaluation purposes, manuscripts may be submitted in print or electronic form (print form is still preferred by most referees), in the latter case preferably as pdf- or zipped ps-files. Lecture Notes volumes are, as a rule, printed digitally from the authors' files. To ensure best results, authors are asked to use the LaTeX2e style files available from Springer's web-pages at

www.springeronline.com

then click on "Mathematics", click on "For Authors" and look for "Macro Packages for books". Macros in LaTeX2.09 and TeX are available on request from: lnm@springer.de. Careful preparation of the manuscripts will help keep production time short besides ensuring satisfactory appearance of the finished book in print and online. After acceptance of the manuscript authors will be asked to prepare the final LaTeX source files (and also the corresponding dvi-, pdf- or zipped ps-files) together with the final printout made from these files. The LaTeX source files are essential for producing the full-text online version of the book

(http://www.springerlink.com/openurl.asp?genre=journal&issn=0075-8434).

The actual production of a Lecture Notes volume takes approximately 8 weeks.

5. Authors receive a total of 50 free copies of their volume, but no royalties. They are entitled to a discount of 33.3 % on the price of Springer books purchased for their personal use, if ordering directly from Springer.

6. Commitment to publish is made by letter of intent rather than by signing a formal contract. Springer-Verlag secures the copyright for each volume. Authors are free to reuse material contained in their LNM volumes in later publications: A brief written (or e-mail) request for formal permission is sufficient.

Addresses:

Professor J.-M. Morel, CMLA,
École Normale Supérieure de Cachan,
61 Avenue du Président Wilson, 94235 Cachan Cedex, France
E-mail: Jean-Michel.Morel@cmla.ens-cachan.fr

Professor F. Takens, Mathematisch Instituut,
Rijksuniversiteit Groningen, Postbus 800,
9700 AV Groningen, The Netherlands
E-mail: F.Takens@math.rug.nl

Professor B. Teissier, Université Paris 7
Institut Mathématique de Jussieu, UMR 7586 du CNRS
Équipe "Géométrie et Dynamique", 175 rue du Chevaleret
75013 Paris, France
E-mail: teissier@math.jussieu.fr

Springer-Verlag, Mathematics Editorial, Tiergartenstr. 17,
69121 Heidelberg, Germany,
Tel.: +49 (6221) 487-8410
Fax: +49 (6221) 487-8355
E-mail: lnm@springer.de

—